System Modeling and Control with Resource-Oriented Petri Nets

AUTOMATION AND CONTROL ENGINEERING

A Series of Reference Books and Textbooks

Series Editors

FRANK L. LEWIS, Ph.D.,
Fellow IEEE, Fellow IFAC

Professor
Automation and Robotics Research Institute
The University of Texas at Arlington

SHUZHI SAM GE, Ph.D.,
Fellow IEEE

Professor
Interactive Digital Media Institute
The National University of Singapore

System Modeling and Control with Resource-Oriented Petri Nets,
Naiqi Wu and MengChu Zhou

Sliding Mode Control in Electro-Mechanical Systems, Second Edition,
Vadim Utkin, Jürgen Guldner, and Jingxin Shi

Optimal Control: Weakly Coupled Systems and Applications,
*Zoran Gajić, Myo-Taeg Lim, Dobrila Skatarić, Wu-Chung Su,
and Vojislav Kecman*

Intelligent Systems: Modeling, Optimization, and Control, *Yung C. Shin
and Chengying Xu*

Optimal and Robust Estimation: With an Introduction to Stochastic Control
Theory, Second Edition, *Frank L. Lewis; Lihua Xie and Dan Popa*

Feedback Control of Dynamic Bipedal Robot Locomotion,
*Eric R. Westervelt, Jessy W. Grizzle, Christine Chevallereau, Jun Ho Choi,
and Benjamin Morris*

Intelligent Freight Transportation, *edited by Petros A. Ioannou*

Modeling and Control of Complex Systems, *edited by Petros A. Ioannou
and Andreas Pitsillides*

Wireless Ad Hoc and Sensor Networks: Protocols, Performance,
and Control, *Jagannathan Sarangapani*

Stochastic Hybrid Systems, *edited by Christos G. Cassandras
and John Lygeros*

Hard Disk Drive: Mechatronics and Control, *Abdullah Al Mamun,
Guo Xiao Guo, and Chao Bi*

Autonomous Mobile Robots: Sensing, Control, Decision Making
and Applications, *edited by Shuzhi Sam Ge and Frank L. Lewis*

Neural Network Control of Nonlinear Discrete-Time Systems,
Jagannathan Sarangapani

Quantitative Feedback Theory: Fundamentals and Applications,
Second Edition, *Constantine H. Houpis, Steven J. Rasmussen,
and Mario Garcia-Sanz*

Fuzzy Controller Design: Theory and Applications, *Zdenko Kovacic
and Stjepan Bogdan*

Chaos in Automatic Control, *edited by Wilfrid Perruquetti
and Jean-Pierre Barbot*

System Modeling and Control with Resource-Oriented Petri Nets

Naiqi Wu

Guangdong University of Technology
Guangzhou, People's Republic of China

MengChu Zhou

New Jersey Institute of Technology
Newark, New Jersey, U.S.A.
and
XiDian University
Xi'an, People's Republic of China

CRC Press
Taylor & Francis Group
Boca Raton London New York

CRC Press is an imprint of the
Taylor & Francis Group, an **Informa** business

CRC Press
Taylor & Francis Group
6000 Broken Sound Parkway NW, Suite 300
Boca Raton, FL 33487-2742

© 2010 by Taylor and Francis Group, LLC
CRC Press is an imprint of Taylor & Francis Group, an Informa business

No claim to original U.S. Government works

Printed in the United States of America on acid-free paper
10 9 8 7 6 5 4 3 2 1

International Standard Book Number: 978-1-4398-0884-9 (Hardback)

Library of Congress Cataloging-in-Publication Data

Wu, Naiqi.
   System modeling and control with resource-oriented Petri nets / Naiqi Wu, MengChu Zhou.
      p. cm. -- (Automation and control engineering)
   Includes bibliographical references and index.
   ISBN 978-1-4398-0884-9 (hardcover : alk. paper)
   1. Flexible manufacturing systems--Mathematical models. 2. Petri nets. 3. Production control--Mathematical models. 4. Discrete-time systems. I. MengChu, Zhou. II. Title. III. Series.

TS155.6.Z493 2010
670.42'7--dc22

2009026763

Visit the Taylor & Francis Web site at
http://www.taylorandfrancis.com

and the CRC Press Web site at
http://www.crcpress.com

Contents

Preface xi

Acknowledgments xvii

The Authors xix

List of Abbreviations xxi

1   Introduction to Petri Net Modeling 1

1.1   The Modeling Process 1
1.2   Automated Manufacturing Systems 2
1.3   Historical Perspective of Petri Nets in Automation 6
1.4   Scope and Objectives 12
1.5   Summary 13
References 14

2   Petri Nets: Basic Concept 15

2.1   Basic Concepts 15
        2.1.1   Definition 15
        2.1.2   Enabling and Firing Rules 16
        2.1.3   Finite Capacity PN 18
        2.1.4   Some Special Structures in PN 18
2.2   Subclass of PN 20
2.3   Properties 22
        2.3.1   Reachability 22
        2.3.2   Boundedness 23
        2.3.3   Incidence Matrix and Conservativeness 24
        2.3.4   Reversibility 24
        2.3.5   Liveness 25
2.4   Timed PN 27
2.5   PN with Inhibitor Arcs 28
2.6   Summary 30
References 30

3   Colored Petri Net 31

3.1   A Simple Example 31
3.2   Definitions of CPN 33
3.3   Transition Enabling and Firing Rules 37
3.4   P-Invariant in CPN 38

3.5   Summary 41
References 41

4   Process-Oriented Petri Net Modeling 43

4.1   Introduction 43
4.2   Modeling Method 44
4.3   Resource Sharing in POPN 47
       4.3.1   Resource Sharing in Part Processing 48
       4.3.2   Resource Sharing in Material Handling 50
4.4   Characteristics of POPN 52
4.5   Summary 54
References 55

5   Resource-Oriented Petri Net Modeling 57

5.1   Introduction 57
5.2   Steps of ROPN Modeling 57
5.3   Modeling Part Production Processes 58
       5.3.1   Subnet Forming 60
       5.3.2   Subnet Merging 60
       5.3.3   Colored ROPN 62
5.4   Modeling Material Handling Processes 65
5.5   Resource Sharing in ROPN 66
5.6   Characteristics of ROPN 68
5.7   Summary 69
References 69

6   Process- vs. Resource-Oriented Petri Nets 71

6.1   Modeling Power and Model Size 71
6.2   Conservativeness 72
6.3   Structure for Liveness 73
6.4   Example 74
6.5   Summary 80
References 81

7   Control of Flexible and Reconfigurable Manufacturing Systems 83

7.1   Introduction 83
7.2   Deadlock in FMS 84
7.3   System Modeling by CROPN 87
7.4   Existence of Deadlock 89
7.5   Deadlock Avoidance Policy 93
       7.5.1   Case 1: Subnet Formed by One PPC 95

7.5.2   Case 2: Interactive Subnet Formed by Two PPCs 95
7.5.3   Case 3: Interactive Subnet Formed by Multiple PPCs...................98
7.6   Liveness of Overall System...102
7.7   Illustrative Example 104
7.8   Implementation 105
7.9   Deadlock Avoidance with Shared Material Handling System...................107
    7.9.1   Deadlock Situations...107
    7.9.2   Deadlock Avoidance with MHS via ROPN Modeling.................109
7.10   Summary...112
References...113

8   Avoiding Deadlock and Reducing Starvation and Blocking 115

8.1   Introduction..115
8.2   A Simple Example 116
8.3   Relaxed Control Policy 118
8.4   Dependent PPCs in Interactive Subnets...121
8.5   Complexity in Applying the Control Law...128
8.6   Performance Improvement through Examples 128
8.7   Summary...131
References...131

9   Control and Routing of Automated Guided Vehicle Systems 133

9.1   Introduction..133
9.2   Control of AGV Systems with Unidirectional Paths 135
    9.2.1   Modeling AGV Systems with Unidirectional Paths
          by CROPN 135
    9.2.2   Deadlock Avoidance Policy...136
    9.2.3   Computational Complexity...139
9.3   Control of AGV Systems with Bidirectional Paths............................140
    9.3.1   Modeling AGV Systems with Bidirectional Paths by CROPN 140
    9.3.2   Deadlock Avoidance for AGV Systems with Cycles...................143
    9.3.3   Deadlock Avoidance in the CROPN...148
    9.3.4   Examples 150
9.4   Routing of AGV Systems Based on CROPN 154
    9.4.1   Problem Description..155
    9.4.2   AGV Rerouting...158
    9.4.3   Route Expansion...162
    9.4.4   Illustrative Examples 163
    9.4.5   Performance Comparison...165
9.5   Summary...169
References...169

10 Control of FMS with Multiple AGVs 171

10.1   Introduction.. 171
10.2   System Modeling with CROPN 173
10.3   Deadlock Avoidance Policy 178
10.4   Illustrative Example 182
10.5   Summary.. 183
References.. 183

11 Control of FMS with Multiple Robots.. 185

11.1   Introduction.. 185
11.2   Motivation through Example 185
11.3   Deadlock Control Policy 186
11.4   Illustrative Example 193
11.5   Summary.. 194
References.. 195

12 Control of Semiconductor Manufacturing Systems 197

12.1   Modeling, Analysis, and Control of Cluster Tools................................ 197
        12.1.1   Cluster Tools... 198
        12.1.2   Analysis by Timed MG 199
        12.1.3   Modeling Cluster Tools by CROPN 203
        12.1.4   Analysis of the Single-Blade Robot Cluster Tool 208
                  12.1.4.1   Deadlock Analysis 209
                  12.1.4.2   Throughput Analysis for the Process without
                             Revisiting 210
                  12.1.4.3   Throughput Analysis of a Process with Revisiting 211
        12.1.5   Analysis of Dual-Blade Robot Cluster Tools.............................. 213
                  12.1.5.1 Deadlock Analysis 213
                  12.1.5.2 Throughput Analysis for the Process without
                           Revisiting 214
                  12.1.5.3 Throughput Analysis of Process with Revisiting............ 215
12.2   Deadlock Avoidance in Track System 217
        12.2.1   Semiconductor Track System 217
        12.2.2   Modeling by ROPN 219
        12.2.3   Deadlock-Free Condition for Strongly Connected Subnet............ 220
        12.2.4   Implementation of the Deadlock-Free Condition 228
        12.2.5   Illustrative Example.. 229
12.3   Deadlock-Free Scheduling of a Track System 230
        12.3.1   Dispatching Rules... 231
        12.3.2   Illustrative Example.. 234
12.4   Summary.. 236
References.. 236

13 Modeling and Control of Assembly/Disassembly Systems.............................239

13.1   Introduction...239
13.2   A Flexible Assembly System 240
13.3   R-Policy..242
13.4   Modeling FAS by CROPN 246
       13.4.1   Models for Resources 246
       13.4.2   Models for Individual Products......................................247
       13.4.3   ROPN for the Whole System..250
13.5   Realizable Resource Requirement 253
13.6   Deadlock Avoidance Control Policy 256
13.7   Illustrative Example 260
13.8   Industrial Case Study 262
13.9   Summary..265
References..266

Bibliography 267
Index..273

Preface

The hardest thing in life is to know which bridge to cross and which to burn.

—David Russell

In the early 1990s, Naiqi Wu was a visiting scholar at the School of Industrial Engineering, at Purdue University, West Lafayette, Indiana, and worked in the field of design, scheduling, and control of automated manufacturing systems (AMS). One day, the paper "Deadlock Prevention and Deadlock Avoidance in Flexible Manufacturing Systems Using Petri Net Models" (Viswanadham et al., *IEEE Transactions on Robotics and Automation*, volume 6, pp. 713–723, 1990) attracted his attention. It deals with a deadlock avoidance problem in flexible manufacturing systems, an important issue in automated manufacturing processes. Wu decided to study the subject. It is known that deadlocks in flexible manufacturing systems are caused by a circular wait for resource competition, caused by limited shared resources. At that time, he knew nothing about the Petri net theory. Then, Wu read Peterson's book *Petri Net Theory and the Modeling of Systems* (Englewood Cliffs, NJ: Prentice-Hall, 1981), Murata's paper "Petri Nets: Properties, Analysis and Applications" (*Proceedings of the IEEE*, volume 77, pp. 541–580, 1989), and some papers written by MengChu Zhou, Dr. Frank DiCesare, and their sizable RPI Petri net research group at Rensselaer Polytechnic Institute. Soon thereafter it was found that process-oriented Petri nets (POPNs) were almost exclusively used as a net model of flexible manufacturing systems in the literature at that time. Although the POPN methodology is powerful in system modeling, the resultant model easily tends to be unmanageably large. Furthermore, it cannot explicitly describe circular wait, the very nature of deadlock in any resource allocation system. Based on this observation, a new way to model a flexible manufacturing system is considered such that resource circuits behind circular wait can be explicitly described. To do so, one can treat a resource in the system as a server and a part to be processed in the system as a customer. The way in which customers (parts/jobs) visit servers (resources) can in fact describe the dynamic behavior of the system. In 1997, Wu and Zhou met at the IEEE International Conference on Robotics and Automation in Albuquerque, New Mexico, and talked with each other regarding the emerging field of Petri nets and their applications to automation. Wu presented his paper "Avoiding Deadlocks in Automated Manufacturing Systems with Shared Material Handling System." This was the first time that he presented the idea of resource-oriented Petri nets and their applications when a material handling system must be considered as a part of an AMS in deadlock resolution. Listening to the presentation, Zhou became deeply interested in the new modeling philosophy. He immediately realized the great significance a modeling paradigm could bring to the solution of difficult deadlock problems, as he had already worked in the process-oriented Petri net modeling field for 10 years and coauthored the first monograph* on the topic with his former advisor, Professor Frank DiCesare. Since that meeting, Wu and Zhou began to communicate with each other and worked together to obtain many new interesting results. Wu spent the summer of 2004 at the New Jersey Institute of Technology to conduct collaborative research with Zhou, upon his invitation. Zhou also paid several visits to Wu and to Wu's laboratory. Some important results gained from this collaboration and related to this monograph are explained as follows:

1. Our first joint paper, "Resource-Oriented Petri Nets for Deadlock Avoidance in Automated Manufacturing" (*Proceedings of 2000 IEEE International Conference on Robotics and Automation*, San Francisco, April 2000, pp. 3377–82), and its extended version, "Avoiding Deadlock and Reducing Starvation and Blocking in Automated Manufacturing Systems Based on a Petri Net Model" (*IEEE Transactions on Robotics and Automation* 17 (2001): 658–69) use resource-oriented Petri nets to address the benefit of applying a "liveness" margin to improve the system productivity. The idea is motivated by the stability margin in traditional feedback control systems. We can also find some real-life examples to explain why it is so important to leave an extra margin for the best performance. For example, consider a circular highway with entrances and exits allowing vehicles to enter and exit freely. When there are too many cars on it, all will be significantly slowed down, and thus the number of cars flowing through any segment in a unit of time will be decreased (called throughput). Yet we find no deadlock, only congestion, as all the cars will be able to exit from the highway. On the contrary, if too few cars are allowed to enter it, the throughput is very low, and each car reaches its highest allowed speed. Clearly this will significantly lower the utilization. Therefore, the best cases are those in which we have enough cars on the highway able to utilize their allowed high speed without yielding congestion. This way we enjoy the highest throughput and highway utilization. This applies to AMS as well. Our work has revealed this important fact: A maximally permissive deadlock control policy may admit some congestion cases that can lower productivity when rule-based or other heuristic scheduling policies apply. As a result, a policy with some margin should be used such that those congestion cases are completely excluded. The contents of these two papers are described in Chapters 7 and 8 of this book.

2. Our second important work is on the use of resource-oriented Petri nets to faithfully model automated guided vehicle (AGV) systems and derive the deadlock control policies suitable for real-time implementation. They can handle both bidirectional and unidirectional paths. The former offer additional flexibility, efficiency, and cost savings when compared with the latter. Yet, they exhibit more challenging AGV management problems. Unlike jobs that can enter and leave an AMS, AGVs always stay in the system. In AGV systems, a lane and a node may hold one—and only one—AGV.

---

* Zhou, M. C., and F. DiCesare, *Petri Net Synthesis for Discrete Event Control of Manufacturing Systems* (Boston: Kluwer, 1993).

This leads to a single-capacity resource allocation system, invalidating many elegant deadlock policies that require multiple-capacity resources. By modeling nodes with places and lanes with transitions, the proposed method can construct Petri net models for changing AGV routes. In addition to a deadlock control policy, our study also leads to a method that finds the shortest time routes while avoiding both deadlock and blocking. The method needs to perform rerouting whenever necessary. As a result, the proposed method can offer better solutions than the existing ones. For some cases, it can find a solution while the existing ones cannot. Our work was published as "Modeling and Deadlock Control of Automated Guided Vehicle Systems" (*IEEE/ASME Transactions on Mechatronics* 9 (2004): 50–57) and "Shortest Routing of Bidirectional Automated Guided Vehicles Avoiding Deadlock and Blocking" (*IEEE/ASME Transactions on Mechatronics* 12 (2007): 63–72). This part of the material is presented in Chapter 9 of this book.

3. The deadlock problems are separately treated for parts in production and transportation. Many techniques are developed for either problem. In general, it is intractable to obtain maximally permissive control policies for either one. Conventional thinking suggests that the combination of two problems would complicate the issue. When we investigated the combined problem for a flexible manufacturing system that adopted multiple AGVs for material handling, we found that they brought the flexibility and opportunity to help resolve deadlocks for parts in production. More surprisingly, we established a novel control policy for deadlock avoidance that was maximally permissive and had only polynomial computational complexity if the complexity for controlling the part transportation by AGVs was limited. Thus, the complexity of deadlock avoidance for the whole system is bounded by the complexity in controlling the AGV system. It is known that most AGV systems in practice seem to be easily controlled because of (1) limited number of AGVs in a system, (2) the prior known configuration, and (3) some special designs. Note that general AGV systems have nonpolynomial complexity for their optimal control. Our work appeared in "Modeling and Deadlock Avoidance of Automated Manufacturing Systems with Multiple Automated Guided Vehicles" (*IEEE Transactions on Systems, Man, and Cybernetics, Part B: Cybernetics* 35 (2005): 1193–202). We have dedicated Chapter 10 to this topic.

4. Multiple AGVs can create opportunities to help resolve those otherwise deadlock situations, partially due to their mobility. Robots in most industrial settings are often fixed while serving one or multiple machines within their reach. The question is whether they can also create some opportunity to ease deadlock issues. Our joint work proves this by extending resource-oriented Petri net models to model both parts in production and their transfers from one machine to another. We are able to eliminate some deadlocks in the corresponding process-oriented models by simply using such models in fact. We have published the results in a paper entitled "Deadlock Resolution in Automated Manufacturing Systems with Robots" (*IEEE Transactions on*

Automation Science and Engineering 4 (2007): 474–80). Chapter 11 discusses the details of the extended models and their application results.

5. The massive need for various chips used in consumer electronics, automobiles, and other engineering systems has greatly promoted the area of semiconductor manufacturing automation. In particular, the development of cluster tools and track systems has helped the industry achieve high yield with high product quality. The modeling, control, and scheduling of these highly integrated flexible automated systems become a very challenging task. Our work has successfully associated time delays with the places and transitions in their resource-oriented Petri net models and derived deadlock-free optimal or near-optimal schedules. The work was published in the following articles: "Real-Time Deadlock-Free Scheduling for Semiconductor Track Systems Based on Colored Timed Petri Nets" (*OR Spectrum* 29 (2007): 421–43), "Deadlock Modeling and Control of Semiconductor Track Systems Using Resource-Oriented Petri Nets" (*International Journal of Production Research* 45 (2007): 3439–56), and "A Petri Net Method for Schedulability and Scheduling Problems in Single-Arm Cluster Tools with Wafer Residency Time Constraints" (*IEEE Transactions on Semiconductor Manufacturing* 21 (2008): 224–37). The majority of these published articles are combined and presented in Chapter 12.

6. Automated flexible assembly systems represent one of the hardest systems in terms of deadlock resolution. This extreme difficulty is due to their particular features that other automated manufacturing systems do not have. For example, deadlocks can take place in both the flows of base components and parts to be mounted onto the former. The assembly operations can also result in a deadlock due to inappropriate allocation of different types of parts or fixtures. The optimal solution to this deadlock avoidance problem is computationally infeasible. Our work uses heuristics to well address this challenge and greatly outperforms the existing method. A short version of this work was presented at the 16th IFAC World Congress, Prague, Czech Republic, and a full version, "Resource-Oriented Petri Net for Deadlock Avoidance in Flexible Assembly Systems," appeared in *IEEE Transactions on Systems, Man, and Cybernetics, Part A: Systems and Humans* (38 (2008): 56–69). Chapter 13 covers this topic with an industrial example.

We have continued our collaboration on the extension of resource-oriented Petri net models to deal with hybrid systems. For example, we have studied the short-term scheduling issues for oil refineries. We have obtained some significant results that we have published in *IEEE Transactions on Automation Science and Engineering*, *IEEE Transactions on Systems, Man, and Cybernetics, Part C: Applications and Reviews*, and *International Journal of Intelligent Control and Systems*. We cannot include the results in this book due to space limitations and the lack of a close relation with the topics of automated manufacturing systems—the focus of this book.

Chapter 1 of this book introduces a modeling process, contemporary automated manufacturing systems, and a historic perspective on Petri net studies as related to automation. Chapter 2 presents the fundamentals of Petri net theory with the definitions and examples known to the Petri net research community. Chapter 3 presents the classical colored Petri nets and their definitions. Chapter 4 presents process-oriented Petri net models and modeling methodologies. Some examples from manufacturing automation are given to illustrate them. We used "Resource-Oriented Petri Nets in Deadlock Prevention and Avoidance," our jointly published chapter in *Deadlock Resolution in Computer-Integrated Systems* (M. C. Zhou and M. P. Fanti, eds. (New York: Marcel Dekker, 2005), 349–406), in Chapter 5. It covers the basics of resource-oriented Petri net models, the colored version, and the related characteristics. In particular, it presents the necessary and sufficient deadlock-free conditions for the class of automated manufacturing systems represented with the proposed models. This significant result was obtained by Naiqi Wu in his 1999 paper "Necessary and Sufficient Conditions for Deadlock-Free Operation in Flexible Manufacturing Systems Using a Colored Petri Net Model" (*IEEE Transactions on Systems, Man, and Cybernetics, Part C: Applications and Reviews* 29 (1999): 192–204). We compare resource- and process-oriented Petri net modeling methodologies and resulting models in Chapter 6. We discuss their advantages and disadvantages in the design of discrete event systems.

Finally, we hope that this monograph will stimulate more engineers and researchers to investigate and apply resource-oriented Petri nets for modeling, analysis, performance evaluation, simulation, control, and scheduling of their particular engineering systems. We hope that the work serves as a beginning step toward the better understanding and modeling of increasingly complex engineering systems and eventually toward the better design of such systems to benefit humankind.

Naiqi Wu, PhD
Gangdong University of Technology, Guangzhou, Gangdong, China
University of Technology of Troyes, Troyes, France

MengChu Zhou, PhD
New Jersey Institute of Technology, Newark, New Jersey, United States
Xidian University, Xi'an, Shanxi, China

Acknowledgments

The following are important books and papers that address the fundamentals of Petri nets and their applications to the field of automation science and engineering:

Murata, T. 1989. Petri nets: Properties, analysis and applications. *Proceedings of the IEEE* 77:541–80.

Viswanadham, N., and Y. Narahari. 1992. *Performance modeling of automated manufacturing systems*. Englewood Cliffs, NJ: Prentice Hall.

Zhou, M. C., and F. DiCesare. 1993. *Petri net synthesis for discrete event control of manufacturing systems*. Boston: Kluwer Academic.

Desrochers, A. A., and R. Y. Al-Jaar. 1995. *Applications of Petri nets in manufacturing systems: Modeling, control, and performance analysis*. Piscataway, NJ: IEEE Press.

Zhou, M. C., ed. 1995. *Petri nets in flexible and agile automation*. Boston: Kluwer Academic Publisher.

Proth, J.-M., and X. L. Xie. 1996. *Petri nets: A tool for design and management of manufacturing systems*. New York: John Wiley & Sons.

Zhou, M. C., and K. Venkatesh. 1998. *Modeling, simulation and control of flexible manufacturing systems: Petri net approach*. Singapore: World Scientific.

Cassandras, C. G., and S. Lafortune. 1999. *Introduction to discrete event systems*. Boston: Kluwer Academic Publishers.

Zhou, M. C., and M. P. Fanti, eds. 2005. *Deadlock resolution in computer-integrated systems*. New York: Marcel Dekker.

Bogdan, S., F. L. Lewis, Z. Kovacic, and J. Mireles. 2006. *Manufacturing systems control design: A matrix based approach*. London: Springer-Verlag.

Hruz, B., and M. C. Zhou. 2007. *Modeling and control of discrete-event dynamic systems*: *With petri nets and other tools*. (Series-Advanced Textbooks in Control and Signal Processing) London: Springer-Verlag.

These publications cover many topics related to this book. The significant difference, however, lies in the fact that they take a Petri net modeling approach from the viewpoint of the processes a job must take, while this book adopts the viewpoint of the resources. As a result, the routes a job takes are not as obviously seen as those in a process-oriented Petri net model. Instead, they are implicitly reflected by the routes through which a job visits resources. Consequently, the graphical modeling complexity is drastically reduced, while the tokens representing jobs have to carry job-related information. From the perspective of deadlock analysis and control, this resource-oriented approach can yield significant benefits.

During our research toward efficient analysis and control of deadlocks in various systems, including flexible manufacturing, material handling, flexible assembly, and semiconductor fabrication tools, we have received much help. The following researchers have contributed to this book through a variety of ways, including guidance, advice, discussion, and critique, and are deserving of our mention: Prof. Peter B. Luh, University of Connecticut; Prof. N. Viswanadham, Indian School of Business; MuDer Jeng, National Taiwan Ocean University; Prof. Y. Narahari, Indian Institute of

Science; Prof. Zhiwu Li, Xidian University; Prof. Changjun Jiang, Tongji University; Prof. Yi-Sheng Huang, National Defense University; Prof. D. Chao, National Cheng Chi University; Prof. Maria Pia Fanti, Politecnico di Bari; Prof. Xiaolan Xie, Ecole Nationale Superieure des Mines de Saint-Etienne; Prof. Feng Chu, University of Technology of Troyes; Prof. Jiacun Wang, Monmouth University; and Prof. Chengbin Chu, Ecole Centrale Paris.

This effort has been made possible through the support of various agencies including the Natural Science Foundation of China grants 69974011, 60574066, 60474018, and 60773001; the National 863 Program grant 2008AA04Z109; and the Ministry of Education of China under the Changjiang Scholars Program.

The Authors

Naiqi Wu received both his MS and PhD degrees in systems engineering from Xi'an Jiaotong University, Xi'an, China, in 1985 and 1988, respectively. From 1988 to 1995, he was with the Chinese Academy of Sciences, Shenyang Institute of Automation, Shenyang, China, and from 1995 to 1998, with Shantou University, Shantou, China. From 1991 to 1992, he was a visiting scholar in the School of Industrial Engineering, Purdue University, West Lafayette, Indiana. In 1999, 2004, and 2007–2008, he was a visiting professor with the Department of Industrial Engineering, Arizona State University, Tempe; the Department of Electrical and Computer Engineering, New Jersey Institute of Technology, Newark; and the Industrial Systems Engineering Department, Industrial Systems Optimization Laboratory, University of Technology of Troyes, Troyes, France. He is currently a professor of industrial and systems engineering in the Department of Industrial Engineering, School of Mechatronics Engineering, Guangdong University of Technology, Guangzhou, China. He is the author or coauthor of many publications in international journals. Professor Wu is an associate editor of the *IEEE Transactions on Systems, Man, and Cybernetics, Part C, IEEE Transactions on Automation, Science, and Engineering*, and editor-in-chief of the *Industrial Engineering Journal*. He was a program committee member of the 2003–2008 IEEE International Conference on Systems, Man, and Cybernetics, a program committee member of the 2005–2008 IEEE International Conference on Networking, Sensing and Control, a program committee member of the 2006 IEEE International Conference on Service Systems and Service Management, a program committee member of the 2007 International Conference on Engineering and Systems Management, and reviewer for many international journals.

MengChu Zhou received his BS degree from Nanjing University of Science and Technology, MS degree from Beijing Institute of Technology, and PhD degree from Rensselaer Polytechnic Institute. He joined New Jersey Institute of Technology (NJIT) in 1990, and is currently a professor of electrical and computer engineering and director of the Discrete-Event Systems Laboratory. His interests are in computer-integrated systems, Petri nets, networks, and manufacturing. He has written more than 300 publications, including 9 books and over 130 journal papers. Professor Zhou was managing editor of *IEEE Transactions on Systems, Man, and Cybernetics* from 2005 to 2008, and associate editor of IEEE *Transactions on Automation Science and Engineering* from 2004 to 2007, *Transactions on Systems, Man, and Cybernetics, Part B* from 2002 to 2005, and *IEEE Transactions on Robotics and Automation* from 1997 to 2000. He is presently editor of *IEEE Transactions on Automation Science and Engineering*; associate editor of IEEE *Transactions on Systems, Man, and Cybernetics: Part A* and *IEEE Transactions on Industrial Informatics*; and editor-in-chief of *International Journal of Intelligent Control and Systems*. He has served as general and program chair of many international conferences.

Dr. Zhou has led or participated in thirty-six research and education projects with budgets totaling over $10M that were funded by the National Science Foundation, the Department of Defense, and industry. He was the recipient of the CIM University–LEAD Award by the Society of Manufacturing Engineers, Harlan J. Perlis Research Award by NJIT, the Humboldt Research Award for U.S. Senior Scientists, and a distinguished lecturer of the IEEE Systems, Man, and Cybernetics Society. Dr. Zhou has been invited to lecture in Australia, Canada, China, France, Germany, Hong Kong, Italy, Japan, Korea, Mexico, Taiwan, and the United States. He was the founding chair of the Discrete Event Systems Technical Committee of the IEEE Systems, Man, and Cybernetics Society, and founding chair of the Semiconductor Manufacturing Automation Technical Committee of IEEE Robotics and Automation Society. Dr. Zhou is currenlty co-chair (founding) of the Enterprise Information Systems Technical Committee of IEEE Systems, Man, and Cybernetics Society and the vice chair of the IFAC Technical Committee on Economic and Business Systems. He is a life member of the Chinese Association for Science and Technology–USA and served as its president in 1999. He is a fellow of the IEEE.

List of Abbreviations

AGV = automated guided vehicle

AMS = automated manufacturing system

$C(p)$ = a set of colors for place p

$C(p_i(A_k))$ = color for a token representing an A part at its kth operation in place p_i

$C(t)$ = a set of colors for transition t

c, c_i = a color for a transition, or a token

CPN = colored Petri net

CROPN = colored resource-oriented Petri net

DES = discrete event system

DPP = a direct place path in a Petri net

FAS = flexible assembly system

FMS = flexible manufacturing system

$I: P \times T \rightarrow N$ = an input function that defines the directed arcs from places to transitions in a Petri net

IBS = ill-behaved siphon in a Petri net

IIT = intercircuit input transition

IOT = intercircuit output transition

$K: P \rightarrow N$ = a capacity function where $K(p)$ represents the maximal number of tokens that place p can hold at a time

$K(v_i)$ = the capacity of PPC v_i

$M: P \rightarrow N$ = the marking vector in a Petri net

M_0 = the initial marking in a Petri net

M_i = the ith marking in a Petri net

$M(p_i)(c_i)$ = the number of tokens with color c_i in p_i at marking M

$M(p_i)(v^n)$ = the number of cycling tokens of subnet v^n in place p_i at marking M

$M(L(v_i))$ = the number of possible leaving tokens in PPC v_i

$M(v)$ = the number of cycling tokens of PPC v in marking M

$M(v^n)$ = the number of cycling tokens in subnet v^n at marking M

MG = marked graph

MHS = material handling system

$\eta(v^n, M)$ = the number of enabled PPCs in subnet v^n

$N = \{0, 1, 2, \ldots\}$, the set of nonnegative integers

$O: P \times T \rightarrow N$ = an output function that defines the directed arcs from transitions to places in a Petri net

P = a set of places in a Petri net

$P(v)$ = a set of places on circuit v

$P(v^n)$ = a set of places in subnet v^n

PME = parallel mutual exclusion

$PN = (P, T, I, O, M_0)$ is a marked Petri net

POPN = process-oriented Petri net

PPC = production process circuit

R = the set of non-negative real numbers

$R(M_0)$ = the set of reachable markings of a PN from initial marking M_0

$R_c(M_0)$ = the set of reachable markings of a CROPN under control from initial marking M_0

RC = resource circuit

RMS = reconfigurable manufacturing system

ROPN = resource-oriented Petri net

S = siphon in a Petri net

$S(v_i)$ = the number of free spaces available in PPC v_i in a CROPN in marking M

$S(v^n)$ = the number of spaces available in subnet v^n in a CROPN in marking M

$S'(v_i)$ = the number of currently potential spaces available in PPC v_i in a CROPN in marking M

SDPP = shared direct place path in ROPN

SM = state machine

SME = sequential mutual exclusion

$\tau\colon P, T \to R$ = time delay in place or transition

T = a set of transitions in a Petri net

$T(v)$ = set of transitions on circuit v

$T(v^n)$ = set of transitions in subnet v^n

$T_I(v^n)$ = set of input transitions of subnet v^n

$T_O(v^n)$ = set of output transitions of subnet v^n

t_{iik} = intercircuit input transition (IIT) external to v_k

t_{iok} = intercircuit output transition (IOT) on v_k

v, v_i = a PPC in an ROPN

v^n = interactive subnet composed of n PPC in an ROPN

Y, Y_i = a cycle in CROPN for AGV system

1 Introduction to Petri Net Modeling

1.1 THE MODELING PROCESS

A modeling process is a method by which people understand a concerned object, system, or phenomenon. Various modeling methods, tools, and models have been developed since people started to design man-made and engineering systems to benefit human beings. It is well known that *mathematical modeling* helps one to understand, analyze, evaluate, simulate, and control the systems under consideration. Traditionally, a resulting mathematical model is a description of a system in terms of equations. These equations are derived based on physical laws, such as Newton's laws, using continuous variables driven by time. *Discrete time models* are driven by clocks such that values of a variable at discrete times are observed, studied, analyzed, and used. The last type of models are *discrete event driven*. In other words, they represent the evolution of system states as driven by events that are often asynchronous. This asynchrony distinguishes the last class of models from the other two types. As a result, differential or difference equations cannot be used to describe them. We have to seek new models in order to well describe asynchronous event-driven dynamics. Additional features include concurrency, choices, and mutually exclusive use of shared resources. In reality, hybrid models may be used to describe the components/subsystems, interactions among components/subsystems, and the entire system.

In ancient times, the Chinese observed the season, sun, and moon changes and developed the lunar calendar, which has greatly helped farmers for more than 1,000 years. A lunar calendar can be viewed as a rough model of the seasonal changes over time in a year. It defines proper times for people to perform all farming activities to ensure a higher likelihood of a successful crop harvest.

A typical electronic circuit consists of resistors, capacitors, inductors, and sources. Each element is characterized by a certain law. For example, a linear resistor satisfies Ohm's law that the voltage across its two ends is proportional to the current flowing through it. A linear capacitor can be described by a model in which the current through it is proportional to the first derivative of the voltage across its two ends. For an electronic circuit containing such elements, one needs to use interconnection laws to characterize the entire circuit, for example, Kirchhoff's voltage law and current law. Modeling such a system takes two steps: First, we build a model for each individual element. Next, we use interconnection laws to build the entire system model. This approach certainly works for discrete event system models as well, which will be discussed in this book.

Another widely used modeling idea is to start and then keep adding to it to make it more complex. Initially, one may ignore most details by just focusing on major aspects of the system. Then, step by step, we add more features to the model to describe more details of a system. This can help one identify some important problems at an early stage without wasting too much time in unnecessary details. Only after the design satisfies the desired properties can one continue with more fine details to make the design better. This modeling strategy will be used when we build Petri net models for the system considered in this book.

In most engineering design, an accurate mathematical model becomes indispensable for designers. It is indeed the first step toward the development of any efficient and sustainable man-made system. This book contributes to the field of discrete event systems by presenting their modeling and analysis using a novel Petri net model.

1.2   AUTOMATED MANUFACTURING SYSTEMS

Conventional manufacturing has such features as mass production, sales from stock, long pipeline, and cost of inventory. Modern manufacturing calls for mass customization or eventually one-of-a-kind production. For example, a computer company now allows customers to select from a range of possibilities to specify their own computer and collects the payment from the customers up front. This minimizes a company's risk in overproducing unwanted or obsolete computers. As a sports wear company, Adidas has a project called miAdidas. It uses laser scanning techniques to create a three-dimensional model of a customer's feet. The model is then used to produce shoes that fit the customer perfectly. It is expensive and used primarily by professionals, but it may become common for average consumers after such value-added service becomes less expensive. For example, if we could use a webcam, available in many personal computers, and related software to catch a precise three-dimensional model, we could then lower the cost to produce such highly individualized products.

As a result, automated manufacturing systems must be designed and developed with new characteristics, i.e., flexibility, agility, and reconfigurability, in mind. Gaining these characteristics requires reconfigurable fixtures, material handling devices and systems, storage space, tools and machines, and more importantly, the use of advanced computer, communication, and management technologies to design reconfigurable control systems. A typical automated manufacturing system (AMS) consists of the following hardware entities:

1. Programmable computer-numerically-controlled components such as machines.
2. Automated material handling system that allows parts to flow freely from one station to another. Programmable robots and automated guided vehicles are often required.
3. Automated storage and retrieval system and buffering spaces where raw, intermediate, and final pieces can be stored until required for further processing.
4. A supervisory control system to monitor, coordinate, and control all the involved entities and release jobs into such an AMS. It has to ensure that the system is deadlock-free and can achieve its highest productivity and best product quality.

Contemporary manufacturing systems have had different goals over the past several decades. These goals are reflected through programmability, flexibility, agility, and reconfigurability, as described below:

1. **Programmability:** The early automated machines were implemented by rigid electromechanical devices, which resulted in so-called wired solutions. Such solutions suffered from many limitations, e.g., bulky size, rigidity in terms of any changes, low scalability, difficulty to debug and maintain, and inability to handle complex functions. Around 1970 the automobile industry realized their limitations and hence developed and adopted programmable logic controllers (PLCs) for their production lines, to keep pace with technical evolution and a growing number of new car models demanded by customers. The invention and use of PLCs greatly moved forward the automation technology and made the control system of a manufacturing system programmable. A PLC can be viewed as a special computing system. Its functionality has been vastly expanded since its early models. Today's PLCs have much computing and communication power, allowing them to accomplish not only control tasks but also communicating tasks with others and the Internet. This further allows engineers to develop remote diagnosis and maintenance capability for automated manufacturing facilities.

2. **Flexibility:** With the advent and wide use of PLCs and computers in control systems, flexible manufacturing is a current reality; its origins can be traced back to the 1960s with the Ingersoll-Rand factory in Roanoke, Virginia. The concept of manufacturing flexibility refers to manufacturing system designs that can adapt when external (likely uncertain) changes occur. According to Browne et al. (1984) and Sethi and Sethi (1990), we have the following eleven types of flexibility:

   - **Machine flexibility:** The different operation types that a machine can perform. It can be partly measured by the number of tool types.
   - **Material handling flexibility:** The ability to move the raw material pieces, parts, and products within a manufacturing facility.
   - **Operation flexibility:** The ability to produce a product in different ways.
   - **Process flexibility:** The set of products that the system can produce.
   - **Product flexibility:** The ability to add new products in the system.
   - **Routing flexibility:** The different routes (through machines and workshops) that can be used to produce a product in the system.
   - **Volume flexibility:** The ease with which the output of an existing system can be profitably increased or decreased.
   - **Expansion flexibility:** The ability to build out the capacity of a system.
   - **Program flexibility:** The ability to run a system automatically.
   - **Production flexibility:** The number of products a system currently can produce.
   - **Market flexibility:** The ability of the system to adapt to market demands.

3. **Agility:** Agile manufacturing was initially proposed to deal with the production of military products, as different technologies often led to very different products. Consequently, the agility to manufacture products from one generation to the next is highly desired. It requires both flexible manufacturing capability and the capability to respond to the rapid changes of customer needs and market demands. For example, Motorola developed an automated factory with the ability to produce physically different pagers on the same production line. Panasonic can manufacture different bicycles from combinations of a group of core parts. Agile manufacturing has been seen as the next step after Lean manufacturing in the evolution of production paradigms. The former is like an athletic person, and the latter a thin one. Thus, agile manufacturing is beneficial if the customer order cycle (the time the customers are willing to wait) is short. Lean manufacturing becomes possible if the supplier has a short lead time. Goldman et al. (1995) suggest that agility has four underlying components:

- Delivering value to the customer
- Being ready for rapid change
- Valuing human knowledge and skills
- Forming virtual partnerships

4. **Reconfigurability:** Reconfigurable manufacturing has become famous due to the a vast amount of work performed by the Engineering Research Center for Reconfigurable Manufacturing Systems (RMSs) at the University of Michigan College of Engineering, which is sponsored by the National Science Foundation and many manufacturing companies, especially those in the automobile industry. Reconfigurable manufacturing systems (Koren et al., 1999; Mehrabi et al., 2000) have the following single goal statement: *exactly the capacity and functionality needed, exactly when needed.* They must be designed at the outset for rapid change in their structure, as well as in their hardware and software components, in order to quickly adjust their production capacity and functionality within a part family in response to sudden market changes or intrinsic system changes. Their characteristics include modularity, integrability, customized flexibility, scalability, convertibility, and diagnosability.

- **Modularity** is the degree to which a system is modularized, e.g., machines, tools, control systems, and material handling systems. Modular components can be replaced or upgraded to better suit new applications.
- **Integrability** is the ability to integrate modules by a set of mechanical, informational, and control interfaces that enable integration and communication. The machine modules are integrated via material handling systems (such as conveyors and gantry robots), and their controllers are integrated into a factory control system.
- **Customized flexibility** is the ability of a system to produce a product or part family, fulfilling the mass customization or one-of-a-kind

production. This characteristic distinguishes reconfigurable manufacturing systems from flexible manufacturing systems (FMS) and allows a reduction in investment cost. Examples of product families are all types of Boeing 747 and all sizes of Adidas tennis shoes. A product family is defined as all products that have similar geometric features and shapes, have the same level of tolerances, require the same processes, and are within the same range of cost.

- **Scalability** is the ability to scale up the system size in terms of the number of machines, robots, and other resources or their processing capabilities in order to produce a larger quantity of products. Scalability is achieved by the additional resources or the improvement of the modules, e.g., conveyor's speed, motor speed, or better tools.

- **Convertibility** is the ability to transform the functionality of existing systems, machines, and controls to suit new production requirements. For example, RMS can switch the production between two members of a product family.

- **Diagnosability** is the ability to identify and use the current and past states of a system for detecting and diagnosing the root cause of output product defects and, subsequently, correcting operational defects quickly. Higher diagnosability means higher capability of detecting machine failure and unacceptable part quality. It requires reconfigurable product quality measurement systems.

It should be noted that flexibility, agility, and reconfigurability share some significant characteristics, such as programmability and modularity. They all emphasize the system's ability to respond to market and demand changes. Their differences, in fact, seem smaller than these terms imply. Consider RMS and FMS. FMS aims at increasing the variety of parts produced. RMS aims at increasing the speed of responsiveness to markets and customers. RMS requires only limited flexibility that is confined only to what is necessary to produce a product family. It is not the general flexibility that FMS offers. RMS tends to offer rapid scalability to the desired volume and convertibility, which are obtained within reasonable cost to manufacturers. Similar goals can be achieved by FMS with its high volume flexibility and expansion flexibility. The best application of an FMS is found in manufacturing small sets of products. With RMS, manufacturers can vary their production volume from small to large for a product family. RMS has introduced a new dimension, i.e., diagnosability, which becomes critically important.

RMS and FMS can only be implemented through a programmable control system, which has to rapidly configure itself when needed. Their design requires a powerful modeling tool that can rigorously reveal their intrinsic discrete event characteristics, such as sequential, concurrent, and mutually exclusive relations among activities, potential conflicts, and deadlock states. This monograph offers a resource-oriented Petri net framework to facilitate the modeling, deadlock analysis and control, and scheduling for them.

1.3   HISTORICAL PERSPECTIVE OF PETRI NETS IN AUTOMATION

A **Petri net** is a graphical tool invented by Carl Adam Petri. Its origins can be traced back to August 1939 when, at the age of 13, Petri created the graphics to describe chemical processes that produced a final compound from various elements through some intermediate compounds. The net-like representations were formalized in his doctoral thesis, "Communication with Automata," at the Technical University of Darmstadt, Germany, in 1962. The rules for transition enabling and firing, also called a token game, were defined. The algebraic aspect of distributed systems was described in detail. His thesis argued that the theory of automata had to be replaced by a new theory that respected the results of modern physics, e.g., the relativity theory and uncertainty principle. The new models were applicable to distributed systems. Thereafter, Petri and his collaborators published a number of papers applying their nets to such areas as economics, mechanics, computer science, logic, organization, biology, and telecommunication protocols, with a purpose to create a net tool for interdisciplinary transfer of structural knowledge. General net theory has been their primary research focus.

Petri net theory and applications were greatly advanced by the Computation Structure Group at MIT in the early 1970s. The first conference related to Petri nets was, in fact, held at MIT in 1975. A more sizable Petri net conference was held in Hamburg, Germany, in October 1979, and was attended by about 135 researchers, mostly from European countries. It included a 2-week advanced course on general net theory of processes and systems. The first Petri net conference proceeding was published by Springer-Verlag in 1980 and contained seventeen papers (Brauer, 1980). The first European Workshop on Applications and Theory of Petri Nets was held in Strasbourg, France, in 1980. Since then, every year there is such a Petri net conference, and selected papers are compiled and published normally in the following year, as edited volumes called *Advances in Petri Nets*. In 1981, Peterson published the first Petri net book, *Petri Net Theory and the Modeling of Systems*, by Prentice-Hall. It has greatly popularized Petri nets as a tool for the modeling of various systems. In particular, it has attracted many noncomputer scientists to consider Petri nets for their specific applications.

Dr. M. Silva of the University of Zaragoza, Spain, has led his group since the early 1980s and produced significant research results related to Petri nets and their applications to automation. His Spanish book *Petri Nets in Automation and Computer Engineering* was published by Editorial AC, Madrid, Spain, in 1985. In the same year, Dr. W. Reisig's 1982 book *Petrinetze*, in German, was translated and published as an English book by Springer-Verlag. *Petri Nets: An Introduction*, Dr. T. Murata's award-winning paper "Petri Nets: Properties, Analysis and Applications" was published in 1989. It is the most cited paper in the area of Petri nets according to SCOPUS—the most extensive database covering all engineering and science research papers in the world. The paper has well documented the key research results prior to 1989, including behavioral and structural properties, marked graphs, free-choice nets, analysis methods (reachability analysis, invariant analysis, and reduction), timed Petri nets, stochastic Petri nets, and high-level Petri nets (predicate/transition nets, colored Petri nets, and nets with individual tokens). Several influential Petri net tools were developed to help various researchers in the 1980s. The theories of generalized stochastic Petri nets (SPNs) and GreatSPN as a simulation and performance analysis tool were developed in the early 1980s by an Italian group (Ajmore Marsan et al., 1995). Stochastic Petri net package (SPNP), another stochastic timed Petri net tool for performance analysis, was generated by the group led by K. S. Trevidi with two key developers, G. Ciardo and J. Dugan, at Duke University. In the late 1980s, CPN was developed by a group led by K. Jensen, with some support from the U.S. government, and has been used to specify and simulate colored Petri nets for the design of complex systems.

The aforementioned publications and tools strongly affected academic researchers and industrial engineers when selecting a powerful model to deal with various issues in the design and implementation of more and more complex man-made systems, especially computer-integrated manufacturing systems. The traditional approaches based on finite state machine or automata were proved inadequate, since the state explosion problems would be met at the beginning of the system design. In addition, any design flaws or mistakes could invalidate the entire system design and the performance analysis results. Any system specification changes could require tremendous effort to modify the design. As result, it is extremely difficult, if not impossible, to use such traditional approaches to design modern manufacturing systems with high flexibility, agility, and reconfigurability.

In the mid-1980s, Prof. N. Viswanadham and Y. Narahari and their group in India made great contributions to the Petri net theory and applications to automated manufacturing systems. Their work dealt with performance modeling and analysis, bottom-up modeling methods, invariant analysis, and deadlock control. They detailed some of their Petri net work in their 1992 book, *Performance Modeling of Automated Manufacturing Systems*, published by Prentice Hall.

In the late 1980s, Dr. Frank DiCesare and Dr. Alan A. Desrochers established a sizable research group at Rensselaer Polytechnic Institute to tackle the design issues in computer-integrated manufacturing systems. They chose Petri nets as their major modeling tool and generated many interesting results in Petri net theory and applications to automated manufacturing. Their research was supported by such industrial firms as IBM, GM, Johnson & Johnson, Sun Microsystems, and Digital Equipment Corporation (now HP) via an 8-year-long Computer Integrated Manufacturing Research Program of the Center for Manufacturing Productivity and Technology Transfer at Rensselaer Polytechnic Institute. The following contributions by the Rensselaer group are summarized:

- Its first doctoral graduate was Dr. Robert Al-Jaar, whose 1989 dissertation was entitled "Performance Evaluation of Automated Manufacturing Systems Using Generalized Stochastic Petri Nets." He used the above-mentioned GreatSPN and SPNP tools to study the transfer lines with varying buffer sizes. He and his advisor, Dr. A. A. Desrochers, published a monograph, "Applications of Petri Nets in Manufacturing Systems: Modeling, Control and Performance," through IEEE Press in 1994.
- The second doctoral graduate was the second author of this book, MengChu Zhou, whose 1990 dissertation was called "A Theory for the Synthesis and Augmentation of Petri Nets in Automation." His dissertation work included

the development of new concepts, e.g., parallel and sequential mutual exclusions for shared resource modeling, formulation of top-down, bottom-up, and hybrid methodologies for net synthesis, Petri net augmentation and its applications in error recovery, Petri net modeling of buffers, and design of discrete event controllers for flexible manufacturing systems. With his advisor, Dr. DiCesare, he published the monograph *Petri Net Synthesis for Discrete Event Control of Manufacturing Systems* through Kluwer Academic Publisher in 1993. Two years later, he edited a volume, "Petri Nets in Flexible and Agile Automation," through the same publisher. In 1998, with his first doctoral graduate, Dr. K. Venkatesh, he published the book *Modeling, Simulation and Control of Flexible Manufacturing Systems: A Petri Net Approach* published by World Scientific.

- Dr. Fei-Yue Wang graduated with his doctoral dissertation, "A Coordination Theory of Intelligent Machines," supervised by Dr. G. N. Saridis, in 1990. He has applied Petri nets to design intelligent machines and has built an intelligent control foundation. He has also contributed to the Petri net modeling and analysis of a communication protocol for manufacturing message specification. He later made a great contribution to the development of modified reachability trees for liveness analysis of unbounded Petri nets.

- Dr. Desrochers' second doctoral graduate, Jagdish S. Joshi, completed his work in the Petri net area in 1991. Dr. Joshi's dissertation title was "Design and Performance Prediction of Computer Resources for Real-Time Computer Integrated Manufacturing Systems." His work dealt with the performance modeling and analysis of computer network and database transactions in a computer-integrated manufacturing environment.

- Dr. I. Koh, the next doctoral graduate of Dr. DiCesare, finished his dissertation, "A Transformation Theory for Petri Nets and Their Applications to Manufacturing Automation," in 1991. His work perfected the bottom-up approach to the synthesis of Petri nets with desired behavioral properties.

- Dr. MuDer Jeng was the third doctoral graduate of Dr. DiCesare in the area of Petri nets and completed his dissertation, "Theory and Applications of Resource Control Petri Nets for Automated Manufacturing Systems," in 1992. He was the primary inventor of a new class of Petri nets called extended resource control net (ERCN)-merged nets. They are still being cited and used by many researchers today. Dr. Jeng has continued his Petri net research. He has made many significant contributions in the areas of Petri net methods for deadlock analysis, discrete event control, scheduling of flexible manufacturing systems, analysis of unbounded Petri nets, and modeling, simulation, and scheduling of semiconductor manufacturing systems.

- In 1992, Dr. DiCesare graduated Alessandro Giua, whose doctoral thesis was entitled "Petri Nets as Discrete Event Models for Supervisory Control." This work fully opened the door to the supervisory control study in a framework of Petri nets when most of supervisory control theory was based on automata. Dr. Giua made many theoretical contributions in the area, from Petri net language to general mutual exclusion constraints. His work has been widely cited and used by other researchers. He and his group have continued to make solid contributions to hybrid Petri nets, optimal control, and identification of discrete event systems. Their applications cover manufacturing, fault detection, and railway networks. He coauthored two research monographs in Italian in 2002 and 2005, respectively.

- Dr. Hauke J. Jungnitz completed his dissertation, "Approximation Methods for Stochastic Petri Nets," under the supervision of Dr. Desrochers in 1992. His work presented flow-equivalent nets for the performance analysis of large generalized stochastic Petri nets.

- In 1992, Dr. Desrochers graduated J. F. Watson III, whose doctoral thesis was entitled "Performance and State-Space Analyses of Systems Using Petri Nets." His work proposed several approximation models for those transitions with nonexponentially distributed delay functions, and thus SPNP software could be used to evaluate the system performance. He also invented a bottom-up methodology to estimate the state-space size of Petri nets. Consequently, one can determine the appropriateness of a particular analysis technique, evaluate the trade-off between model detail and solution complexity, and provide data for state-space size reduction algorithms.

- Dr. Doo Yong Lee graduated in 1993 with his dissertation, "Scheduling and Supervisory Control of Flexible Manufacturing Systems Using Petri Nets and Heuristic Search," under Dr. DiCesare's supervision. His work proposed to use heuristic search to perform the scheduling of flexible manufacturing systems in a framework of Petri nets. It has motivated many Petri net–based scheduling studies. His later work successfully applied Petri nets to deadlock analysis and scheduling of wafer production systems such as photolithography equipment and integrated cluster tools in semiconductor fabrication.

- Dr. Arthur C. Sanderson graduated Tiehua Cao, whose 1993 dissertation was entitled "Task Planning with Uncertainty for Robotic Systems." Their work combined fuzzy set theory and Petri nets and developed fuzzy Petri nets for intelligent task planning in robotic systems. The reasoning algorithms and their applications were presented. The research led to their coauthored 1996 book, *Intelligent Task Planning Using Petri Nets*, in the Intelligent Control and Intelligent Automation series of World Scientific Publisher.

- In 1993, Dr. Desrochers graduated Chengche Feng with his doctoral thesis "Sensitivity Analysis of Discrete Event Dynamic Systems by a Petri Net-Based Perturbation Method." Dr. Feng's work conducted the Petri net sensitivity analysis with respect to a structural change of discrete event dynamic systems.

- Dr. Jongwook Kim received his doctoral degree in 1995 with his dissertation "The Modeling, Analysis and Simulation of a Discrete Event Dynamic System Using Time Petri Net Models." He was also advised by Dr. Desrochers. His work found the eigenvalue bounds of a stochastic Petri net and proposed a stochastic Petri net synthesis method based on such knowledge. He applied Petri nets to semiconductor manufacturing plants, task planning, and project management.

Some good industrial applications in automation emerged in the late 1980s and 1990s. Some of them were compiled into the 1995 volume *Petri Nets in Flexible and Agile Automation*, edited by MengChu Zhou and published by Kluwer Academic Publisher. In the 1990s and thereafter, there were numerous groups who worked and made their contributions in the areas of Petri nets and their applications. Notably, Dr. Frank L. Lewis's group at the University of Texas–Arlington invented a matrix-based approach to the solution of many discrete event system problems. They were able to automatically generate a Petri net model and related design for analysis, control, and simulation of flexible manufacturing systems given the manufacturing specifications expressed in a bill of materials, assembly tree, task sequencing matrix, and resource requirement matrix. They also made in-depth contributions to deadlock resolution problems. Their work has been comprehensively presented in their 2006 book, *Manufacturing Systems Control Design: A Matrix Based Approach* (S. Bogdan, F. L. Lewis, Z. Kovacic, and J. Mireles (London: Springer-Verlag)).

Another group in the area of Petri nets and automation research was led by Prof. J.-M. Proth and X. Xie in France. They have been the primary contributors to the theory of event graphs (also called marked graphs—a subclass of Petri nets) and the application of Petri nets to management, planning, and scheduling of automated manufacturing systems. Some of their research appeared in their 1996 book entitled *Petri Nets: A Tool for Design and Management of Manufacturing Systems* (New York: John Wiley & Sons). Dr. Xie has also made great contributions to the mathematical programming approach to deadlock analysis and control problems in Petri nets and various applications to semiconductor manufacturing and the healthcare service industry.

It is clear that there have been many great studies in the areas of Petri nets and discrete event systems. We cannot enumerate them here. Authors can enter "Petri nets" as a search item in a database, e.g., the previously mentioned SCOPUS. Based on the snapshot of our March 7, 2009 research result (Figure 1.1), 11,655 papers that have "Petri net" as a term in their title, keywords, or abstract are published in journals and conference papers. It should be noted that SCOPUS does not include such publications as books, book chapters, theses, and technical reports.

In this figure, we also find the most productive researchers in the area of Petri nets in the world so far. The top ten are

1. MengChu Zhou (total 147, with 102 under Zhou, M., and 45 under Zhou, M. C.)
2. Alessandro Giua (72)
3. MuDer Jeng (total 68, listed evenly under Jeng, M., and Jeng, M. D.)
4. Chuan Lin (63)
5. Manuel Silva (58)
6. Kishor S. Trivedi (57)
7. Zhiwu Li (51)
8. Wil M.P. Van Der Aalst (46)
9. Jonathan Billington (46)
10. Tadao Murata (46)

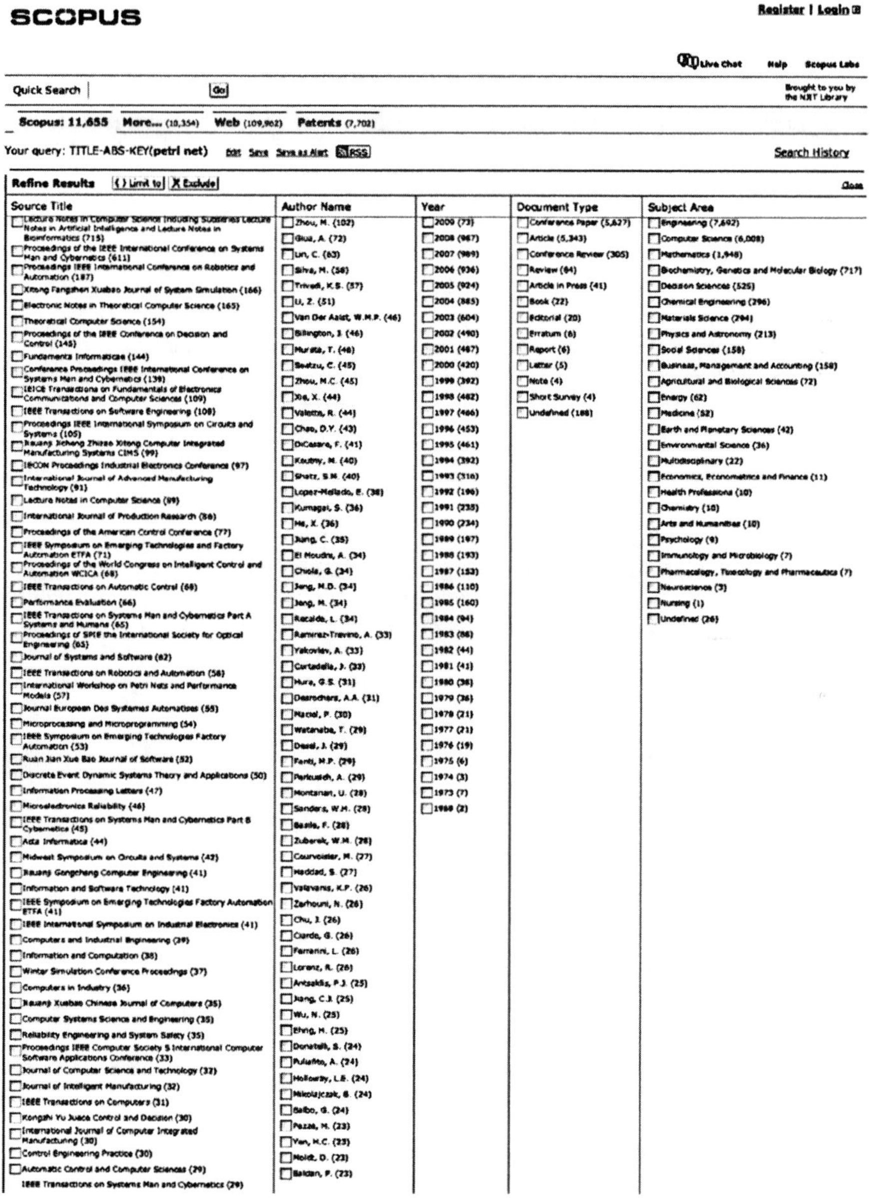

SCOPUS

Live chat   Help   Scopus Labs

Quick Search | [Go]   Brought to you by the NIT Library

Scopus: 11,655   More... (10,354)   Web (109,902)   Patents (7,702)

Your query: TITLE-ABS-KEY(petri net)   Edit Save Save as Alert RSS   Search History

Refine Results { } Limit to | X Exclude   Close

Source Title	Author Name	Year	Document Type	Subject Area
Lecture Notes in Computer Science Including Subseries Lecture Notes in Artificial Intelligence and Lecture Notes in Bioinformatics (715)	Zhou, M. (102)	2009 (73)	Conference Paper (5,827)	Engineering (7,692)
Proceedings of the IEEE International Conference on Systems Man and Cybernetics (611)	Giua, A. (72)	2008 (967)	Article (5,343)	Computer Science (6,008)
Proceedings IEEE International Conference on Robotics and Automation (187)	Lin, C. (63)	2007 (989)	Conference Review (305)	Mathematics (1,948)
Xitong Fangzhen Xuebao Journal of System Simulation (166)	Silva, M. (56)	2006 (936)	Review (84)	Biochemistry, Genetics and Molecular Biology (717)
Electronic Notes in Theoretical Computer Science (165)	Trivedi, K.S. (57)	2005 (924)	Article in Press (41)	Decision Sciences (525)
Theoretical Computer Science (154)	Li, Z. (51)	2004 (865)	Book (22)	Chemical Engineering (296)
Proceedings of the IEEE Conference on Decision and Control (145)	Van Der Aalst, W.M.P. (46)	2003 (604)	Editorial (20)	Materials Science (294)
Fundamenta Informaticae (144)	Billington, J. (46)	2002 (490)	Erratum (6)	Physics and Astronomy (213)
Conference Proceedings IEEE International Conference on Systems Man and Cybernetics (139)	Murata, T. (46)	2001 (467)	Report (6)	Social Sciences (158)
IEICE Transactions on Fundamentals of Electronics Communications and Computer Sciences (109)	Seatzu, C. (45)	2000 (420)	Letter (5)	Business, Management and Accounting (158)
IEEE Transactions on Software Engineering (108)	Zhou, M.C. (45)	1999 (392)	Note (4)	Agricultural and Biological Sciences (72)
Proceedings IEEE International Symposium on Circuits and Systems (105)	Xie, X. (44)	1998 (462)	Short Survey (4)	Energy (62)
Jisuanji Jicheng Zhizao Xitong Computer Integrated Manufacturing Systems CIMS (99)	Valette, R. (44)	1997 (466)	Undefined (168)	Medicine (52)
IECON Proceedings Industrial Electronics Conference (97)	Chen, D.Y. (43)	1996 (453)		Earth and Planetary Sciences (42)
International Journal of Advanced Manufacturing Technology (91)	DiCesare, F. (41)	1995 (461)		Environmental Science (36)
Lecture Notes in Computer Science (89)	Koutny, M. (40)	1994 (392)		Multidisciplinary (22)
International Journal of Production Research (86)	Shatz, S.M. (40)	1993 (316)		Economics, Econometrics and Finance (11)
Proceedings of the American Control Conference (77)	Lopez-Mellado, E. (38)	1992 (196)		Health Professions (10)
IEEE Symposium on Emerging Technologies and Factory Automation ETFA (71)	Kumagai, S. (36)	1991 (235)		Chemistry (10)
Proceedings of the World Congress on Intelligent Control and Automation WCICA (71)	He, X. (36)	1990 (234)		Arts and Humanities (10)
IEEE Transactions on Automatic Control (68)	Jiang, C. (35)	1989 (197)		Psychology (9)
Performance Evaluation (66)	El Moudni, A. (34)	1988 (193)		Immunology and Microbiology (7)
IEEE Transactions on Systems Man and Cybernetics Part A Systems and Humans (65)	Chiola, G. (34)	1987 (153)		Pharmacology, Toxicology and Pharmaceutics (7)
Proceedings of SPIE the International Society for Optical Engineering (65)	Jeng, M.D. (34)	1986 (110)		Neuroscience (3)
Journal of Systems and Software (62)	Jeng, M. (34)	1985 (160)		Nursing (1)
IEEE Transactions on Robotics and Automation (58)	Recalde, L. (34)	1984 (94)		Undefined (26)
International Workshop on Petri Nets and Performance Models (57)	Ramirez-Trevino, A. (33)	1983 (86)		
Journal European Des Systemes Automatises (55)	Yakovlev, A. (33)	1982 (44)		
Microprocessing and Microprogramming (54)	Cortadella, J. (33)	1981 (41)		
IEEE Symposium on Emerging Technologies Factory Automation (53)	Mura, G.S. (31)	1980 (36)		
Ruan Jian Xue Bao Journal of Software (52)	Desrochers, A.A. (31)	1979 (36)		
Discrete Event Dynamic Systems Theory and Applications (50)	Racal, P. (30)	1978 (21)		
Information Processing Letters (47)	Watanabe, T. (29)	1977 (21)		
Microelectronics Reliability (46)	Desel, J. (29)	1976 (19)		
IEEE Transactions on Systems Man and Cybernetics Part B Cybernetics (45)	Fanti, M.P. (29)	1975 (6)		
Acta Informatica (44)	Perkusich, A. (29)	1974 (4)		
Midwest Symposium on Circuits and Systems (43)	Montanari, U. (29)	1973 (7)		
Jisuanji Gongcheng Computer Engineering (41)	Sanders, W.H. (28)	1968 (2)		
Information and Software Technology (41)	Basile, F. (28)			
IEEE Symposium on Emerging Technologies Factory Automation ETFA (41)	Zuberek, W.M. (28)			
IEEE International Symposium on Industrial Electronics (41)	Courvoisier, M. (27)			
Computers and Industrial Engineering (39)	Haddad, S. (27)			
Information and Computation (38)	Valavanis, K.P. (26)			
Winter Simulation Conference Proceedings (37)	Zarhouni, N. (26)			
Computers in Industry (36)	Chu, J. (26)			
Jisuanji Xuebao Chinese Journal of Computers (35)	Ciardo, G. (26)			
Computer Systems Science and Engineering (35)	Ferranti, L. (26)			
Reliability Engineering and System Safety (35)	Lorenz, R. (26)			
Proceedings IEEE Computer Society's International Computer Software Applications Conference (33)	Antsaklis, P.J. (25)			
Journal of Computer Science and Technology (32)	Jiang, C.J. (25)			
Journal of Intelligent Manufacturing (32)	Wu, N. (25)			
IEEE Transactions on Computers (31)	Ehrig, H. (25)			
Kongzhi Yu Juece Control and Decision (30)	Donatelli, S. (24)			
International Journal of Computer Integrated Manufacturing (30)	Pulaflto, A. (24)			
Control Engineering Practice (30)	Holloway, L.E. (24)			
Automatic Control and Computer Sciences (29)	Nikolajzak, B. (24)			
IEEE Transactions on Systems Man and Cybernetics (29)	Balbo, G. (24)			
	Pezze, M. (23)			
	Yen, H.C. (23)			
	Moldt, D. (23)			
	Baldan, P. (23)			

FIGURE 1.1 The SCOPUS search results.

As an interesting note, the top three were Dr. Frank DiCesare's students and received their doctoral degrees from the Electrical, Computer, and Systems Engineering Department, Rensselaer Polytechnic Institute, in the early 1990s. Naiqi Wu ranks in the top fifty.

In addition, we find that the top three sources of the papers collected by SCOPUS are

1. Lecture Notes in Computer Science, including subseries Lecture Notes in Artificial Intelligence and Lecture Notes in Bioinformatics (715)
2. Proceedings of the IEEE International Conference on Systems, Man, and Cybernetics (611)
3. Proceedings of the IEEE International Conference on Robotics and Automation (187)

From Figure 1.2, we can see the growth of the number of Petri net papers. Years 2007 and 2008 have 989 and 987 papers, respectively. Year 1968 saw two papers, and none were found in the years 1969 to 1972 in SCOPUS.

Another source of Petri net activities is Petri Nets World. It aims to provide a variety of online services for the international Petri nets community. The services mainly include information on the International Conferences on Application and Theory of Petri Nets, mailing lists, bibliographies, tool databases, newsletters, and addresses. The Petri Nets Steering Committee supervises these activities, and the site is maintained by the TGI group at the University of Hamburg, Germany. One can freely join its mailing list. Its website is http://www.informatik.uni-hamburg.de/TGI/PetriNets/.

1.4 SCOPE AND OBJECTIVES

The scope of this book is confined to Petri nets—in particular, the class of resource-oriented Petri nets. To distinguish different jobs, colors are introduced, thereby resulting in colored resource-oriented Petri nets. We focus on their modeling applications

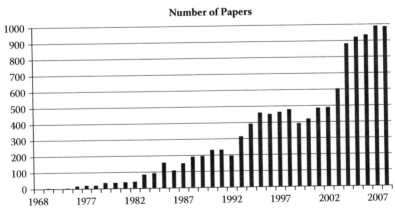

FIGURE 1.2 The number of Petri net papers published each year over the past 40 years.

to manufacturing automation. This monograph intends to achieve the following objectives:

1. Present the definitions of Petri nets, colored Petri nets, and their properties related to manufacturing automation.
2. Present both process- and resource-oriented Petri nets and modeling methodologies and their modeling examples.
3. Present a comparative study between process- and resource-oriented Petri nets to illustrate their pros and cons.
4. Establish the necessary and sufficient conditions of deadlock in the framework of resource-oriented Petri nets.
5. Derive a deadlock control policy that can achieve better throughput and utilization by avoiding congestion arising under a control policy in an expert-rule-based scheduling environment.
6. Develop a resource-oriented Petri net model for automated guided vehicle (AGV) systems, a deadlock-free control policy, and shortest routing methodology.
7. Develop an efficient deadlock avoidance policy for automated manufacturing systems with multiple AGVs, where multiple AGVs can be effectively used to help resolve the deadlocks that would be inevitable in part production processes without them.
8. Develop a similar deadlock avoidance policy for automated manufacturing systems with multiple robots.
9. Develop resource-oriented Petri net models for deadlock control and scheduling of single- and dual-robot cluster tools and track systems that are core pieces of equipment in today's semiconductor manufacturing processes.
10. Develop resource-oriented Petri net models and related deadlock resolution methods for automated flexible assembly systems—one of the hardest systems in terms of deadlock resolution.

1.5 SUMMARY

State machines have been among the most important models that can describe discrete-event-driven dynamics. They have been widely used in academia and industry. However, they suffer from many problems when we design automated complex distributed systems. Since they describe a system at the level of states, any minor changes lead to a change in their model, often in a drastic way. Thus, state-machine-based approaches cannot deal well with those systems subject to frequent changes, and thus cannot achieve high system flexibility and agility as required by today's customers. Another major issue that such approaches face is state explosion. The increasing problem size can easily enlarge the solution to an unmanageable scale. Furthermore, any design flaws or unconsidered cases may invalidate the entire system design and often requires extremely time-consuming recovery.

This book introduces Petri nets as a powerful modeling tool for discrete event systems that are often distributed and subject to changes, in order to ensure the high-level system flexibility, agility, and reconfigurability. In particular, we confine ourselves to resource-oriented models that have been proved to be highly compact and useful in deadlock analysis, control, and scheduling.

We explain flexible manufacturing, agile manufacturing, and reconfigurable manufacturing and the need for a powerful modeling tool for discrete event systems. We also offer some interesting perspectives about Petri nets and their evolution in the application field of automation.

REFERENCES

Ajmore Marsan, M., G. Balbo, G. Conte, S. Donatelli, and G. Franceschinis. 1995. *Modelling with generalized stochastic Petri nets*. New York: John Wiley & Sons.

Brauer, W., ed. 1980. *Net theory and applications*. Lecture Notes in Computer Science (LNCS), vol. 84. New York: Springer-Verlag. Out of print.

Browne J., D. Dubois, K. Rathmill, S. P. Sethi, and K. E. Stecke. 1984. Classification of flexible manufacturing systems. *The FMS Magazine* 2:114–17.

Goldman, L., R. L. Nagel, and K. Preiss. 1995. *Agile competitors and virtual organizations— Strategies for enriching the customer*. New York: Van Nostrand Reinhold.

Koren, Y., F. Jovane, U. Heisel, T. Moriwaki, G. Pritschow, G. Ulsoy, and H. VanBrussel. 1999. Reconfigurable manufacturing systems, a keynote paper. *CIRP Annals* 48:6–12.

Mehrabi, M., G. Ulsoy, and Y. Koren. 2000. Reconfigurable manufacturing systems: Key to future manufacturing. *Journal of Intelligent Manufacturing* 11:403–19.

Murata, T. 1989. Petri nets: Properties, analysis and applications. *Proceedings of the IEEE* 77:541–80.

Peterson, J. L. 1981. *Petri net theory and the modeling of systems*. Englewood Cliffs, NJ: Prentice-Hall. Out of print.

Proth, J.-M., and X. Xie. 1996. *Petri nets: A tool for design and management of manufacturing systems*. New York: John Wiley & Sons.

Sethi, A. K., and S. P. Sethi. 1990. Flexibility in manufacturing: A survey. *International Journal of Flexible Manufacturing Systems* 2:289–328.

Zhou, M. C., and F. DiCesare. 1993. *Petri net synthesis for discrete event control of manufacturing systems*. Boston: Kluwer Academic Publications.

Zhou, M. C., and K. Venkatesh. 1998. *Modeling, simulation and control of flexible manufacturing systems: Petri net approach*. Singapore: World Scientific.

2 Petri Nets
Basic Concept

This chapter defines Petri net (PN) and introduces its basic concepts and properties that are used in this book. For details, see Murata (1989), Zhou and DiCesare (1993), Zhou and Venkatesh (1998), and Proth and Xie (1996).

2.1 BASIC CONCEPTS

2.1.1 DEFINITION

A Petri net is a particular kind of directed bipartite graph (or digraph) together with an initial state called the initial marking. It contains two types of nodes, places, and transitions. In graphical representation, places are drawn as circles, and transitions as bars or boxes. Arcs are either from a place to a transition or from a transition to a place. Arcs are labeled with their weights (positive integers). Labels for unity weight are usually omitted. A marking (state) assigns to each place a nonnegative integer k; we say that p is marked with k tokens. Pictorially, we place k black dots (tokens) in place p. If k is large, one can simply write the number k inside p to represent k tokens. A marking is denoted by M, an m-vector, where m is the total number of places. The number of tokens in p is denoted by $M(p)$.

As an illustration, consider the PN shown in Figure 2.1a. It contains three places p_{1-3} and a transition t. The initial marking $M_0 = (2, 2, 0)^T$. The weights for arcs (p_1, t), (p_2, t), and (t, p_3) are 2, 1, and 2, respectively.

Definition 2.1: A PN is a five-tuple PN = (P, T, I, O, M_0), where

1. $P = \{p_1, p_2, \ldots, p_m\}$ is a finite set of places.
2. $T = \{t_1, t_2, \ldots, t_n\}$ is a finite set of transitions, $P \cup T \neq \emptyset$, $P \cap T = \emptyset$.
3. $I: P \times T \to N$ is an input function that defines the directed arcs from places to transitions, where $N = \{0, 1, 2, \ldots\}$.
4. $O: P \times T \to N$ is an output function that defines the directed arcs from transitions to places.
5. $M: P \to N$ is a marking representing the numbers of tokens in places with M_0 denoting the initial marking.

In modeling, often places represent conditions and transitions represent events. A transition (event) has a certain number of input and output places representing the preconditions and postconditions, respectively. The presence of a token in a place is interpreted as holding the truth of the condition associated with the place. Some typical interpretations of transitions and their input and output places are given in Table 2.1.

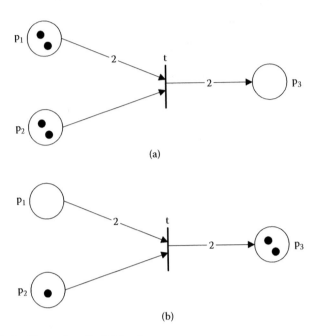

FIGURE 2.1 An illustration of a Petri net.

2.1.2 ENABLING AND FIRING RULES

Definition 2.2: A PN is said to be an ordinary PN if for any arc in the net its weight is 1.

The *preset* of transition t is the set of all input places to t, i.e., $^\bullet t = \{p: p \in P$ and $I(p, t) > 0\}$. The *postset* of t is the set of all output places from t, i.e., $t^\bullet = \{p: p \in P$ and $O(p, t) > 0\}$. Similarly, p's preset $^\bullet p = \{t \in T: O(p, t) > 0\}$ and postset $p^\bullet = \{t \in T: I(p, t) > 0\}$. Given a set $S \subset P \cup T$, S's preset and postset are defined as $^\bullet S = \cup_{x \in S}{}^\bullet x$ and $S^\bullet = \cup_{x \in S}x^\bullet$, respectively.

Definition 2.3: A transition $t \in T$ in PN is enabled in marking M if for all $p \in {}^\bullet t$,

$$M(p) \geq I(p, t) \tag{2.1}$$

TABLE 2.1
Some Typical Interpretations of Transitions and Places

Input Places	Transition	Output Places
Preconditions	Event	Postconditions
Input data	Computation step	Output data
Input signals	Signal processor	Processed results
Resource needed	Task or job	Resource released
Conditions	Clause in logic	
Buffers		

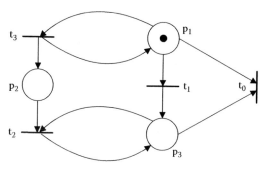

FIGURE 2.2 A Petri net.

If a transition is enabled, it can fire. Firing an enabled transition t in marking M yields

$$M' = (p) \begin{cases} M(p) - I(p, t), \ p \in {}^\bullet t \\ M(p) + O(p, t), \ p \in t^\bullet \\ M(p), \text{ otherwise} \end{cases} \tag{2.2}$$

For example, transition t in Figure 2.1a is enabled; after firing t, two tokens are taken away from p_1, and one from p_2, and at the same time, two tokens are added into p_3. Hence, $M' = (0, 1, 2)^T$, as shown in Figure 2.1b. If firing t changes marking M_1 to M_2, it can be denoted as $M_1[t > M_2$. Similarly, after firing a sequence of transitions $\sigma = t_1 t_2 \ldots t_k$, the marking is changed from M_1 to M_2. This fact can be denoted as $M_1[\sigma > M_2$. The set of all markings reachable from M_0 is denoted by $R(M_0)$.

Consider the Petri net shown in Figure 2.2. We have $M_0 = (1, 0, 0)^T$, $M_0[t_1 > M_1$, and $M_1 = (0, 0, 1)^T$. At M_1, no transition is enabled. $M_0[\sigma_1 > M_2$, where $\sigma_1 = t_3 t_3 \ldots t_3$, i.e., firing α times of t_3, and $M_2 = (1, \alpha, 0)^T$. $M_2[t_1 > M_3$, and $M_3 = (0, \alpha, 1)^T$. $M_3[\sigma_2 > M_4$, $\sigma_2 = t_2 t_2 \ldots t_2$, i.e., firing β times of t_2, and $M_4 = (0, \alpha\text{-}\beta, 1)^T$. Because α, β, and $\alpha\text{-}\beta$ can be arbitrary integers with constraint $\beta \le \alpha$, M_2, M_3, and M_4 can be denoted as $M_2 = (1, \omega, 0)^T$, $M_3 = M_4 = (0, \omega, 1)^T$. Thus, for this PN, $R(M_0) = \{M_0, M_1, M_2, M_3\}$ and the relation is shown in Figure 2.3. Note that t_0 is never enabled starting at M_0.

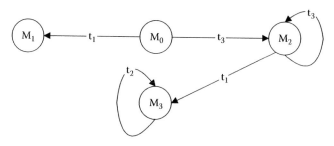

FIGURE 2.3 The reachable marking relation of a PN.

2.1.3 FINITE CAPACITY PN

For Definition 2.3, it is assumed that each place can hold an unlimited number of tokens. Such assumption may not be true for some physical systems. It is natural to set an upper limit to the number of tokens that each place can hold. Such a PN is referred to as *finite capacity* PN. $K(p)$ is used to denote the maximal number of tokens that p can hold at a time. The previously defined PN is sometimes called infinite capacity PN. We have the following enabling and firing rule for finite capacity PN.

Definition 2.4: A transition $t \in T$ in a finite capacity PN is enabled in marking M if for all $p \in P$,

$$M(p) \geq I(p, t) \tag{2.3}$$

and

$$K(p) \geq M(p) - I(p, t) + O(p, t) \tag{2.4}$$

If a transition is enabled, it can fire. Firing an enabled transition t in marking M yields for all $p \in P$

$$M'(p) = M(p) - I(p, t) + O(p, t) \tag{2.5}$$

2.1.4 SOME SPECIAL STRUCTURES IN PN

Some special structures in a PN affect the behavior of a PN. They are source places and transitions, sink places and transitions, p-path, elementary circuits, self-loops, siphons, and traps.

Definition 2.5: A place (transition) in a PN is said to be a source place (transition) if it has no input transitions (places), i.e., $^\bullet p = \varnothing$ ($^\bullet t = \varnothing$). A place (transition) in a PN is said to be a sink place (transition) if it has no output transitions (places), i.e., $p^\bullet = \varnothing$ ($t^\bullet = \varnothing$).

In Figure 2.1, both p_1 and p_2 are source places, but p_3 is a sink place. Transition t_1 in Figure 2.4 is a source transition, but t_3 is a sink transition.

Definition 2.6: A subnet in a PN is called a p-path if it contains $k \geq 2$ places, starts with p_1 and ends with p_k, and there is a transition t_i connecting p_i and p_{i+1}, $i = 1, \ldots,$ $k - 1$, such that $^\bullet t_i = \{p_i\}$ and $t_i^\bullet = \{p_{i+1}\}$. The path capacity is $\sum_{j=1}^{k} K(p_j)$.

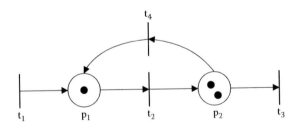

FIGURE 2.4 PN with source and sink transitions.

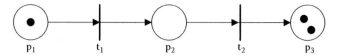

FIGURE 2.5 A p-path.

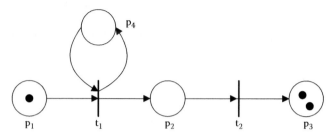

FIGURE 2.6 A PN contains a self-loop.

Sometimes a p-path can be merged to a single place in applications. A p-path with three places is shown in Figure 2.5.

Definition 2.7: (p, t) in a PN is said to be a self-loop if $p \in {}^\bullet t$ and $p \in t^\bullet$, or $t \in {}^\bullet p$ and $t \in p^\bullet$.

In other words, p is an input and output place of t. (p_4, t_1) in Figure 2.6 is a self-loop. Later we can see that if certain processes and their resources are modeled by self-loops, some undesired behavior can be avoided just by modeling.

Definition 2.8: A circuit in a PN is said to be elementary if it goes from one node, through a series of nodes, back to this node, such that no node is repeated.

In Figure 2.7, circuits $\{p_1, t_2, p_2, t_1, p_1\}$ and $\{p_2, t_4, p_3, t_3, p_2\}$ are two elementary ones, but $\{p_1, t_2, p_2, t_4, p_3, t_3, p_2, t_1, p_1\}$ is not, for p_2 is repeated.

Definition 2.9: A PN is said to be strongly connected if from any node there exists a directed path that leads to any other node in the net.

A strongly connected PN implies no existence of source place and transition, and sink place and transition in it. For example, the PN shown in Figure 2.7 is a strongly connected one.

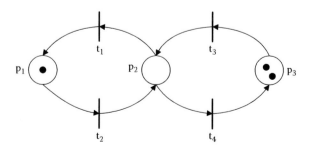

FIGURE 2.7 A PN contains circuits.

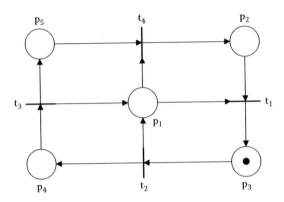

FIGURE 2.8 A PN contains siphons.

Definition 2.10: A set of places S in a PN is called a *siphon* if $^\bullet S \subseteq S^\bullet$, i.e., any input transition of S is also an output transition of S. A siphon is said to be minimal if it does not contain other siphons. It is called a trap if $S^\bullet \subseteq {^\bullet S}$.

In the PN shown in Figure 2.8, there are four siphons: $S_1 = \{p_2 - p_5\}$, $S_2 = \{p_1, p_3, p_4\}$, $S_3 = \{p_1, p_3, p_4, p_5\}$, and $S_4 = \{p_1 - p_5\}$. Among them, S_1 and S_2 are minimal ones. A siphon may make a PN nonlive. A siphon may be marked at initial marking M_0, but may be emptied after firing sequences of transitions. In Figure 2.8, $\{p_2 - p_5\}$, $\{p_1 - p_3\}$, $\{p_1 - p_4\}$, and $\{p_1 - p_5\}$ are traps.

2.2 SUBCLASS OF PN

Until now, there was no efficient technique for the analysis of a general PN. However, such techniques are available to analyze a subset of PN. In this section, such subclasses of PN are introduced.

A state machine (SM) is an ordinary PN such that each transition t has exactly one input place and one output place, i.e.,

$$|{^\bullet t}| = |t^\bullet| = 1 \text{ for all } t \in T$$

A marked graph (MG) is an ordinary PN such that each place p has exactly one input transition and one output transition, i.e.,

$$|{^\bullet p}| = |p^\bullet| = 1 \text{ for all } p \in P$$

A free-choice (FC) net is an ordinary PN such that every arc from a place is either a unique outgoing arc or a unique incoming arc to a transition, i.e.,

$$\text{for all } p \in P, |p^\bullet| \leq 1 \text{ or } {^\bullet(p^\bullet)} = \{p\}$$

equivalently, for all $p_1, p_2 \in P$, $p_1^\bullet \cap p_2^\bullet \neq \varnothing \Rightarrow |p_1^\bullet| = |p_2^\bullet| = 1$

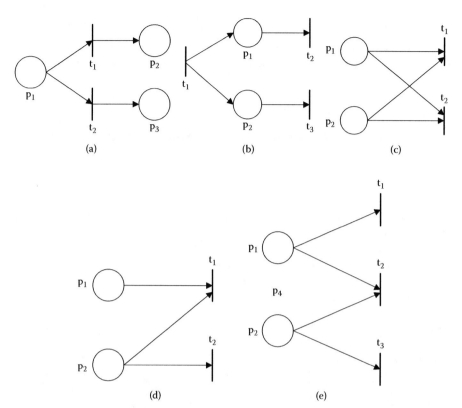

FIGURE 2.9 Key structures for subclasses of PN: (a) SM, $\overline{\text{MG}}$, (b) MG, $\overline{\text{SM}}$, (c) EFC, $\overline{\text{FC}}$, (d) AC, $\overline{\text{EFC}}$, and (e) PN, $\overline{\text{AC}}$.

An extend free-choice (EFC) net is an ordinary PN such that

$$p_1{}^\bullet \cap p_2{}^\bullet \neq \varnothing \Rightarrow p_1{}^\bullet = p_2{}^\bullet \text{ for all } p_1{}^\bullet, p_2{}^\bullet \in P$$

An asymmetric choice (AC) net is an ordinary PN such that

$$p_1{}^\bullet \cap p_2{}^\bullet \neq \varnothing \Rightarrow p_1{}^\bullet \subseteq p_2{}^\bullet \text{ or } p_1{}^\bullet \supseteq p_2{}^\bullet \text{ for all } p_1, p_2 \in P$$

The PN structures shown in Figure 2.9 are the keys for the subclasses. Those shown in Figure 2.9a and b are SM and MG, respectively. FCs are a generalization of the structures common to both SM and MG. They allow the conflict structure of SM and the synchronization structure of MG, but exclude the structure shown in Figure 2.9c. SM admits no synchronization and MG admits no conflicts. Thus, SM can describe the asynchronous and conflicting processes in manufacturing systems. MG is suitable for the deterministic, concurrent, and decision-free processes.

2.3 PROPERTIES

The properties in a PN are very important in system analysis and control. This section introduces the main properties in a PN.

2.3.1 REACHABILITY

Reachability is a fundamental basis for studying the dynamic properties of a system. The reachability problem for PN is to find if a given M_n is in $R(M_0)$. By solving the reachability problem, we can check if the desired markings are reachable and if there are deadlocks in the system.

A coverability tree method can be used to analyze reachability for PN. From M_0, we can obtain many new markings, and from each new marking, again reach more markings. This process results in a tree representation of markings. Nodes represent markings, and each arc represents a transition firing. The tree will grow infinitely large if the net is unbounded. To solve this problem, a special symbol, ω, is introduced. It can be thought of as infinity, with the properties that for each finite integer n, $\omega > n$, $\omega \pm n = \omega$, and $\omega \geq \omega$. A marking M_2 covers M_1 if and only if $\forall p \in P$, $M_2(p) \geq M_1(p)$ and $\exists p \in P \ni M_2(p) > M_1(p)$. During the evolution of the reachability tree, if M_2 covers M_1, we often use ω to replace $M_2(p)$, where $M_2(p) > M_1(p)$.

Consider the PN shown in Figure 2.2. Initially, $M_0 = (1, 0, 0)^T$ and t_1 and t_3 are enabled. Firing t_1 transforms M_0 to $M_1 = (0, 0, 1)^T$, which is a dead-end node. Firing t_3 at M_0 results in a marking $(1, 1, 0)^T$, which covers M_0. Therefore, the new marking is $M_2 = (1, \omega, 0)^T$. At M_2, t_1 and t_3 are again enabled. Firing t_1 leads to $M_3 = (0, \omega, 1)^T$, from which t_2 can fire, and it results in M_3. Firing t_3 at M_2 results in the old marking M_2. Thus, we have the coverability tree shown in Figure 2.10, and the reachable marking relation is shown in Figure 2.3. The coverability tree generation algorithm can be found in Murata (1989) and Zhou and Venkatesh (1998).

For bounded PN, the reachability problem is decidable, but the complexity is exponential for a general PN. However, for some special PN, the properties can be obtained based on the special structure of the PN studied.

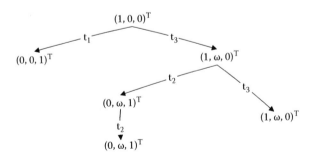

FIGURE 2.10 The coverability tree for the PN shown in Figure 2.2.

2.3.2 BOUNDEDNESS

A PN is said to be k-bounded or simply bounded if the number of tokens in each place does not exceed a finite number k for any marking reachable from M_0. It is said to be safe if it is 1-bounded.

Consider the PN shown Figure 2.11a. It is a PN for an automated manufacturing flow shop and contains two machines. Places p_{1-2} and p_{4-5} together with transitions t_{1-6} model a machine prone to failures, p_3 models a buffer between two machines, and p_6 and t_6 model a reliable machine. The number of raw parts to be released to the system at a time is unlimited. It is easy to verify by coverability tree analysis that places p_{1-2} and p_{4-6} are safe, but p_3 is unbounded. Thus, the PN is not bounded. However, if only two raw parts are allowed to be released into the system at a time, the system can be modeled by the PN shown in Figure 2.11b. It can be verified that this time the PN shown in Figure 2.11b is bounded.

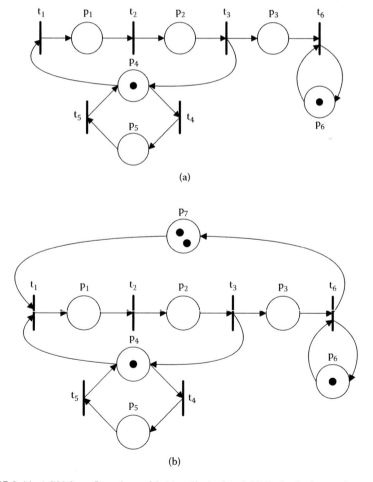

(a)

(b)

FIGURE 2.11 A PN for a flow shop with (a) unlimited and (b) limited release of raw parts.

Property 2.1: A PN is bounded if and only if the markings related to the nodes of the converability tree do not contain the symbol ω.

2.3.3 INCIDENCE MATRIX AND CONSERVATIVENESS

A PN is said to be (partially) conservative if there exists a positive integer $y(p)$ for every (some) place p such that the weighted sum of tokens $M^T y = M_0^T y = $ a constant, for every $M \in R(M_0)$ and any initial marking M_0. Clearly, if a PN is conservative and marked with finite tokens at M_0, the PN is bounded. Thus, for some PN, we can study the boundedness by analyzing its conservativeness.

For a PN with m places and n transitions, the incidence matrix $A = [a_{ij}]$ is an $m \times n$ matrix of integers and its entry is given by

$$a_{ij} = a_{ij}^+ - a_{ij}^-$$

where $a_{ij}^+ = O(p_j, t_i)$ and $a_{ij}^- = I(p_j, t_i)$. For example, the incidence matrix A for the PN shown in Figure 2.12 is

$$
A = \begin{array}{c} \\ p_1 \\ p_2 \\ p_3 \\ p_4 \\ p_5 \\ p_6 \\ p_7 \end{array}
\begin{array}{c}
\begin{array}{ccccccc} t_1 & t_2 & t_3 & t_4 & t_5 & t_6 & t_7 \end{array} \\
\left[\begin{array}{ccccccc}
-1 & -1 & 0 & 0 & 0 & 0 & 1 \\
1 & 0 & -1 & 0 & 0 & 0 & 0 \\
0 & 0 & 0 & -1 & 0 & 0 & 0 \\
0 & 0 & 0 & 0 & -1 & 0 & 0 \\
0 & 1 & 0 & 0 & 0 & -1 & 0 \\
0 & 1 & 1 & 0 & 1 & 0 & -1 \\
0 & 0 & 0 & 1 & 0 & 1 & -1
\end{array} \right]
\end{array}
$$

Based on the incidence matrix A, we have the following property.

Property 2.2: A PN is (partially) conservative if and only if there exists a vector y of positive integers such that $A^T y = 0$, $y \neq 0$ and y is called P-invariant.

Clearly, positive y can be found by solving $A^T y = 0$ if existing. For example, for the PN shown in Figure 2.12, solving $A^T y = 0$ leads to one such solution, i.e., $y = (2, 1, 1, 1, 1, 1, 1)^T$.

2.3.4 REVERSIBILITY

A PN is said to be reversible if for each marking $M \in R(M_0)$, M_0 is reachable from M. Thus, in a reversible PN one can always get back to the initial marking or state. For manufacturing systems, we often require such property to be held. In some applications, it is not necessary to get back to the initial state as long as one can get back to

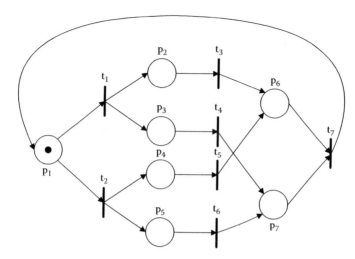

FIGURE 2.12 A PN for illustration of P-invariant.

some state that is termed a home state. Thus, the reversibility condition is relaxed and a home state is defined. Formally, M' is said to be a home state if for each marking $M \in R(M_0)$, M' is reachable from M.

2.3.5 LIVENESS

The concept of liveness is closely related to the deadlock-free operation in a system. Thus, it is a basic requirement for many systems. However, for some systems it is too strong to require such a condition. For liveness in PN we have the following definitions:

1. Dead (L0-live): If transition t can never fire in any firing sequence from M_0.
2. L1-live: If t can fire at least once in some firing sequence from M_0.
3. L2-live: If, given any integer k, t can fire at least k times in some firing sequence from M_0.
4. L3-live: If t appears infinitely in some firing sequence from M_0.
5. L4 live (live): t is L1 live for every marking $M \in R(M_0)$.

A PN is said to be Lk-live if every transition in the PN is Lk-live, $k = 1, 2, 3, 4$. Thus, a PN is said to be live if every transition is live. Consider the PN shown in Figure 2.13 that models a manufacturing system composed of two manufacturing resources and two products to be processed. The production processes of the two products are represented by the elementary circuits $\{p_1, t_1, p_2, t_2, p_3, t_3, p_1\}$ and $\{p_4, t_4, p_5, t_5, p_6, t_6, p_4\}$ with initial marking $M_0 = (1\ 0\ 0\ 1\ 0\ 0\ 1\ 1)^T$. Firing t_1 and t_4 results in $M = (0\ 1\ 0\ 0\ 1\ 0\ 0\ 0)^T$. Then no transition is enabled, or the system is deadlocked. Thus, the PN is not live.

The PN shown in Figure 2.14 is strictly L1-live since each transition can fire exactly once in the order of t_2, t_4, t_5, t_1, and t_3. For applications in manufacturing

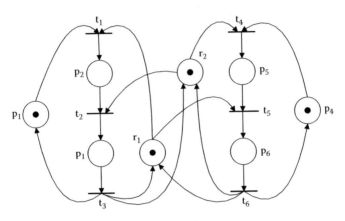

FIGURE 2.13 A PN for an automated manufacturing system.

systems, sometimes L1-live is enough to guarantee the correct operations. For example, we require the PN model for automated guided vehicle (AGV) systems to be L1-live. Transitions t_0, t_1, t_2, and t_3 in Figure 2.2 are strictly L0, L1, L2, and L3-live, respectively.

To analyze the liveness for a general PN, a coverability tree may be used, but it is prohibited for a large PN because of the exponential computation complexity. Furthermore, the liveness analysis via the coverability tree method is not applicable for unbounded PN in general. However, for some special PN we can analyze the liveness based on the special structure of the PN. We have the following properties:

Property 2.3: If a PN is live and safe, then it is free of source/sink place/transition.

Property 2.3 implies that if a PN is live and safe, it must be strongly connected. Thus, when we model systems by PN, we desire a strongly connected PN.

Property 2.4: An infinite capacity SM is live if and only if it is strongly connected and M_0 has at least one token.

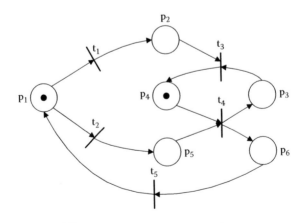

FIGURE 2.14 An L1-live PN.

In manufacturing systems, we can assume that there are always jobs to be processed, or at M_0 there are tokens in the system. The problem is whether we can model the system by an infinite capacity SM. We will see later that it is not the case for many applications in manufacturing systems due to the limited resources and their processing capacity. Thus, control policies are developed to guarantee the liveness of the system.

Property 2.5: For an MG, the token count in a circuit is invariant under any firing.

Property 2.5 implies that if there is no token in a circuit, it remains token-free and the transitions on the circuit are never enabled.

Property 2.6: An MG is live if and only if M_0 places at least one token in each circuit.

Because of Properties 2.5 and 2.6, MG is widely used in the modeling of concurrent choice-free manufacturing systems for performance evaluation.

Siphons in PN play an important role in its liveness because of the following properties:

Property 2.7: A free-choice PN is live if and only if every siphon contains a marked trap.

Property 2.8: If a siphon S is free of tokens at a marking, it remains token-free, and all the transitions connected to S are not live.

Therefore, to make a PN live, one should develop a control policy to keep every siphon to be marked at any marking.

Definition 2.11: An initially marked siphon may never be emptied no matter what marking $M \in R(M_0)$ is reached. These are called well-behaved siphons. However, some initially marked siphons can be emptied after a marking $M \in R(M_0)$ is reached and will never be marked again. These are called ill-behaved siphons (IBSs).

2.4   TIMED PN

The PN that we have introduced up to now is called a regular PN. To analyze performance associated with time, time is introduced into the PN, and such a PN is called a timed PN. It is an extension of a regular PN. Time can be associated with places, transitions, or both places and transitions. If time is associated with places, then when a token enters a place, the token has to remain in this place for a specified amount of time. If time is associated with transitions, there is a specified time delay in firing the transition.

Assume that time is associated with transitions, transition t begins firing at time τ_0, and the time delay for firing t is θ. Then the marking changes are given by

$$M'(p) = \begin{cases} M(p) - I(p, t), \ p \in {}^\bullet t, \ \text{in time interval } (\tau_0, \ \tau_0 + \theta) \\ M(p) + O(p, t), \ p \in t^\bullet, \ \text{at time} = \tau_0 + \theta \end{cases}$$

The marking change for a timed PN is illustrated in Figure 2.15. In the time interval between τ_0 and $\tau_0 + \theta$ the tokens are supposed to remain in transition t.

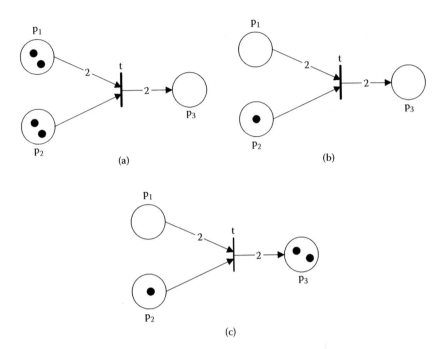

FIGURE 2.15 Time delay in timed PN: (a) marking at time τ_0, (b) marking in time interval between τ_0 and $\tau_0 + \theta$, and (c) marking at time $\tau_0 + \theta$.

An important use of a timed PN is to calculate the cycle time in an MG. Let v denote a circuit in an MG, and $d(t)$ and $d(p)$ denote the time delay associated with transition t and place p, respectively. If transition t (place p) is not a timed transition (place), $d(t)$ $(d(p)) = 0$. Further, let $P(v)$ denote the place set on v and $T(v)$ the transition set on v, and let $\pi(v) = (\Sigma_{p_i \in p(v)} d(p_i) + \Sigma_{t_j \in T(v)} d(t_j)) / M(v)$ be the time delay of the circuit, where $M(v)$ is the tokens in circuit v. Assume that $V = \{v_1, v_2, \ldots, v_k\}$ is the set of circuits in the MG; then

$$\pi^* = \max(\pi(v_1), \pi(v_2), \ldots, \pi(v_k)) \tag{2.6}$$

is the cycle time of the MG.

The PN shown in Figure 2.16 is an MG and there are two circuits: $v_1 = \{p_1, t_2, p_2, t_3, p_3, t_4, p_5, t_1, p_1\}$ and $v_2 = \{p_3, t_4, p_5, t_5, p_6, t_3, p_3\}$, and $M(v_1) = 3$, $M(v_2) = 5$. We have $d(v_1) = (d(p_1) + d(t_2) + d(p_2) + d(t_3) + d(p_3) + d(t_4) + d(p_4) + d(t_1))/3$, $d(v_2) = (d(p_3) + d(t_4) + d(p_5) + d(t_5) + d(p_6) + d(t_3))/5$ and $\pi^* = \max(\pi(v_1), \pi(v_2))$.

2.5 PN WITH INHIBITOR ARCS

An inhibitor arc connects a place to a transition. Pictorially, it ends with a small circle, not an arrow. If (p, t) is an inhibitor arc and there are tokens in p, then t cannot fire. The introduction of inhibitor arcs can increase the power of PN. For example, by

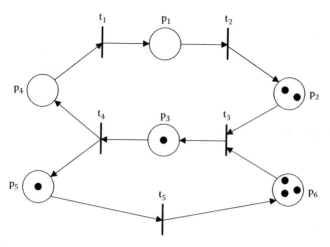

FIGURE 2.16 An MG.

using an inhibitor arc we can model priority in a manufacturing system. Consider the manufacturing process represented by Figure 2.17a. A machine processes two types of jobs, and job J_1 has a higher priority than J_2. The PN in Figure 2.17b describes this process. In the model, t_1 and t_2 model the arrival of J_1 and J_2, tokens in p_1 and p_2 represent jobs of J_1 and J_2 in buffers b_1 and b_2 waiting for processing, and p_3 and p_5 represent

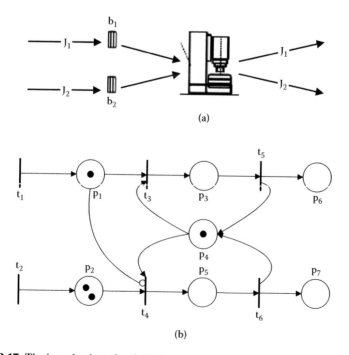

FIGURE 2.17 The introduction of an inhibitor arc.

the processing of J_1 and J_2 by the machine. By adding an inhibitor arc from p_1 to t_4, as shown in Figure 2.17b, the priority is exactly described and indicates that J_1 has a higher priority to be processed.

2.6 SUMMARY

PN is an effective tool in modeling, analysis, performance evaluation, and control for discrete event systems (DES). This chapter briefly introduces the basic concepts, terminology, and properties of a PN. In this book, we are concerned with the modeling, analysis, and control of automated manufacturing systems (AMS) by using PN. Deadlock-free operation in AMS is the basic requirement. We mainly discuss how an AMS should be modeled and controlled via PN so that the system is deadlock-free.

REFERENCES

Murata, T. 1989. Petri nets: Properties, analysis and applications. *Proceedings of the IEEE* 77:541–80.

Proth, J.-M., and X. L. Xie. 1996. *Petri nets: A tool for design and management of manufacturing systems.* New York: John Wiley & Sons.

Zhou, M. C., and F. DiCesare. 1993. *Petri net synthesis for discrete event control of manufacturing systems.* Boston: Kluwer Academic Publications.

Zhou, M. C., and K. Venkatesh. 1998. *Modeling, simulation and control of flexible manufacturing systems: Petri net approach.* Singapore: World Scientific.

3 Colored Petri Net

As pointed out in Jensen (1986), in modeling by regular Petri net (PN), it is often necessary to have several identical subsets. This is because a folding into a single subnet would destroy the possibility of distinguishing different processes. Thus, the size of PN becomes large in modeling those practical systems. To overcome this problem, colors are introduced into the regular PN, and the regular PN is extended into colored PN (CPN) (Jensen, 1981, 1986; Viswanadham and Narahari, 1987). A CPN is compact, and the mathematical theory for it is well developed. Thus, it is widely used in modeling, analysis, simulation, and control. This chapter introduces CPN and some CPN analysis methods.

3.1 A SIMPLE EXAMPLE

Like timed PN, CPN is an extension of regular PN by introducing colors into the regular PN. Before we present the formal definition of CPN, we present a simple example to show how CPN works.

Example 3.1

Consider a manufacturing system with two machines m_1 and m_2 processing two part types J_1 and J_2. Each part type goes through one stage of operation, and this operation can be performed on either m_1 or m_2.

This system can be modeled by a regular PN, as shown in Figure 3.1. A token in p_1 (p_4) represents that J_1 (J_2) is waiting for processing. A token in p_7 (p_8) represents that machine m_1 (m_2) is idle for performing operations. Thus, at M_0, there are a number of jobs of J_1 and J_2 waiting for processing, and both machines m_1 and m_2 are idle. Transition t_1 represents that m_1 starts processing a J_1. A token in p_2 represents that m_1 is processing a J_1. Transition t_3 represents that m_1 completes the processing of J_1. The other places and transitions are interpreted in Table 3.1.

This system can also be modeled by a CPN, as shown in Figure 3.2. A token in p_1 indicates that there is a job waiting for processing. A token in p_2 indicates that a job is being processed on a machine. A token in p_3 indicates that a machine is available. Transition t_1 (t_2) represents that a machine starts (completes) processing of a job. The places and transitions are interpreted in Table 3.2. However, in this way, confusion exists. For example, when a token is in p_1, one does not know whether it is J_1 or J_2, and by firing t_1, we need to know which machine is used to process which job. To distinguish tokens in places and the type of transition firing, colors are introduced for places and transitions.

There are two types of jobs, so to distinguish the tokens in p_1, two colors are needed and are named J_1 and J_2, denoted by ① and ②, respectively. For p_3, there are two machines, so two colors are needed and are named m_1 and m_2, denoted by ❶ and ❷, respectively. For p_2, there are four possible cases: (1) m_1 processes J_1, (2) m_1 processes J_2, (3) m_2 processes J_1, or (4) m_2 processes J_2. Thus, four colors are

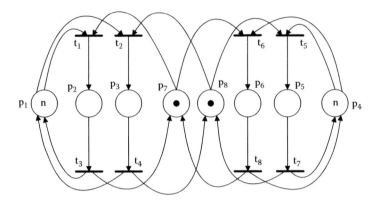

FIGURE 3.1 Regular PN model for Example 3.1.

TABLE 3.1
Interpretation of Places and Transitions in Figure 3.1

Places	Transitions
p_1: J_1 is waiting for processing	t_1: m_1 starts processing J_1
p_2: J_1 is being processed on m_1	t_2: m_2 starts processing J_1
p_3: J_1 is being processed on m_2	t_3: m_1 completes the processing of J_1
p_4: J_2 is waiting for processing	t_4: m_2 completes the processing of J_1
p_5: J_2 is being processed on m_2	t_5: m_2 starts processing J_2
p_6: J_2 is being processed on m_1	t_6: m_1 starts processing J_2
p_7: m_1 is available	t_7: m_2 completes the processing of J_2
p_8: m_2 is available	t_8: m_1 completes the processing of J_2

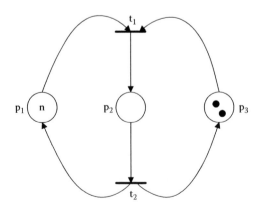

FIGURE 3.2 Colored PN model for Example 3.1.

TABLE 3.2

Interpretation of Places and Transitions in Figure 3.2

Places	Transitions
p_1: Jobs are waiting for processing	t_1: Start of processing a job
p_2: Jobs are processing	t_1: Completion of processing a job
p_3: Machine is available	

needed and named (m_1, J_1), (m_1, J_2), (m_2, J_1), and (m_2, J_2), and denoted by ■, ●, ✱, and □, respectively. Similarly, four colors are needed to distinguish the firing types for transitions, and they are denoted by ✚, ✕, ✦, and ○, respectively.

With the colors defined, the relation of input and output by firing a transition with respect to a color can be given. For example, when there is a token with color ① in p_1 and at the same time a token with color ❶ in p_3, then transition t_1 with respect to color ✚ is enabled and can fire. After firing t_1 with respect to color ✚, a token with color ① and a token with color ❶ are taken away from p_1 and p_2, respectively, and at the same time a token with color ■ is put into p_2. The input and output relation of firing transitions t_1 and t_2 with respect to different colors is presented in Table 3.3. In this way, the dynamics of the system are accurately modeled. However, the structure of the CPN for this example is much simpler than the regular PN model. Note that the simple structure works only after colors are introduced for tokens, places, and transitions.

3.2 DEFINITIONS OF CPN

Now we present the formal definitions of CPN.

Definition 3.1: A CPN is a six-tuple CPN = (P, T, I, O, C, M_0), where

1. $P = \{p_1, p_2, \ldots, p_m\}$, $m > 0$, is a finite set of places.
2. $T = \{t_1, t_2, \ldots, t_n\}$, $n > 0$, is a finite set of transitions such that $P \cup T \neq \emptyset$ and $P \cap T = \emptyset$.

TABLE 3.3

The Input and Output of Firing a Transition of the CPN in Figure 3.2

Transition	Firing Color	Input			Output		
		P_1	P_2	P_3	P_1	P_2	P_3
t_1	✚	①	—	❶	—	■	—
	✕	②	—	❶	—	●	—
	✦	①	—	❷	—	✱	—
	○	②	—	❷	—	□	—
t_2	✚	—	■	—	①	—	❶
	✕	—	●	—	②	—	❶
	✦	—	✱	—	①	—	❷
	○	—	□	—	②	—	❷

3. $C(p)$ and $C(t)$ are the sets of colors associated with place $p \in P$ and transition $t \in T$. We let

$$C(p_i) = \{a_{i1}, a_{i2}, \ldots, a_{iu_i}\}, u_i = |C(p_i)|$$

$$C(t_j) = \{b_{j1}, b_{j2}, \ldots, b_{jv_j}\}, v_j = |C(t_j)|$$

4. $I(p, t): C(p) \times C(t) \to N$ is an input function where $N = \{0, 1, 2, \ldots, \}$ is the set of nonnegative integers.
5. $O(p, t): C(p) \times C(t) \to N$ is an output function.
6. $M: P \to N$ is a marking representing the numbers of tokens in places. $M = (M(p_1), M(p_2), \ldots, M(p_m))^T$. $M(p_i)$ denotes the number of tokens in p_i regardless of the colors of the tokens, and $M(p_i, a_{ij})$ denotes the number of tokens with color a_{ij} in p_i. M_0 denotes the initial marking.

Example 3.1 (continued)

For the CPN shown in Figure 3.2, according to the definition above we have the sets of colors for places and transitions as follows:

$C(p_1) = \{J_1, J_2\} = \{①, ②\}$

$C(p_3) = \{m_1, m_2\} = \{❶, ❷\}$

$C(p_2) = \{(m_1, J_1), (m_1, J_2), (m_2, J_1), (m_2, J_2)\} = \{■, ●, ＊, □\}$

and

$C(t_1) = C(t_2) = \{✚, ✖, ✦, ○\}$

From Table 3.3 we can present the input and output functions as follows:

$I(p_1, t_1)(①, ✚) = 1$ $I(p_1, t_1)(①, ✖) = 0$

$I(p_1, t_1)(①, ✦) = 1$ $I(p_1, t_1)(①, ○) = 0$

$I(p_1, t_1)(②, ✚) = 0$ $I(p_1, t_1)(②, ✖) = 1$

$I(p_1, t_1)(②, ✦) = 0$ $I(p_1, t_1)(②, ○) = 1$

$O(p_1, t_1) = 0$

$I(p_3, t_1)(❶, ✚) = 1$ $\qquad\qquad$ $I(p_3, t_1)(❶, ✗) = 1$

$I(p_3, t_1)(❶, ✦) = 0$ $\qquad\qquad$ $I(p_3, t_1)(❶, ○) = 0$

$I(p_3, t_1)(❷, ✚) = 0$ $\qquad\qquad$ $I(p_3, t_1)(❷, ●) = 0$

$I(p_3, t_1)(❷, ✦) = 1$ $\qquad\qquad$ $I(p_3, t_1)(❷, ○) = 1$

$O(p_3, t_1) = 0$

$I(p_2, t_1) = 0$

$O(p_2, t_1)(■, ✚) = 1$

$O(p_2, t_1)(■, ✗) = O(p_2, t_1)(■, ✦) = O(p_2, t_1)(■, ○) = 0$

$O(p_2, t_1)(●, ✗) = 1$

$O(p_2, t_1)(●, ✚) = O(p_2, t_1)(●, ✦) = O(p_2, t_1)(●, ○) = 0$

$O(p_2, t_1)(✽, ✦) = 1$

$O(p_2, t_1)(✽, ✚) = O(p_2, t_1)(✽, ✗) = O(p_2, t_1)(✽, ○) = 0$

$O(p_2, t_1)(□, ○) = 1$

$O(p_2, t_1)(□, ✚) = O(p_2, t_1)(□, ✗) = O(p_2, t_1)(□, ✦) = 0$

$I(p_2, t_2)(■, ✚) = 1$

$I(p_2, t_2)(■, ✗) = I(p_2, t_2)(■, ✦) = I(p_2, t_2)(■, ○) = 0$

$I(p_2, t_2)(●, ✗) = 1$

$I(p_2, t_2)(●, ✚) = I(p_2, t_2)(●, ✦) = I(p_2, t_2)(●, ○) = 0$

$I(p_2, t_2)(✽, ✦) = 1$

$I(p_2, t_2)(✽, ✚) = I(p_2, t_2)(✽, ✗) = I(p_2, t_2)(✽, ○) = 0$

$I(p_2, t_2)(□, ○) = 1$

$I(p_2, t_2)(□, ✚) = I(p_2, t_2)(□, ✗) = I(p_2, t_2)(□, ✦) = 0$

$O(p_2, t_2) = 0$

$I(p_1, t_2) = 0$ $O(p_1, t_2)(①, ✚) = 1$

$O(p_1, t_2)(①, ✕) = 0$ $O(p_1, t_2)(①, ✦) = 1$

$O(p_1, t_2)(①, ○) = 0$ $O(p_1, t_2)(②, ✚) = 0$

$O(p_1, t_2)(②, ✕) = 1$ $O(p_1, t_2)(②, ✦) = 0$

$O(p_1, t_2)(②, ○) = 1$

$I(p_3, t_2) = 0$ $O(p_3, t_2)(❶, ✚) = 1$

$O(p_3, t_2)(❶, ✕) = 1$ $O(p_3, t_2)(❶, ✦) = 0$

$O(p_3, t_2)(❶, ○) = 0$ $O(p_3, t_2)(❷, ✚) = 0$

$O(p_3, t_2)(❷, ●) = 0$ $O(p_3, t_2)(❷, ✦) = 1$

$O(p_3, t_2)(❷, ○) = 1$

These functions can be written in matrix form as follows:

$$I(p_1, t_1) = \begin{bmatrix} 1 & 0 & 1 & 0 \\ 0 & 1 & 0 & 1 \end{bmatrix}$$

$$O(p_1, t_1) = [0]_{2 \times 4}$$

$$I(p_3, t_1) = \begin{bmatrix} 1 & 1 & 0 & 0 \\ 0 & 0 & 1 & 1 \end{bmatrix}$$

$$O(p_3, t_1) = [0]_{2 \times 4}$$

$$I(p_2, t_1) = [0]_{4 \times 4} = O(p_2, t_2)$$

$$O(p_2, t_1) = I(p_2, t_2) = \begin{bmatrix} 1 & 0 & 0 & 0 \\ 0 & 1 & 0 & 0 \\ 0 & 0 & 1 & 0 \\ 0 & 0 & 0 & 1 \end{bmatrix}$$

$I(p_1, t_2) = I(p_3, t_2) = [0]_{2 \times 4}$

$$O(p_1, t_2) = \begin{bmatrix} 1 & 0 & 1 & 0 \\ 0 & 1 & 0 & 1 \end{bmatrix}$$

$$O(p_3, t_2) = \begin{bmatrix} 1 & 1 & 0 & 0 \\ 0 & 0 & 1 & 1 \end{bmatrix}$$

Assume that at the initial state two machines m_1 and m_2 are both idle, and there are n_1 J_1-parts and n_2 J_2-parts. Then the initial marking is

$$M_0 = \begin{bmatrix} n_1 + n_2 \\ 0 \\ 2 \end{bmatrix}$$

$M_0(p_1, ①) = n_1$, $M_0(p_1, ②) = n_2$, $M_0(p_3, ❶) = M_0(p_3, ❷) = 1$, and $M_0(p_2, ■) = M_0(p_2, ●) = M_0(p_2, ✱) = M_0(p_2, □) = 0$.

3.3 TRANSITION ENABLING AND FIRING RULES

Based on the definitions of colors, markings, and the input and output functions, we can present the transition enabling and firing rules.

Definition 3.2: A transition $t_j \in T$ is said to be enabled with respect to color b_{jk} at marking M if and only if

$$M(p_i, a_{ih}) \geq I(p_i, t_j)(a_{ih}, b_{jk}), \ \forall p_i \in p \text{ and } a_{ih} \in C(p_i) \tag{3.1}$$

When a transition is enabled it can fire. Firing an enabled transition t_j with respect to color b_{jk} at marking M changes the marking from M to M' according to the following equation:

$$M'(p_i, a_{ih}) = M(p_i, a_{ih}) - I(p_i, t_j)(a_{ih}, b_{jk}) + O(p_i, t_j)(a_{ih}, b_{jk}), \ \forall p_i \in p \text{ and } a_{ih} \in C(p_i) \tag{3.2}$$

Example 3.1 (continued)

Assume that $n_1 = n_2 = 1$. Then at the initial marking M_0 transition t_1 is enabled with respect to colors ✚, ✗, ✦, and ◯. It can fire individually with respect to any one of the colors. It can also fire concurrently with respect to colors ✚ and ◯, or ✗ and ✦. If t_1 fires with respect to color ✚, then M_0 changes into M, and $M(p_1) = M(p_2) = M(p_3) = 1$ with $M(p_1, ①) = M(p_3, ❶) = M(p_2, ●) = M(p_2, ✱) = M(p_2, □) = 0$ and $M(p_1, ②) = M(p_3, ❷) = M(p_2, ■) = 1$. The marking changes due to firing t_1 with respect to colors ➕, ✖, ✦, ○, ➕, and ○, and ✖ and ✦ are given in Table 3.4, where ①, ②, ❶, ❷, ■, ●, ✳, and □ are used to denote $M(p_1, ①)$, $M(p_1, ②)$, $M(p_3, ❶)$, $M(p_3, ❷)$, $M(p_2, ■)$, $M(p_2, ●)$, $M(p_2, ✳)$, and $M(p_2, □)$, respectively.

3.4   P-INVARIANT IN CPN

Because of the relation between P-invariant and conservativeness in PN, P-invariant plays an important role in PN analysis. The theory of P-invariant for CPN has been well developed. For P-invariant in CPN, the incidence matrix for CPN should be defined first.

Definition 3.3: The incidence matrix A of a CPN is a $\sum_{i=1}^{m} u_i \times \sum_{i=1}^{n} v$; matrix. It is defined by the $m \times n$ block matrix $A = [W_{ij}]$, $i = 1, 2, \ldots, m$, $j = 1, 2, \ldots, n$, where $W_{ij} = O(p_i, t_j) - I(p_i, t_j)$.

Recall that the P-invariant for a regular PN is determined by a vector $y > 0$ with a nonzero entry corresponding to a place. For CPN, there are multiple colors for each place, and thus a matrix is needed to determine the P-invariant in CPN.

Definition 3.4: Let Q be a nonempty set of colors with cardinality $|Q| = q$. A set of places with respect to Q is a function X defined on P such that $X(p)$: $Q \times C(p) \to Z$, where Z is the set of integers.

The set of places can also be represented by a block matrix $X = [X_1, X_2, \ldots, X_m]^T$, where X_i, $i = 1, 2, \ldots, m$, is a matrix of integers of dimension $q \times u_i$. In fact, the set of places treats each color in CPN as a place, like in regular PN.

TABLE 3.4
The Marking Changes after Firing t_1 with Respect to Different Color

Color	$M(p_1)$		$M(p_3)$		$M(p_2)$			
	①	②	❶	❷	■	●	✳	□
➕	1		1			1		
	0	1	0	1	1	0	0	0
✖	1		1				1	
	1	0	0	1	0	1	0	0
✦	1		1			1		1
	0	1	1	0	0	0	1	0
○	1		1			1		
	1	0	1	0	0	0	0	1
➕ and ○	0		0			2		
	0	0	0	0	1	0	0	1
✖ and ✦	0		0			2		
	0	0	0	0	0	1	1	0

Definition 3.5: A weighted matrix for a set of places $X = [X_1, X_2, \ldots, X_m]^T$ is called a P-invariant of a CPN if and only if

$$A^T X = 0 \tag{3.3}$$

Let $M(C) = (M(p_1, a_{11}), M(p_1, a_{12}), \ldots, M(p_1, a_{1u1}), M(p_2, a_{21}), \ldots, M(p_2, a_{2u2}), \ldots,$
$M(p_m, a_{m1}), \ldots, M(p_m, a_{mum}))^T$. If X is a P-invariant of a CPN, then we have

$$M(C)^T X = M_0(C)^T X, \forall M \in R(M_0) \tag{3.4}$$

This implies the CPN is conservative or partially conservative. Notice that a color represents a type of token. Hence, if for every color the CPN is conservative, the CPN must be conservative in the sense of a regular PN. Therefore, if the entries with values greater than zero in X contain all places corresponding to the colors, then for every color the CPN must be conservative in the sense of a regular PN. Thus, there exists a positive integer vector $y = (y_1, y_2, \ldots, y_m)^T$ such that

$$M^T y = M_0^T y, \forall M \in R(M_0) \tag{3.5}$$

Example 3.1 (continued)

Based on the input and output functions, incidence A can be calculated as follows:

$$W_{11} = O(p_1, t_1) - I(p_1, t_1) = \begin{bmatrix} -1 & 0 & -1 & 0 \\ 0 & -1 & 0 & -1 \end{bmatrix}$$

$$W_{12} = O(p_1, t_2) - I(p_1, t_2) = \begin{bmatrix} 1 & 0 & 1 & 0 \\ 0 & 1 & 0 & 1 \end{bmatrix}$$

$$W_{21} = O(p_2, t_1) - I(p_2, t_1) = \begin{bmatrix} 1 & 0 & 0 & 0 \\ 0 & 1 & 0 & 0 \\ 0 & 0 & 1 & 0 \\ 0 & 0 & 0 & 1 \end{bmatrix}$$

$$W_{22} = O(p_2, t_2) - I(p_2, t_2) = \begin{bmatrix} -1 & 0 & 0 & 0 \\ 0 & -1 & 0 & 0 \\ 0 & 0 & -1 & 0 \\ 0 & 0 & 0 & -1 \end{bmatrix}$$

$$W_{31} = O(p_3,t_1) - I(p_3,t_1) = \begin{bmatrix} -1 & -1 & 0 & 0 \\ 0 & 0 & -1 & -1 \end{bmatrix}$$

$$W_{32} = O(p_3,t_2) - I(p_3,t_2) = \begin{bmatrix} 1 & 1 & 0 & 0 \\ 0 & 0 & 1 & 1 \end{bmatrix}$$

Thus, we obtain the incidence matrix:

$$A = \begin{bmatrix} -1 & 0 & 1 & 0 & 0 & 0 & -1 & 0 \\ 0 & -1 & 0 & 1 & 0 & 0 & -1 & 0 \\ -1 & 0 & 0 & 0 & 1 & 0 & 0 & -1 \\ 0 & -1 & 0 & 0 & 0 & 1 & 0 & -1 \\ 1 & 0 & -1 & 0 & 0 & 0 & 1 & 0 \\ 0 & 1 & 0 & -1 & 0 & 0 & 1 & 0 \\ 1 & 0 & 0 & 0 & -1 & 0 & 0 & 1 \\ 0 & 1 & 0 & 0 & 0 & -1 & 0 & 1 \end{bmatrix}$$

Consider that the colors for p_2 are dependent on the colors for p_1 and p_3, and there are two colors for both p_1 and p_3. Thus, $q = 4$ is the largest number of colors to be considered in determining the P-invariant. Let

$$X = \begin{bmatrix} 1 & 0 & 0 & 0 \\ 0 & 1 & 0 & 0 \\ 1 & 0 & 1 & 0 \\ 0 & 1 & 1 & 0 \\ 1 & 0 & 0 & 1 \\ 0 & 1 & 0 & 1 \\ 0 & 0 & 1 & 0 \\ 0 & 0 & 0 & 1 \end{bmatrix}$$

It is easy to verify that $AX = 0$, or X is a P-invariant for the CPN shown in Figure 3.2. In fact, in X, $x_1 = (1, 0, 1, 0, 1, 0, 0, 0)^T$, $x_2 = (0, 1, 0, 1, 0, 1, 0, 0)^T$, $x_3 = (0, 0, 1, 1, 0, 0, 1, 0)^T$, and $x_4 = (0, 0, 0, 0, 1, 1, 0, 1)^T$ correspond to the P-invariant for colors ① (J_1), ② (J_2), ❶ (m_1), and ❷ (m_2), respectively. This implies that the number of tokens for each type is a constant. Further, $x_1 + x_2 + x_3 + x_4 = (1, 1, 2, 2, 2, 2, 1, 1)^T$, every component of which is greater than zero. Thus, if each color is thought of as a place, then X contains all places. Thus, according to Equation 3.5, $M^Ty = M_0^Ty$ with $y = (1, 2, 1)^T$ must hold, or the CPN is conservative.

3.5 SUMMARY

CPN is an extension of regular PN. In modeling a practical discrete event system (DES) by regular PN, the size of the resulting model is very large. This gives rise to a problem for system analysis and control by using the PN model. By introducing colors into a regular PN, the resulting CPN model is very compact. Thus, CPN is widely applied. In this book, we will use CPN as a modeling tool to model AMS for performance evaluation and control.

REFERENCES

Jensen, K. 1981. Colored Petri nets and the invariant method. *Theoretical Computer Science* 14:317–36.

Jensen, K. 1986. *Colored Petri nets*, 248–99. Lecture Notes in Computer Science, vol. 254, part 1. Berlin: Springer-Verlag.

Viswanadham, N., and Y. Narahari. 1987. Colored Petri net models for automated manufacturing systems. Paper presented at Proceedings of IEEE International Conference on Robotics and Automation, Raleigh, NC.

4 Process-Oriented Petri Net Modeling

4.1 INTRODUCTION

In operating an automated manufacturing system (AMS), it is very important that its flexibility should be exploited sufficiently. This requires the effective application of control and management techniques and theories to model and analyze the behavior of the systems. Thus, during the last several decades, the modeling, analysis, simulation, and control of AMS have become emerging topics and been extensively studied. A model is needed for analysis, simulation, and control of AMS.

Because Petri nets (PNs) can model concurrency well, they are widely used to model AMS for analysis, simulation, and control (Zhou and Venkatesh, 1998). In earlier studies, PNs were used for performance evaluation (Viswanadham and Narahari, 1987; Narahari and Viswanadham, 1985; Al-Jaar and Desrochers, 1990). Zhou and colleagues (1991, 1992, 1993) used Petri nets to synthesize a conservative, reversible, and live model. PNs are also used to model AMS for deadlock prevention and deadlock avoidance (Banaszak and Krogh, 1990; Xing et al., 1996; Ezpeleta et al., 1995; Chu and Xie, 1997; Li and Zhou, 2004; Roszkowska, 2004). In all the applications mentioned, process-oriented Petri nets (POPNs) are used. They are obtained via process-oriented modeling methods. One major difficulty in modeling AMS lies in the method of dealing with the shared resources (Zhou and DiCesare, 1991; Zhou et al., 1992; Jeng and DiCesare, 1995). POPN can do that straightforwardly. Such methods use operation places to describe the operation sequences of every part to be processed in the system, and then resource places to describe the resource requirements for all the operations. POPN can describe manufacturing processes clearly and easily and are widely applicable. This chapter presents the modeling method of POPN and discusses the properties of its resulting models. These properties will be compared with another modeling method later in this book.

The complexity of AMS is mainly from resource sharing. Based on resource acquisition for each operation, AMS are classified into three classes: (1) single resource allocation system—each operation requires a single unit of a single resource (Lawley, 1999); (2) assembly/disassembly systems (Fanti et al., 2002; Roszkowska, 2004; Wu et al., 2008); and (3) conjunctive and disjunctive resource allocation system—each operation can require an arbitrary number of units from an arbitrary resource set (Park and Reveliotis, 2002). Here, we confine the discussion to the first class of AMS.

4.2   MODELING METHOD

In modeling discrete event systems (DES) by PN, often places and tokens are used for conditions, and transitions are used for the occurrence of events. In modeling AMS by POPN, the following interpretations for places, transitions, and tokens are employed:

1. A place represents a resource or job order status or an operation. When it represents the former, the initial number of tokens can be a constant, e.g., the number of machines, assuming that the plant is fixed. If it represents the job order status, the number of initial tokens may be a variable, e.g., the number of jobs to be processed in the system.
2. If a place represents a resource status, one or more tokens in it indicate that the resource is available, and no token indicates that it is not available. If it represents job order status, tokens in it indicate that there are jobs waiting to be processed, and no token indicates that there is no job for processing. If it represents an operation, a token in it shows that an operation is being executed, and no token shows that it is not being performed.
3. A transition represents either the start or completion of an event or operation process.

With the interpretation of places, transitions, and tokens above, the POPN models an AMS according to the following steps:

1. Identify the activities and resources required for the production of one item of each product.
2. Order the activities by the precedence relation as given in the process plans.

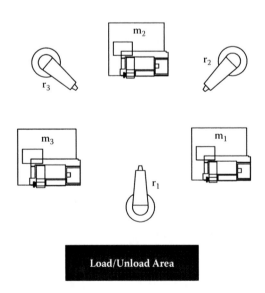

FIGURE 4.1 An example of AMS.

3. For each activity, in order: Create and label a place to represent the status of that activity, add a transition (start activity) with an output arc(s) to the place(s), and add a transition (stop activity) with an input arc(s) from the activity place(s). In general, the stop transition for one activity is the same as the start transition for the next activity. When the net is executed, a token in an activity place indicates the activity is taking place. Multiple tokens indicate the activity occurring in multiplicity. For example, in a buffer space, two tokens might represent two parts being stored at the same time. The firing of the start transition represents starting the activity or process, and the firing of the stop transition represents the completion of the activity and may also represent the start of the next activity.

4. For each item of product, create and label a place that represents job order status with an output arc(s) to the start transition(s) of the first activity of the item and an input arc(s) from the stop transition(s) of the last activity of the item. The tokens in this place indicate the number of job orders for the item to be performed in the system.

5. For each activity, in order: If such a place has not already been created, create and label a place for each resource that must be available to start the activity. Connect all appropriate resource availability places such that there is an input arc from each resource to the starting transition for the activity. Create output arcs to connect the stop transition, following the activity to any resource places representing resources that become available (are released) upon completion of the activity.

6. Specify the initial marking for the system.

To illustrate the POPN modeling method, take the simple AMS as shown in Figure 4.1 as an example. It is comprised of three machines, m_1, m_2, and m_3; three robots, r_1, r_2, and r_3; and a load/unload area. Robot r_1 is used for loading and unloading parts and for part transfer between machines m_1 and m_3. Robots r_2 and r_3 are used for part transfer between machines m_1 and m_2, and m_2 and m_3, respectively. We assume that there is a perennial supply of raw workplaces. The parts to be processed undergo three stages of operations with route $m_1 \rightarrow m_2 \rightarrow m_3$.

Following the POPN modeling method above, the modeling process for the example is as follows:

1. The activities required for each item are loading, machining, transferring, and unloading. The resources are m_1, m_2, m_3, r_1, r_2, and r_3.

2. The order of the activities is as follows:

p_1: Robot r_1 loads a part from the load/unload area onto m_1.

p_2: Machine m_1 performs the first operation.

p_3: After completion of the first operation by m_1, r_2 transfers a part from m_1 to m_2.

p_4: Machine m_2 performs the second operation.

p_5: After completion of the second operation by m_2, r_3 transfers a part from m_2 to m_3.

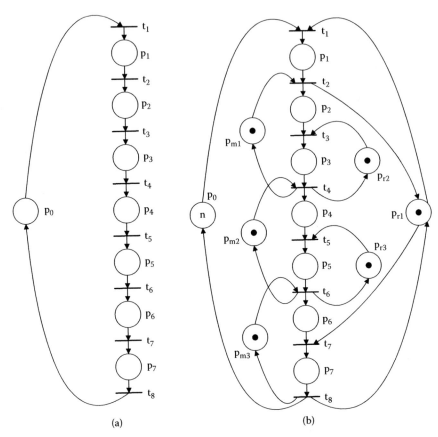

(a)   (b)

FIGURE 4.2 Illustration of modeling method by POPN. (a) PN graph of the activity sequence with a place representing job order status added for the AMS example. (b) PN with the resource places added and connected and the initial marking.

p_6: Machine m_3 performs the third operation.

p_7: After completion of the third operation by m_3, r_1 unloads a part from m_3 to the load/unload area.

3. As shown in Figure 4.2a, places p_{1-7} are created to model the activity sequence for one part. Transition t_1 models the start of activity p_1; t_2 models the stop of activity p_1 and start of activity p_2; t_3, the stop of activity p_2 and start of activity p_3; t_4, the stop of activity p_3 and start of activity p_4; t_5, the stop of activity p_4 and start of activity p_5; t_6, the stop of activity p_5 and start of activity p_6; t_7, the stop of activity p_6 and start of activity p_7; and t_8, the stop of activity p_7.

4. As shown in Figure 4.2a, place p_0 representing a job order status is created and added into the model with an output arc to t_1 and an input arc from t_8.

5. Consider t_1. Its enabling requires that raw material and robot r_1 be available. There is already an input arc from p_0 representing a job order available.

Another input arc must be added from p_{r1} that represents that r_1 is available. Next consider t_2, the stop of activity p_1 and start of p_2. After the completion of p_1, r_1 should be released. However, as the start transition of p_2, it requires that m_1 be available. Thus, an output arc is added to p_{r1}, and at the same time an input arc is added from p_{m1} representing that m_1 is available. Consider t_3, the stop transition of p_2 machining the first operation by m_1. After completion of the activity, m_1 cannot be released since m_1 is not available before it is unloaded. Transition t_3 requires that r_2 be available, an input arc is added from p_{r2} representing that r_2 is available. Consider t_4, the stop transition of activity p_3 and start transition of p_4. After completion of p_3, m_1 and r_2 can be released and, at the same time, it requires m_2. An output arc is added to p_{m1} and p_{r2}, respectively, and an input arc is added from p_{m2} representing m_2's availability. Consider t_5, the stop transition of activity p_4 and start transition of p_5. After the completion of p_4, m_2 cannot be released, but t_5 requires r_3. Thus, an input arc is added from p_{r3}. Consider t_6, the stop transition of activity p_5 and start transition of p_6. After the completion of p_5, both m_2 and r_3 can be released, but it requires m_3. Thus, an output arc is added to both p_{m2} and p_{r3}, respectively, and at the same time an input arc is added from p_{m3} representing m_3. Consider t_7, the stop transition of activity p_6 and start transition of p_7. After the completion of p_6, m_3 cannot be released yet. Enabling t_7 requires r_1. Thus, an input arc is added from p_{r1}. Consider t_8, the stop transition of the last activity p_7. After the completion of p_7, both m_3 and r_1 should be released and no other resources are required. Thus, an output arc to both p_{m3} and p_{r1} is added. In this way, the modeling for resource requirements is completed and the resulting net is shown in Figure 4.2b.

6. The initial marking is formulated for system start-up as shown in Figure 4.2b. There are n job orders in p_0, implying that it is a variable and there are always jobs for processing. Places p_{m1}, p_{m2}, p_{m3}, p_{r1}, p_{r2}, and p_{r3} are all marked with a token, implying that all the resources are available and single. All the places for activities are not marked, implying that no activity is being executed. At this marking, only t_1 is enabled.

From this simple example, it can be seen that the modeling process is straightforward and easy to understand.

From the POPN modeling method, it is known that a POPN model of AMS contains three types of places: (1) places for operations, (2) places for resources, and (3) places representing job order status. We name the sets of the three types of places as P_A, P_B, and P_C, respectively.

4.3   RESOURCE SHARING IN POPN

Resource competition or resource sharing is the main characteristic of AMS and reconfigurable manufacturing systems (RMS). It can lead to deadlocks that disable the operation of the whole system. Thus, how to deal with resource sharing in modeling an AMS and RMS is very important. Zhou and colleagues (1991, 1992, 1993) identified two types of resource sharing modes in RMS, named parallel mutual

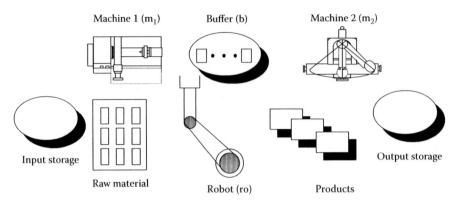

FIGURE 4.3 A simple manufacturing cell.

exclusion (PME) and sequential mutual exclusion (SME), that reveal the essential properties of dynamics in RMS. In RMS, there are mainly two types of resources: resources for part production and resources for part delivery. Because of the difference between these two types of resources in the operation of RMS, we discuss resource sharing for them separately.

4.3.1 RESOURCE SHARING IN PART PROCESSING

To illustrate resource sharing in part processing, consider the simple AMS shown in Figure 4.3. It contains machines m_1 and m_2, and robot ((r)). Two different situations are discussed as Examples 4.1 and 4.2.

Example 4.1

Assume that there are two types of parts, A and B, to be processed in the system shown in Figure 4.3, and both have two operations. For A, operation 1 requires m_1 to process, and operation 2 requires m_2. B requires m_2 to process first, then m_1. Here, we do not consider the operations of the robot. The intermediate buffer is not used.

Following the POPN modeling method we can obtain the POPN model for this example (Figure 4.4 and Table 4.1). We use p_{10} and p_{20} to represent A-parts and B-parts in the input and output storage. The production processes of A-part and B-part are modeled by $p_{10} \rightarrow t_{11} \rightarrow p_{11} \rightarrow t_{12} \rightarrow p_{12} \rightarrow t_{13} \rightarrow p_{10}$ and $p_{20} \rightarrow t_{21} \rightarrow p_{21} \rightarrow t_{22} \rightarrow p_{22} \rightarrow t_{23} \rightarrow p_{20}$, respectively. The usage of resources m_1 and m_2 is described by $p_{m1} \rightarrow t_{11} \rightarrow p_{11} \rightarrow t_{12} \rightarrow p_{m1}$, $p_{m1} \rightarrow t_{22} \rightarrow p_{22} \rightarrow t_{23} \rightarrow p_{m1}$, $p_{m2} \rightarrow t_{21} \rightarrow p_{21} \rightarrow t_{22} \rightarrow p_{m2}$, and $p_{m2} \rightarrow t_{12} \rightarrow p_{12} \rightarrow t_{13} \rightarrow p_{m2}$. The number n in p_{10} and p_{20} represents that there are n tokens in the places (implying the number of A- and B-parts to be processed as variables). In this model, because of the sharing of two machines, $(p_{m1}, (t_{11}, t_{12}), (t_{22}, t_{23}))$ and $(p_{m2}, (t_{12}, t_{13}), (t_{21}, t_{22}))$ are two PME structures.

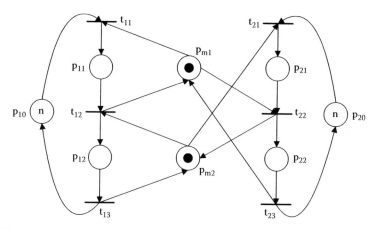

FIGURE 4.4 Process-oriented PN for Example 4.1's part processing.

TABLE 4.1

Interpretation of Places and Transitions for Figure 4.4

Places		Transitions	
p_{10}	A-part waiting in storage	t_{11}	Load A-part to m_1 for operation 1
p_{11}	Operation 1 of A-part being processed by m_1	t_{12}	Load A-part to m_2 for operation 2
p_{12}	Operation 2 of A-part being processed by m_2	t_{13}	Move a finished part to storage
p_{20}	B-part waiting in storage	t_{21}	Load B-part to m_2 for operation 1
p_{21}	Operation 1 of B-part being processed by m_2	t_{22}	Load B-part to m_1 for operation 2
p_{22}	Operation 2 of B-part being processed by m_1	t_{23}	Move a finished part to storage
p_{m1}	m_1 ready		
p_{m2}	m_2 ready		

Example 4.2

Assume that there is only one type of parts with three operations to be processed in the system. It requires m_1 to perform operation 1, m_2 for operation 2, and again m_1 for operation 3. Like Example 4.1, we do not consider the robot operations, and the buffer is not used.

Similarly, we can obtain the POPN for this example, and it is shown in Figure 4.5. The places and transitions are interpreted in Table 4.2. The production process is modeled by $p_{10} \rightarrow t_{11} \rightarrow p_{11} \rightarrow t_{12} \rightarrow p_{12} \rightarrow t_{13} \rightarrow p_{13} \rightarrow t_{14} \rightarrow p_{10}$. The usage of resources m_1 and m_2 is described by $p_{m1} \rightarrow t_{11} \rightarrow p_{11} \rightarrow t_{12} \rightarrow p_{m1}, p_{m1} \rightarrow t_{13} \rightarrow p_{13} \rightarrow t_{14} \rightarrow p_{m1}$, and $p_{m2} \rightarrow t_{12} \rightarrow p_{12} \rightarrow t_{12} \rightarrow p_{m2}$. Because of different resource sharing modes, this time $(p_{m1}, (t_{11}, t_{12}), (t_{13}, t_{14}))$ forms an SME structure.

In fact, these two situations correspond to the parallel and sequential mutual exclusions (Zhou and DiCesare, 1991; Zhou et al., 1992), respectively.

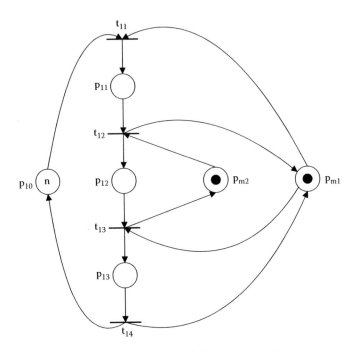

FIGURE 4.5 Process-oriented PN for Example 4.2's part processing.

TABLE 4.2

Interpretation of Places and Transitions for Figure 4.5

Places		Transitions	
P_{10}	Part waiting in storage	t_{11}	Load a part to m_1 for operation 1
P_{11}	Operation 1 being processed by m_1	t_{12}	Load a part to m_2 for operation 2
P_{12}	Operation 2 being processed by m_2	t_{13}	Load a part to m_1 for operation 3
P_{13}	Operation 3 being processed by m_1	t_{21}	Move a finished part to storage
P_{m1}	m_1 ready		
P_{m2}	m_2 ready		

4.3.2 Resource Sharing in Material Handling

In modeling AMS by POPN, a material handling device in the system, as a kind of resource, is treated just like a machine. We make some additional effort to discuss its modeling to facilitate the comparative studies in the later chapters.

Consider AMS shown in Figure 4.3. Assume that parts can be loaded into m_1 and m_2 automatically, and the robot is used to unload parts and final products from them.

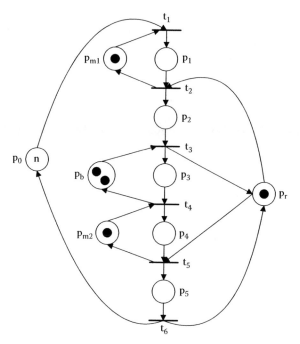

FIGURE 4.6 Process-oriented PN for Example 4.3.

Example 4.3

Assume that only one type of parts is to be processed in the system, and the buffer can hold two parts at a time. This part has two operations. Operation 1 should be performed by m_1 and operation 2 by m_2.

The POPN obtained for this example is shown in Figure 4.6 and Table 4.3. Note that because it models the material handling device just as it models a machine,

TABLE 4.3
Interpretation of Places and Transitions in Figure 4.6

Places		Transitions	
p_0	Part waiting in storage	t_1	Load a part to m_1
p_1	Operation 1 at m_1	t_2	Pick a part from m_1 by robot
p_2	Robot moving a part	t_3	Put a part into buffer by robot
p_3	Part waiting in the buffer	t_4	Load a part to m_2
p_4	Operation 2 at m_2	t_5	Pick a part from m_2 by robot
p_5	Robot moving a part	t_6	Move a finished part to storage by robot
p_{m1}	m_1 ready		
p_{m2}	m_2 ready		
p_b	Buffer slot available		
p_r	Robot available		

places p_2 and p_5 are used to model the robot moving a part from m_1 and m_2, respectively. p_b models the storage of the intermediate buffer. The production process is modeled by $p_0 \rightarrow t_1 \rightarrow p_1 \rightarrow t_2 \rightarrow p_2 \rightarrow t_3 \rightarrow p_3 \rightarrow t_4 \rightarrow p_4 \rightarrow t_5 \rightarrow p_5 \rightarrow t_6 \rightarrow p_0$. The usage of machines and buffer is described by $p_{m1} \rightarrow t_1 \rightarrow p_1 \rightarrow t_2 \rightarrow p_{m1}$, $p_{m2} \rightarrow t_4 \rightarrow p_4 \rightarrow t_5 \rightarrow p_{m2}$ and $p_b \rightarrow t_3 \rightarrow p_3 \rightarrow t_4 \rightarrow p_b$. The usage of robot is described by $p_r \rightarrow t_2 \rightarrow p_2 \rightarrow t_3 \rightarrow p_r$ and $p_r \rightarrow t_5 \rightarrow p_5 \rightarrow t_6 \rightarrow p_r$. It can be seen that this time with resource p_r, an SME (p_r, (t_2, t_3), (t_5, t_6)) is formed.

If a PN model obtained by process-oriented modeling satisfies $|{}^\bullet p| = |p^\bullet| = 1$ for all the places other than the resource places, it is called an augmented marked graph (Chu and Xie, 1997). If the part operations (due to flexible routes) are modeled as a state machine and then resource places are added, the resulting POPN is called an extended resource control net ERCN-merged net (Xie and Jeng, 1999).

4.4   CHARACTERISTICS OF POPN

From the examples, we know that POPN can model the production processes straightforwardly. Hence, it is easy to understand. Furthermore, POPN can model the system in detail. However, when the number of part types or the number of operations for each part is large, the resulting POPN is large. This in general increases the complexity for analysis and control.

It can be shown that the POPN obtained above is conservative. Notice that when a token is in an operation place, this implies that a job is processed by a resource. Thus, let $y = (y_1, y_2, \ldots, y_m)^T$ be a vector corresponding to the place set such that $y_i = 2$ if y_i corresponds to an operation place; otherwise, $y_i = 1$. Then $M^T y = M_0^T y$ for any $M \in R(M_0)$. For example, for the POPN obtained in Figures 4.3 to 4.6, let the entries in vector y that correspond to the operation places be 2, and the others be 1. Then $M^T y$ is a constant. In Figure 4.5, if the place vector is (p_{10}, p_{11}, p_{12}, p_{13}, p_{m1}, p_{m2}), let $y = (1, 2, 2, 2, 1, 1)^T$, and then $M^T y = n + 2$ holds for any marking $M \in R(M_0)$. It can be shown that if the resulting POPN is live, then it is also reversible (Xie and Jeng, 1999). Thus, an important issue is how we can analyze the liveness of the resulting models to be discussed next.

It is well known that the difficulty in operating an AMS comes from the limited resources. The resources in the system are shared by a number of jobs and used in a mutually exclusive way. Consider the set of places $S_1 = \{p_{m1}, p_{m2}, p_{12}, p_{22}\}$ in the PN shown in Figure 4.4. It follows from the definition of siphon that S_1 is a siphon. It is easy to verify that after firing t_{11} and t_{21}, S_1 is emptied. The system is deadlocked. Hence, this siphon is an ill-behaved siphon (IBS). Now let us examine the POPN in Figure 4.5. When the system reaches a marking M such that places p_{11} and p_{12} are both marked, and places p_{m1}, p_{m2}, and p_{13} are emptied, places p_{m1}, p_{m2}, and p_{13} will never be marked again. It can be verified that $S_2 = \{p_{m1}, p_{m2}, p_{13}\}$ is an IBS. Similarly, in Figure 4.6, we can verify that $S_3 = \{p_r, p_b, p_{m2}, p_5\}$ is also an IBS.

The IBS in POPN can be formed in different and complex ways. For example, the PN shown in Figure 4.7 describes an AMS that contains three machines and two part types with five operations. It is easy to check that there is an IBS $S_4 = \{p_{m1}, p_{m2}, p_{m3}, p_{13}, p_{22}\}$. Now consider the PN in Figure 4.8 for an AMS with three machines and three part types; $S_5 = \{p_{m1}, p_{m2}, p_{m3}, p_{12}, p_{22}, p_{32}\}$ is also an IBS.

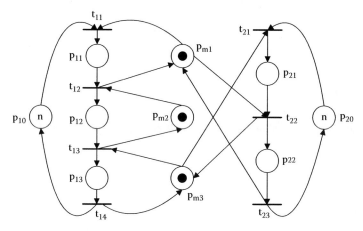

FIGURE 4.7 Process-oriented PN with two processes and three PMEs.

The IBS in POPN is closely related to a circular structure (Xie and Jeng, 1999). The circular structures corresponding to S_{1-5} are $RC_1 = \{p_{m1}, t_{22}, p_{m2}, t_{12}, p_{m1}\}$, $RC_2 = \{p_{m1}, t_{13}, p_{m2}, t_{12}, p_{m1}\}$, $RC_3 = \{p_r, t_5, p_{m2}, t_4, p_b, t_3, p_r\}$, $RC_4 = \{p_{m1}, t_{22}, p_{m3}, t_{13}, p_{m2}, t_{12}, p_{m1}\}$, and $RC_5 = \{p_{m1}, t_{32}, p_{m3}, t_{22}, p_{m2}, t_{12}, p_{m1}\}$. Note that in each of these circular structures, all the places are resource places in a POPN; we call such a circular structure a resource circuit (RC).

Note that both place sets P_A and P_B do not contain p_{i0}, representing the central storage (representing job order status). Further, assume that $p_i \in P_B$, and let $T_{ai} = \{t_{ai} | t_{ai} \in p_i{}^\bullet\}$ and $T_{bi} = \{t_{bi} | t_{bi} \in {}^\bullet p_i\}$. Then for $p_i \in P_B$, $\forall t_{ai} \in {}^\bullet p_i$, $\exists t_{bi} \in p_i{}^\bullet$, such that

$$t_{ai}{}^\bullet \cap {}^\bullet t_{bi} \subset P_A \qquad (4.1)$$

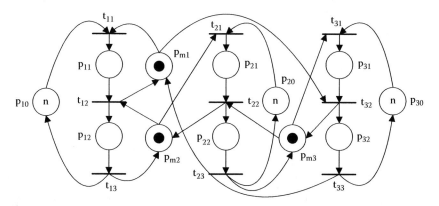

FIGURE 4.8 Process-oriented PN with three processes and three PMEs.

Because resources are conservative and each operation requires one resource, we further have:

$$|t_{ai}{}^{\bullet} \cap {}^{\bullet}t_{bi}| = 1 \tag{4.2}$$

Assume $\{p_1, t_k, p_k, t_{k-1}, \ldots, p_{i+1}, t_i, p_i, \ldots, t_1, p_1\}$ is an RC, or places p_{1-k} are resource places and $t_1 \in {}^{\bullet}p_1$ and $t_1 \in p_2{}^{\bullet}$, $t_2 \in {}^{\bullet}p_2$ and $t_2 \in p_3{}^{\bullet}$, $\ldots$, $t_i \in {}^{\bullet}p_i$ and $t_i \in p_{i+1}{}^{\bullet}$, $\ldots$, $t_k \in {}^{\bullet}p_k$ and $t_k \in p_1{}^{\bullet}$. Let $T_k = \{t_1, t_2, \ldots, t_k\}$, $P_I = ({}^{\bullet}T_k - T_k{}^{\bullet}) \cap P_A$ and $P_S = (T_k{}^{\bullet} - {}^{\bullet}T_k) \cap P_A$. Then we have the following property:

Property 4.1: Assume that in a POPN, RC $= \{p_1, t_k, p_k, t_{k-1}, \ldots, p_{i+1}, t_i, p_i, \ldots, t_1, p_1\}$ is a resource circuit, $M_0(p) = 0$ for $\forall p \in P_A$, and each place p_{j0} is sufficiently marked at marking M_0. Then $S = \{p_1, p_2, \ldots, p_k\} \cup P_S$ forms an IBS.

Proof: First we show that S is a siphon. Because RC is a circuit and $P_m = \{p_1, p_2, \ldots, p_k\} \subset S$, $\{t_1, t_2, \ldots, t_k\} \subset {}^{\bullet}P_m \subset {}^{\bullet}S$ and $\{t_1, t_2, \ldots, t_k\} \subset P_m{}^{\bullet} \subset S^{\bullet}$. We also have $P_S{}^{\bullet} \subset {}^{\bullet}P_m \subset {}^{\bullet}S$ and $P_S{}^{\bullet} \subset S^{\bullet}$ due to 4.1. At the same time, ${}^{\bullet}P_I \subset P_m{}^{\bullet} \subset S^{\bullet}$ according to Expressions 4.1 and 4.2, but P_I is not in S, or ${}^{\bullet}P_I$ is not in ${}^{\bullet}S$. Thus, we have ${}^{\bullet}S = {}^{\bullet}P_m = P_S{}^{\bullet} \cup \{t_1, t_2, \ldots, t_k\}$, and $P_S{}^{\bullet} \subset S^{\bullet}$, $\{t_1, t_2, \ldots, t_k\} \subset S^{\bullet}$, ${}^{\bullet}P_I \subset S^{\bullet}$, or ${}^{\bullet}S \subset S^{\bullet}$. Thus, S is a siphon.

Let $T_b = \{t_{bi}\}$ and $T_a = \{t_{ai}\}$. Then $T_k \subset T_b$. Because $\{p_1, t_k, p_k, t_{k-1}, \ldots, p_{i+1}, t_i, p_i, \ldots, t_1, p_1\}$ is a circuit, $\forall t_i \in {}^{\bullet}p_i$ and $t_i \notin {}^{\bullet}(P - p_i)$ hold. Then, for $\forall p_i, t_i \notin T_b$. Hence, $\exists t_{ai} \in T_a, t_{ai} \notin T_k$, and ${}^{\bullet}t_i \cap t_{ai} = 1$. Thus, let $T_{am} = \{t_{ami}\}$ and $P_F = T_{am}{}^{\bullet} \cap {}^{\bullet}T_k$; we have $|P_F| = |T_{am}| = k$. Assume that at some moment the system is in initial marking, or every resource place is marked and n in p_0 is large enough to consume all the resources. At this time there exists a firing sequence by firing the transitions in T_{am} such that all the k places in P_F become marked and at the same time empty the k resource places $p_1, p_2, \ldots,$ and p_k. Notice that any place $p \in P_S$ is not in P_F, and remains unmarked. In this way, place set S is empty. Thus, S is an IBS. ∎

To illustrate Property 4.1, consider the POPN shown in Figure 4.7. There is a circuit $\{p_{m1}, t_{22}, p_{m3}, t_{13}, p_{m2}, t_{12}, p_{m1}\}$ with all the places being resource places. We have $P_I = \{p_{11}, p_{12}, p_{21}\}$ and $P_S = \{p_{13}, p_{22}\}$. After firing transitions t_{11}, t_{21}, t_{12}, and t_{11}, all the places in P_I are marked, and $S = \{p_{m1}, p_{m2}, p_{m3}, p_{13}, p_{22}\}$ is emptied and can never be marked again.

From Property 4.1 we have seen the important role that an RC plays. We know that if there exists such an RC, there exists an IBS if Property 4.1's condition is met. It is obvious that in a POPN, if there is no such RC, then after the PN reaches any marking, the resources can be finally released. Hence, every transition is live, or there is no IBS. This implies that if IBS exists, then there must be an RC.

4.5   SUMMARY

This chapter presents the POPN modeling method for AMS. It is shown that a POPN is a straightforward way to model an AMS. In a POPN model, resource sharing can be described by PME and SME. The absence and existence of IBS in a POPN model determine the liveness of the model. The relationship between IBS and RC is also given. Later, the POPN modeling method will be compared with another modeling method, resource-oriented PN (ROPN) modeling.

REFERENCES

Al-Jaar, R. Y., and A. A. Desrochers. 1990. Performance evaluation of automated manufacturing systems using generalized stochastic Petri nets. *IEEE Transactions on Robotics and Automation* 6:621–39.

Banaszak, Z. A., and B. H. Krogh. 1990. Deadlock avoidance in flexible manufacturing systems with concurrently competing process flows. *IEEE Transactions on Robotics and Automation* 6:724–34.

Chu, F., and X. L. Xie. 1997. Deadlock analysis of Petri nets using siphons and mathematical programming. *IEEE Transactions on Robotics and Automation* 13:793–804.

Ezpeleta, J., J. M. Colom, and J. Martinez. 1995. A Petri net based deadlock prevention policy for flexible manufacturing systems. *IEEE Transactions on Robotics and Automation* 11:171–84.

Fanti, M. P., G. Maione, and B. Turchiano. 2002. Design of supervisors to avoid deadlock in flexible assembly system. *International Journal of Flexible Manufacturing Systems* 14:157–75.

Jeng, M. D., and F. DiCesare. 1995. Synthesis using resource control nets for modeling shared-resource systems. *IEEE Transactions on Robotics and Automation* 11:317–27.

Lawley, M. A. 1999. Deadlock avoidance for production systems with flexible routing. *IEEE Transactions on Robotics and Automation* 15:1–13.

Li, Z. W., and M. C. Zhou. 2004. Elementary siphons of Petri nets and their application to deadlock prevention in flexible manufacturing systems. *IEEE Transactions on Systems, Man, and Cybernetics A* 34:38–51.

Narahari, Y., and N. Viswanadham. 1985. A Petri net approach to modeling and analysis of FMS. *Annals of Operations Research* 3:449–72.

Park, J., and S. A. Reveliotis. 2002. Liveness-enforcing supervision of resource allocation systems with uncontrollable behavior and forbidden states. *IEEE Transactions on Robotics and Automation* 18:234–39.

Roszkowska, E. 2004. Supervisory control for deadlock avoidance in compound processes. *IEEE Transactions on System, Man, and Cybernetics A* 34:52–64.

Viswanadham, N., and Y. Narahari. 1987. Colored Petri net models for automated manufacturing systems. Paper presented at Proceedings of IEEE International Conference on Robotics and Automation, Raleigh, NC.

Wu, N. Q., M. C. Zhou, and Z. W. Li. 2008. Resource-oriented Petri net for deadlock avoidance in flexible assembly systems. *IEEE Transactions on System, Man, and Cybernetics A* 38:56–69.

Xie, X., and M. D. Jeng. 1999. ERCN-merged nets and their analysis using siphons. *IEEE Transactions on Robotics and Automation* 15:692–703.

Xing, K. Y., B. S. Hu, and H. X. Chen. 1996. Deadlock avoidance policy for Petri net modeling of flexible manufacturing systems with shared resources. *IEEE Transactions on Robotics and Automation* 41:289–95.

Zhou, M. C., and F. DiCesare. 1991. Parallel and sequential mutual exclusions for Petri net modeling of manufacturing systems with shared resources. *IEEE Transactions on Robotics and Automation* 7:515–27.

Zhou, M. C., and F. DiCesare. 1993. *Petri net synthesis for discrete event control of manufacturing systems*. Boston: Kluwer Academic Publications.

Zhou, M., F. DiCesare, and A. Desrochers. 1992. A hybrid methodology for synthesis of Petri nets for manufacturing systems. *IEEE Transactions on Robotics and Automation* 18:350–61.

Zhou, M. C., and K. Venkatesh. 1998. *Modeling, simulation and control of flexible manufacturing systems: Petri net approach*. Singapore: World Scientific.

5 Resource-Oriented Petri Net Modeling

5.1 INTRODUCTION

When studying the deadlock avoidance problem in automated manufacturing systems (AMS), Wu (1997, 1999) and Wu and Zhou (2001a, 2001b) first proposed resource-oriented Petri nets (ROPNs) to model AMS. Such nets are then used to study the control problem in AGV systems and semiconductor manufacturing systems (Wu and Zhou, 2001a, 2001b, 2002, 2003, 2004, 2007a; Wu and Zeng, 2002). ROPN views a part manufacturing process as a part visiting the resources one by one according to its prescribed processing routes. In this net, each resource is modeled by a single place, and the part manufacturing processes are modeled by the paths on which the parts visit the resources. Since each part manufacturing process is modeled in an implicit way, the model is very compact, and some useful structural characteristics can be clearly revealed. This chapter presents the ROPN modeling method and analyzes the characteristics of the ROPN model for AMS.

5.2 STEPS OF ROPN MODELING

Different from POPN modelers, ROPN modelers think that part production processes in an AMS are a process of dynamic resource allocation, and a resource can be thought of as a server. Thus, they model each resource in the system by just a single place. The parts to be processed by the system come to the resources competing for them. By describing the part flow, the production processes can be modeled too. ROPN modelers treat machines (workstations) and their buffers as the same type resource. There is no difference between a part being processed by a machine and waiting in a buffer in the sense of resource occupation. However, they treat the material handling devices in a different way (the reason will be given later). Because the resources in manufacturing systems are limited, a finite capacity PN, together with its corresponding transition enabling and firing rules is used. The ROPN is a finite capacity PN, and the interpretations for places, transitions, and tokens are as follows:

1. In the ROPN model for an AMS, every place represents a resource. In an AMS there are various types of resources, such as machines, fixtures, tools, buffers, and material handling devices. However, in practice, the allocation of fixtures and tools is done at the planning level. When an operation is initiated at a machine for processing a part, the fixture and tools required for the operation must already be in place. Thus, one can mainly consider

only three types of resources in the modeling, i.e., machines, buffers, and material handling devices.

2. Machines and buffers are treated as the same type of resources and are named H-resources. Material handling devices are treated as another type of resources and are named G-resources.

3. If a place in an ROPN model represents the status of H-resources, a token in it indicates that a job utilizes the resource, i.e., a job is being processed or just sitting there; no token in it indicates that the resource is free.

4. If a place in an ROPN model represents the status of H-resources, its availability is determined by the number of the free spaces available in it. The free spaces are defined as the capacity of the place minus the number of tokens in it. If the number of free spaces is greater than zero, this indicates that the resource is available; no free space implies that it is not available. The capacity of such a place is a constant.

5. A place represents the central storage in an AMS as a special H-resource; its capacity is considered infinite. Thus, such a resource is always available. Tokens in such a place indicate that job orders are waiting for processing. No token in it indicates that there is no job for processing.

6. Consider a place that represents the status of H-resource. Its being not marked, implies that it is free and available. However, the place representing the central storage is marked.

7. If a place represents the status of G-resources, the initial number of tokens in it is a constant. A token in it indicates that the resource is available; no token in it indicates that it is not available.

8. A transition in an ROPN model represents that a job is moved from a resource to another resource.

With the interpretation of places and transitions above, modeling an AMS by ROPN can be done in two phases: (1) model the part production processes regardless of the material transportation, and (2) based on the model from phase 1, model the material delivery processes. A simple example will be used to illustrate the modeling method.

5.3   MODELING PART PRODUCTION PROCESSES

According to the interpretation of places above, only H-resources are involved in modeling part production processes by ROPN. An H-resource is modeled by an H-place, as shown in Figure 5.1. Place p's capacity represents the number of parts that can be held by the buffer or machine. A token in p means that there is a part in the buffer or machine. p's multiple input transitions represent the sharing of the machine or buffer by multiple jobs. Its multiple output transitions represent multiple jobs routed to different machines or buffers. Place p_i denotes the ith resource in the system.

Unlike the POPN modeling method that uses operation places to describe the activity sequence, ROPN modelers use no operation places but only resource places. Thus, the key is how to describe an activity sequence. The following method is used in modeling part production processes by ROPN:

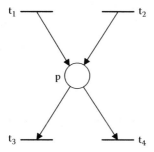

FIGURE 5.1 The PN model for an H-resource.

1. Identify the operations and H-resources required for the production of each route of one item of each production.
2. Order the operations and corresponding H-resources required. Assume that there are L operations for the production of a route of one item. Name the resource required for the first operation, γ_1, the resource required for the ith operation, γ_i, and for the last, γ_L. It is possible that the ith and jth operations with $i \neq j$ require the same resource. In this case, it is said that $\gamma_i = \gamma_j$.
3. Create a place that represents the central storage and label it p_0. Create a place that represents resource γ_1 and label it p_1. Add a transition between p_0 and p_1 with an arc from p_0 to the transition and an arc from the transition to p_1. Let $\Gamma = \{\gamma_1\}$ be a set of resources and $i = j = 1$. Form a subnet PN for each single route of an item as follows:
   a. Increase i by 1.
   b. Check whether γ_i is already in Γ. If not, go to step c; otherwise, do the following. Find the places that correspond to γ_{i-1} and γ_i, and assume they are p_h and p_x. Check whether there is a transition between p_h and p_x with an arc from p_h to the transition and an arc from the transition to p_x. If yes, go to step d; otherwise, add a transition between p_h and p_x with an arc from p_h to the transition and an arc from the transition to p_x. Then go to step d.
   c. Let $\Gamma = \Gamma + \{\gamma_i\}$ and $j = j + 1$. Create a place and label it p_j, and find the place that corresponds to γ_{i-1}. Assume the place corresponding to γ_{i-1} is p_h, and add a transition between p_h and p_j with an arc from p_h to the transition and an arc from the transition to p_j. Then go to step d.
   d. If $i = L$, go to the next step; otherwise, go to step a.
   e. Find the place that corresponds to γ_L. Assume the place corresponding to γ_L is p_h, and add a transition between p_h and p_0 with an arc from p_h to the transition and an arc from the transition to p_0. Then specify the initial marking and go to step 4.
4. Merge the subnets to form the whole ROPN according to the definition given later.
5. Define colors for the places and transitions in the model, and input and output functions to form a colored ROPN (CROPN) according to definitions given later.

To illustrate the ROPN modeling method, we use the AMS shown in Figure 4.1 as an example.

Example 5.1

The AMS is shown in Figure 4.1. Assume that there are two types of parts to be processed in the system: A-part and B-part. A-part has only one route: $m_1 \rightarrow m_2 \rightarrow m_3$. B-part has two routes: $m_1 \rightarrow m_2$ and $m_1 \rightarrow m_3$. Assume that the number of parts to be processed for A-part and B-part in the system is n_1 and n_2, respectively.

5.3.1 SUBNET FORMING

We first show how to form the subnets by using this example. For A-part, first create a place representing the central storage and label it p_0. Next, consider the resource for the first operation, i.e., m_1. Model it as p_1, and add a transition t_1 between p_0 and p_1 with an arc from p_0 to t_1 and an arc from t_1 to p_1. Consider the resource for the second operation, i.e., m_2. Model it as p_2, and add a transition t_2 between p_1 and p_2 with an arc from p_1 to t_2 and an arc from t_2 to p_2. Consider the resource for the third operation, i.e., m_3. Model it as p_3, and add a transition t_3 between p_2 and p_3 with an arc from p_2 to t_3 and an arc from t_3 to p_3. Since operation 3 is the last, add a transition t_4 between p_3 and p_0 with an arc from p_3 to t_4 and an arc from t_4 to p_0. Then mark p_0 as n_1. In this way, the subnet for A-part is formed and is shown in Figure 5.2a.

Similarly, the subnets for the two routes of B-part are obtained and shown in Figure 5.2b and c, respectively, where p_0 in Figure 5.2b is marked by n_{21} and p_0 in Figure 5.2c is marked by n_{22}, with $n_{21} + n_{22} = n_2$.

Note that if there is revisiting in a route for a production process, a revisited resource is modeled by only one H-place. Such a subnet is shown in Figure 5.3. The route for this subnet is $m_1 \rightarrow m_2 \rightarrow m_3 \rightarrow m_2 \rightarrow m_3$. In modeling such a production process, when we consider the second and third operations, p_2 and p_3 are created, and transition t_3 is added with an arc from p_2 to t_3 and an arc from t_3 to p_3. Then, when we consider the fourth operation, the resource required is m_2 and place p_2 corresponding to m_2 is already in the model. This time we do not create any place and just add transition t_4 and the corresponding arcs. When we consider the fifth operation, the resource required is m_3 and place p_3 corresponding to m_3 is already in the model. Further, there is already a transition (t_3) that connects p_2 and p_3 with the same direction from p_2 to p_3. Thus, this time, we neither create any places nor add any transitions and arcs. The model is resource oriented.

5.3.2 SUBNET MERGING

Based on the subnets obtained above, ROPN for the whole system can be obtained by merging the subnets. To merge the subnets, we define the union of two PNs as follows:

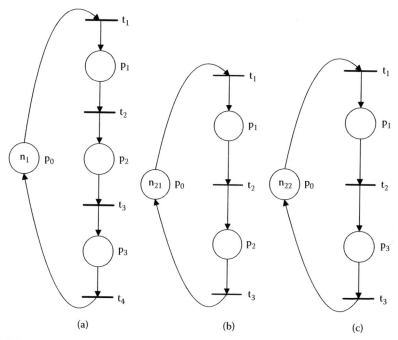

FIGURE 5.2 Subnets for Example 5.1: (a) subnet for A-part with route $m_1 \to m_2 \to m_3$; (b) subnet for route 1 of B-part with route $m_1 \to m_2$; and (c) subnet for route 2 of B-part with route $m_1 \to m_3$.

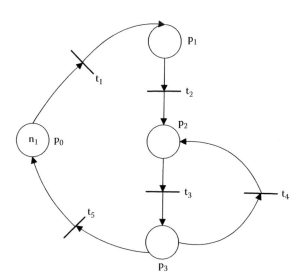

FIGURE 5.3 Subnet with a revisiting route $m_1 \to m_2 \to m_3 \to m_2 \to m_3$.

Definition 5.1: PN = (P, T, I, O, M_0) for an AMS and is said to be the union of two subnets, $PN_i = (P_i, T_i, I_i, O_i, M_{i0})$ and $PN_j = (P_j, T_j, I_j, O_j, M_{j0})$, for part types i and j if

$$P = P_i \cup P_j, \text{ and } T = T_i \cup T_j \tag{5.1}$$

$$M_0(p) = \begin{cases} M_{i0}(p), p \in P_i - P_j \\ M_{j0}(p), p \in P_j - P_i \\ M_{i0}(p) + M_{j0}(p), p \in P_i \cap P_j \end{cases} \tag{5.2}$$

$$I(p,t) = \begin{cases} I_i(p,t), p \in P_i, t \in T_i \\ I_j(p,t), p \in P_j, t \in T_j \\ 0, \text{ otherwise} \end{cases} \tag{5.3}$$

$$O(p,t) = \begin{cases} O_i(p,t), p \in P_i, t \in T_i \\ O_j(p,t), p \in P_j, t \in T_j \\ 0, \text{ otherwise} \end{cases} \tag{5.4}$$

If $p_1, p_2 \in P_i \cap P_j$, the union of PN_i and PN_j merges two transitions, $p_1 \rightarrow p_2$ ($p_2 \rightarrow p_1$) in PN_i and $p_1 \rightarrow p_2$ ($p_2 \rightarrow p_1$) in PN_j, as one. Thus, there is at most one transition between two places along the same direction in the union PN. For example, in each subnet shown in Figure 5.2, there is a transition from p_0 to p_1, a transition from p_1 to p_2 in both subnets a and b, and a transition from p_3 to p_0 in both a and c. Such transitions will be merged into one. By defining the union of multiple subnets in the same way, we obtain the ROPN model for the whole system. The union PN of the three subnets in Figure 5.2 is shown in Figure 5.4. Place p_0 is marked with $M(p_0) = n = n_1 + n_2$. Notice that the obtained PN is structurally a state machine. Hence, it is structurally simple. Since each transition has one and only one input and output place in an ROPN, and every job's production is modeled by a circuit, it is graphically a strongly connected state machine PN.

5.3.3　COLORED ROPN

In a free-choice PN, a token in a place enables all the output transitions of that place, or any one of the output transitions can be chosen to fire. In ROPN for AMS, however, a token in a place does not enable all the output transitions of that place. For example, in Figure 5.4, a token representing B-part in place p_1 enables both transitions t_2 and t_3, but a token representing A-part in that place enables only t_2. A token in p_2 representing A-part enables t_4, and if representing B-part, it enables t_5. This implies that when a token is in a place, only some of the output transitions of that place can be chosen to fire; this is the difference between the ROPN and free-choice PN. To model this characteristic, we introduce colors into the ROPN, and the ROPN becomes colored ROPN (CROPN).

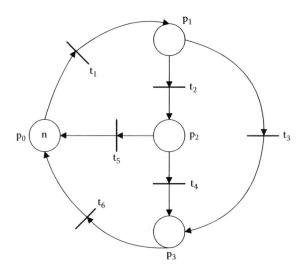

FIGURE 5.4 The ROPN for Example 5.1 with $n = n_1 + n_2$.

A simple way to introduce colors to the system is to name a color for each job, and each job keeps the color all the time, as is done in Chapter 3. However, doing so in the ROPN will lead to confusion when there exists revisiting for a production route. Consider the net shown in Figure 5.3; when a token in p_3 is being processed for the third operation, it enables t_4. However, if it is being processed for the last operation, it enables t_5. In these two cases, if the token has the same color, the net execution is problematic. To solve the problem, we first define colors for transitions.

Definition 5.2: $C(t_i) = \{c_i\}$ is the only color of transition $t_i \in T$.

This means that each transition in CROPN has a unique color, and different transitions have different colors. Based on Definition 5.2, a token's color is defined as follows:

Definition 5.3: If transition $t_j \in p_i^{\bullet}$, the tokens in place p_i enabling t_j have the same color c_j as t_j does. The number of tokens with color c_j in p_i at marking M is denoted by $M(p_i, c_j)$.

With Definitions 5.2 and 5.3 the enabling and firing rules for finite capacity PN and colored PN can be applied to the CROPN. Note that the colors in CROPN are not used to distinguish part types as other models do. They are used to describe the enabling relation, and the color of a token may change when it moves from one place to another. For example, in Figure 5.3, a token representing A-part in p_1 has color c_2 and a token representing A-part in p_2 has color c_4. The goal here is to describe where a token in a place goes next.

Because of the routing flexibility, a token in place p may enable more than one transition simultaneously. Hence, the token has a color set $\{c_1, \ldots, c_k\}$ corresponding to transition set $\{t_1, \ldots, t_k\}$ with $k \geq 1$. This implies that this token can be used to fire any output transitions of place p with color c_i, $i = 1, \ldots, k$. For example, a token representing B-part in place p_1 of Figure 5.4 has a set of colors $\{c_2, c_3\}$. In this way, we complete the modeling of the part production processes.

Now we are ready to define the colors and input and output functions for Example 5.1. According to Definition 5.2, $C(t_i) = \{c_i\}$ for transitions t_{1-6}. Let $C(p_i, A_j)$ and $C(p_i, B_j)$ denote the colors of tokens of A-part and B-part in p_i for the processing of the jth operation with the 0th operation as the raw part, respectively. Then, $C(p_0, A_0) = C(p_0, B_0) = \{c_1\}$, $C(p_1, A_1) = \{c_2\}$, $C(p_1, B_1) = \{c_2, c_3\}$, $C(p_2, A_2) = \{c_4\}$, $C(p_2, B_2) = \{c_5\}$, and $C(p_3, A_3) = C(p_3, B_2) = \{c_6\}$. Based on colors for the places and transitions, the input and output functions in matrix format are as follows:

$$I(p_0, t_1) = \begin{bmatrix} 1 \\ 1 \end{bmatrix}, \quad O(p_0, t_1) = \begin{bmatrix} 0 \\ 0 \end{bmatrix}, \quad I(p_0, t_2) = O(p_0, t_2) = \begin{bmatrix} 0 \\ 0 \end{bmatrix}$$

$$I(p_0, t_3) = O(p_0, t_3) = \begin{bmatrix} 0 \\ 0 \end{bmatrix}, \quad I(p_0, t_4) = O(p_0, t_4) = \begin{bmatrix} 0 \\ 0 \end{bmatrix}$$

$$I(p_0, t_5) = \begin{bmatrix} 0 \\ 0 \end{bmatrix}, \quad O(p_0, t_5) = \begin{bmatrix} 0 \\ 1 \end{bmatrix}$$

$$I(p_1, t_1) = \begin{bmatrix} 0 \\ 0 \end{bmatrix}, \quad O(p_1, t_1) = \begin{bmatrix} 1 \\ 1 \end{bmatrix}, \quad I(p_1, t_2) = \begin{bmatrix} 1 \\ 1 \end{bmatrix}, \quad O(p_1, t_2) = \begin{bmatrix} 0 \\ 0 \end{bmatrix}$$

$$I(p_1, t_3) = \begin{bmatrix} 0 \\ 1 \end{bmatrix}, \quad O(p_1, t_3) = \begin{bmatrix} 0 \\ 0 \end{bmatrix}, \quad I(p_1, t_4) = O(p_1, t_4) = \begin{bmatrix} 0 \\ 0 \end{bmatrix}$$

$$I(p_1, t_5) = O(p_1, t_5) = \begin{bmatrix} 0 \\ 0 \end{bmatrix}, \quad I(p_1, t_6) = O(p_1, t_6) = \begin{bmatrix} 0 \\ 0 \end{bmatrix}$$

$$I(p_2, t_1) = O(p_2, t_1) = \begin{bmatrix} 0 \\ 0 \end{bmatrix}, \quad I(p_2, t_2) = \begin{bmatrix} 0 \\ 0 \end{bmatrix}, \quad O(p_2, t_2) = \begin{bmatrix} 1 \\ 1 \end{bmatrix}$$

$$I(p_2, t_3) = O(p_2, t_3) = \begin{bmatrix} 0 \\ 0 \end{bmatrix}, \quad I(p_2, t_4) = \begin{bmatrix} 1 \\ 0 \end{bmatrix}, \quad O(p_2, t_4) = \begin{bmatrix} 0 \\ 0 \end{bmatrix}$$

$$I(p_2, t_5) = \begin{bmatrix} 0 \\ 1 \end{bmatrix}, \quad O(p_2, t_5) = \begin{bmatrix} 0 \\ 0 \end{bmatrix}, \quad I(p_2, t_6) = O(p_2, t_6) = \begin{bmatrix} 0 \\ 0 \end{bmatrix}$$

$$I(p_3, t_1) = O(p_3, t_1) = \begin{bmatrix} 0 \\ 0 \end{bmatrix}, \quad I(p_3, t_2) = O(p_3, t_2) = \begin{bmatrix} 0 \\ 0 \end{bmatrix}$$

$$I(p_3, t_3) = \begin{bmatrix} 0 \\ 0 \end{bmatrix}, \quad O(p_3, t_3) = \begin{bmatrix} 0 \\ 1 \end{bmatrix}, \quad I(p_3, t_4) = \begin{bmatrix} 0 \\ 0 \end{bmatrix}, \quad O(p_3, t_4) = \begin{bmatrix} 1 \\ 0 \end{bmatrix}$$

$$I(p_3, t_5) = O(p_3, t_5) = \begin{bmatrix} 0 \\ 0 \end{bmatrix}, \quad I(p_3, t_6) = \begin{bmatrix} 0 \\ 0 \end{bmatrix}, \quad O(p_3, t_6) = \begin{bmatrix} 1 \\ 1 \end{bmatrix}$$

Up to now, we have completed the CROPN modeling for part production processing. It is clear that by introducing colors into ROPN, the part flows are accurately described. Further, for each transition, only a single color is needed. Thus it is easy to analyze. The next step is to model the material handling process.

5. MODELING MATERIAL HANDLING PROCESSES

In modeling the production processes by ROPN, a resource containing a machine or buffer is modeled by a single place. A token in it represents that the machine is processing a part, or the buffer is holding a part. Thus, such a place plays the role of both resource and operation. At the same time, all the transitions in an ROPN represent the transfer of materials, or they model the transfer operations. Thus, if a place that represents a material handling device acts as both a resource and transfer operation, there will be a conflict. Therefore, it is better to treat the material handling resource differently. The problem is how to model such devices.

To solve this problem, we consider the model shown in Figure 5.5a. It is the way in which POPN models the behavior of a material handling device. Transition t_1 represents that the device picks a part up, while t_2 means that the device loads the part into the destination place, and p_{rm} models that the robot moves from one place to another place while carrying a part. This model focuses on how a robot transfers a part from one place to another. Consider the ROPN obtained without considering the material handling process; each transition, in fact, represents the transfer of a part from one place to another. Thus, if a material handling device is treated just as a machine, then each transition in an ROPN must be replaced by a subnet given

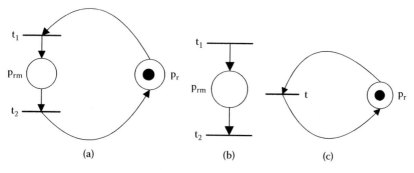

FIGURE 5.5 The modeling of a material handling device.

in Figure 5.5b. This violates the principle that ROPN models each resource by just a single place. Consider the model shown in Figure 5.5a again. It is seen that t_1, p_{rm}, and t_2 together just means that a part is transferred from one place to another (including pickup, moving, and unloading). Thus, it may be enough to model this process by just a single transition t that models the same process as a transition in ROPN with material handling process unconsidered. Therefore, a material handling device can be modeled by a PN shown in Figure 5.5c. In this way, each material handling device can be modeled by places. Thus the ROPN's principle is not violated and the multiple transitions connected from this place imply the sharing of the device among multiple transfer jobs. Therefore, unlike the models of machines and buffers where operations are embedded in the places, we use a place to model the availability of the material handling device, and its operations are associated with transitions that exist already in ROPN. In this way, there is still one-to-one mapping each of H and G resources and each of the places.

Based on the modeling method for a material handling process, the ROPN model is completed and shown in Figure 5.6. It can be seen that for the material handling process, the self-loops (p_{r1}, t_1), (p_{r1}, t_3), (p_{r1}, t_5), (p_{r1}, t_6), (p_{r2}, t_2), and (p_{r3}, t_4) are added into the ROPN for the production processes shown in Figure 5.4. Notice that we use a bidirectional arc to represent a self-loop for simplicity.

5.5 RESOURCE SHARING IN ROPN

Based on the modeling method of ROPN, we can analyze resource sharing in ROPN models. For the goal of comparison with POPN, we use Examples 4.1, 4.2, and 4.3 in Chapter 4.

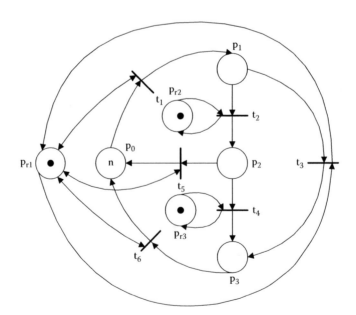

FIGURE 5.6 The ROPN for Example 5.1 with the material handling process added.

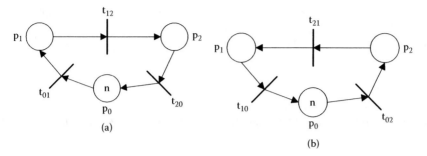

FIGURE 5.7 The subnets for Example 4.1.

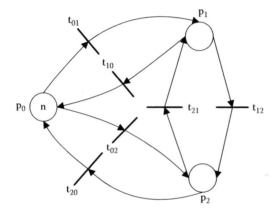

FIGURE 5.8 The ROPN for Example 4.1's part processing.

Following the modeling method of ROPN, the subnets for Example 4.1 are obtained as shown in Figure 5.7. The two subnets are then merged into the ROPN shown in Figure 5.8. The interpretation of places and transitions is given in Table 5.1. The colors for places and transitions, and the input and output functions, can be defined accordingly.

TABLE 5.1
Interpretation of Places and Transitions for Figure 5.8

	Places		Transitions
p_0	Parts in storage	t_{01}	Load a part to m_1
p_1	A part being processed by m_1	t_{10}	Move a finished part to storage
p_2	A part being processed by m_2	t_{02}	Load a part to m_2
		t_{20}	Move a finished part to storage
		t_{12}	Load a part to m_2
		t_{21}	Load a part to m_1

The ROPN models for Examples 4.2 and 4.3 can be obtained, as shown in Figures 5.9 and 5.10, by using the modeling method. It is interesting to note that the structure of the ROPN model is different from that of POPN. In ROPN, there is no PME or SME structure that describes the resource sharing in a POPN model. However, there are circuits in ROPN that model the resource sharing in ROPN.

It should be noted that the resulting PN model, by adding the material handling system, is no longer a state machine and is called an augmented state machine, but the liveness of the original PN is still held. At the first view, such a change in modeling may not be meaningful. Later, it is shown that such modeling can avoid certain IBSs that exist in a process-oriented model.

5.6 CHARACTERISTICS OF ROPN

Observing the ROPNs in Figures 5.8 and 5.9, both models have a circuit $\{p_1, t_{12}, p_2, t_{21}, p_1\}$. Such a circuit is special in ROPN and greatly affects the behavior of the system. We call such an elementary circuit that excludes p_0 a production process circuit (PPC) (Wu, 1999).

Definition 5.4: A circuit in an ROPN is called a production process circuit (PPC), denoted by v, if it does not contain place p_0 representing the load/unload station.

A place p at marking M is said to be full if $M(p) = K(p)$. Let $P(v)$ denote the set of places on PPC v. A PPC v at M is said to be full if $\sum M(p_i) = \sum K(p_i) = K(v)$, $p_i \in P(v)$. It is easy to verify that when deadlock occurs, some PPCs are full. The system can be deadlocked even if only one PPC is full. Thus, we have the following results:

Lemma 5.1: If an ROPN is not live, then there exists a PPC in the net.

Lemma 5.2: The following two statements are equivalent:

1. An ROPN is not live.
2. An ROPN is not deadlock-free.

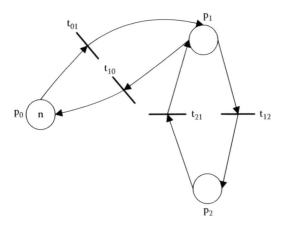

FIGURE 5.9 The ROPN for Example 4.2's part processing.

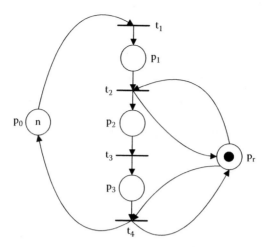

FIGURE 5.10 The ROPN model for Example 4.3, including material handling.

Lemmas 5.1 and 5.2 show that in a CROPN, if deadlocks occur, then they occur in some PPCs. Thus, we have the following result:

Theorem 5.1: The necessary condition for the existence of potential deadlocks in a CROPN is that there exists at least one PPC v in the net such that $M_0(p_0) \geq K(v)$.

In the augmented state machine we have the following result:

Theorem 5.2: If the original CROPN is live, then the augmented CROPN, by adding the model of the material handling process, is also live.

5.7   SUMMARY

This chapter presents the modeling methodology of ROPN for AMS. The resulting ROPN model for AMS is a kind of colored and finite capacity PN. It uses different concepts of colors as defined in Chapter 3. The method to define colors and the input and output functions is also presented. Because theory for CPN is well developed, CROPN can be an effective tool for modeling, analysis, and control of AMS. In this book, this method is used to model AMS, automated guided vehicle systems, and semiconductor manufacturing systems for analysis and control. The RORN models will be compared with their corresponding POPN models in Chapter 6.

REFERENCES

Wu, N. Q. 1997. Avoiding deadlocks in automated manufacturing systems with shared material handling system. In *Proceedings of 1997 IEEE International Conference on Robotics and Automation*, Albuquerque, NM, 2427–33.
Wu, N. Q. 1999. Necessary and sufficient conditions for deadlock-free operation in flexible manufacturing systems using a colored Petri net model. *IEEE Transactions on Systems, Man, and Cybernetics C* 29:192–204.
Wu, N. Q., and W. Q. Zeng. 2002. Deadlock avoidance in AGV system using colored Petri net model. *International Journal of Production Research* 40:223–38.

Wu, N. Q., and M. C. Zhou. 2001a. Avoiding deadlock and reducing starvation and blocking in automated manufacturing systems. *IEEE Transactions on Robotics and Automation* 17: 657–68.

Wu, N. Q., and M. C. Zhou. 2001b. Resource-oriented Petri nets in deadlock avoidance of AGV systems. In *Proceedings of 2001 IEEE International Conference on Robotics and Automation*, Seoul, Korea, 64–69.

Wu, N. Q., and M. C. Zhou. 2002. Deadlock avoidance in semiconductor track systems. In *Proceedings of 2002 IEEE International Conference on Robotics and Automation*, Washington, DC, 193–98.

Wu, N. Q., and M. C. Zhou. 2003. AGV routing for conflict resolution in AGV systems. In *Proceedings of 2003 IEEE International Conference on Robotics and Automation*, Taipei, Taiwan, 1428–33.

Wu, N. Q., and M. C. Zhou. 2004. Modeling and deadlock control of automated guided vehicle systems. *IEEE Transactions on Mechatronics* 9:50–57.

Wu, N. Q., and M. C. Zhou. 2007. Shortest routing of bi-directional automated guided vehicles avoiding deadlock and blocking. *IEEE/ASME Transactions on Mechatronics* 12: 63–72.

6 Process- vs. Resource-Oriented Petri Nets

Petri nets are widely used as a tool for modeling, simulation, performance evaluation, and control of automated manufacturing systems (AMS). Because of the complexity of AMS, regular PNs must be extended for their various applications. Thus, various extensions to regular PN have been made. For different goals, different extended PNs are used. Thus, it is meaningful to analyze the differences and applicability among different extended PNs.

Among different PNs, process-oriented Petri net (POPN) and resource-oriented Petri net (ROPN) have been effectively applied in modeling, analysis, and control of AMS. They were introduced in Chapters 4 and 5, respectively. A POPN models the dynamics of an AMS based on the activity sequence and resources sharing. Thus, part flows can be described straightforwardly. ROPN modelers see AMS from another viewpoint. They view a resource in an AMS as a server, and jobs visit servers for processing in a prescribed sequence. In this way, part flows in an AMS can also be modeled. Based on Wu et al. (2007), this chapter makes comparisons between them, bridges the relation between them, and also points out the differences between them.

6.1 MODELING POWER AND MODEL SIZE

As for the modeling power, the POPN is more powerful than ROPN. POPN models the production processes in a very intuitive way, can model a system in detail, and deals with complex resource requirements and other conditions. Although ROPN can model the production processes, colors should be introduced into it. Consequently, it is not in such an intuitive way. There are AMS where operations require multiple resources. POPN can model such a system with no difficulty, but ROPN cannot model such a system in any straightforward way.

In modeling a system intuitively, POPN uses much more places and transitions, so the model is large in size. ROPN uses only one place for a resource, leading to a very simple and compact model. A well-known drawback is that a PN model is usually large and complex, since if a PN is large in size, it is difficult to analyze and control. Thus, it is very important to make the PN model simple and concise. Thus, in the sense of model complexity, ROPN is better.

Assume that k is the number of job types, j is the average number of steps (operations), including the operations of material handling processes of all job types, and r is the total number of resources. Then it is easy to show the following result:

Theorem 6.1: The number of places $|P|$ and transitions $|T|$ for POPN and ROPN are

$$|P| = \begin{cases} O(kj+r), \text{for POPN} \\ O(r), \text{for ROPN} \end{cases}$$

and

$$|T| = \begin{cases} O(kj), \text{for POPN} \\ O(r^2), \text{for ROPN} \end{cases}$$

6.2 CONSERVATIVENESS

Conservativeness is another property to be required for PN modeling, since matter in an AMS must be conservative. We have shown in Chapter 4 that POPN is conservative. Now we show that ROPN is conservative too. Notice that a token representing a part never goes into a G-place, and a token representing a material handling device never goes into an H-place. Thus, conservativeness of a CROPN is independent of G-places. Hence, to show the conservativeness of a CROPN, one needs to show the conservativeness for the H-places in the CROPN. Assume that there are u H-places and k types of parts that are being processed in the system. Then, the incidence A must be an $n \times m$ matrix, with n being the number of transitions in CROPN and $m = u \times k$, because both $I(p_i, t_j)$ and $O(p_i, t_j)$ are a $k \times 1$ matrix. Assume that the dth row in $I(p_i, t_j)$ and $O(p_i, t_j)$ corresponds to the enabling and firing behavior of part type d. Let $X_i = (x_{i1}, x_{i2}, \ldots, x_{im})^T, i \in \{1, 2, \ldots, k\}$ such that

$$x_{if} = \begin{cases} 1, \text{if } f = kj+i, j = 0,1,\ldots,u-1 \\ 0, \text{otherwise} \end{cases}$$

and

$$X = [X_1, X_2, \ldots, X_k]$$

Then, we have the following result:

Theorem 6.2: $AX = 0$.

Proof: Let $A_d = (a_{d1}, a_{d2}, \ldots, a_{dm})$ be the dth row of A for $\forall d \in \{1, 2, \ldots, n\}$ and a_{df} with $f = kj + i$ for an $i \in \{1, 2, \ldots, k\}$, and $j = 0, 1, \ldots, u - 1$ be u elements in A_d. Then, with the structure of ROPN, we have that either all the a_{df} are zero or there are two elements $a_{d(kg+i)}$ and $a_{d(kh+i)}$ such that one is -1 and the other is 1, and the other a_{df}'s are zero. Hence, we have $A_d X_i = 0 \Rightarrow AX_i = 0 \Rightarrow AX = 0$. ∎

Notice that $X_1 + X_2 + \ldots, X_k = (1, 1, \ldots, 1)^T$. Thus, Theorem 6.2 implies that a CROPN is conservative.

Consider Example 5.1. From the definition of colors and the input and output functions induced for the CROPN, according to the theory for CPN given in Chapter 3, the incidence matrix A can be calculated as

$$
A^T = \begin{bmatrix}
-1 & -1 & 1 & 1 & 0 & 0 & 0 & 0 \\
0 & 0 & -1 & -1 & 1 & 1 & 0 & 0 \\
0 & 0 & 0 & -1 & 0 & 0 & 0 & 1 \\
0 & 0 & 0 & 0 & -1 & 0 & 1 & 0 \\
0 & 1 & 0 & 0 & 0 & -1 & 0 & 0 \\
1 & 1 & 0 & 0 & 0 & 0 & -1 & -1
\end{bmatrix}
$$

We can define a matrix X as

$$
X = \begin{bmatrix}
1 & 0 \\
0 & 1 \\
1 & 0 \\
0 & 1 \\
1 & 0 \\
0 & 1 \\
1 & 0 \\
0 & 1
\end{bmatrix}
$$

by letting $x_1 = (1, 0, 1, 0, 1, 0, 1, 0)^T$ and $x_2 = (0, 1, 0, 1, 0, 1, 0, 1)^T$, as given above. It is easy to verify that $A^T X = 0$. This implies that X is a P-invariant. Since each element in $x_1 + x_2 = (1, 1, 1, 1, 1, 1, 1, 1)^T$ is greater than zero, the PN is conservative. In fact, let $y = (1, 1, 1, 1)^T$; then $M^T y = M_0^T y$ for any $M \in R(M_0)$. Thus, there is no difference between them in the sense of conservativeness analysis.

6.3   STRUCTURE FOR LIVENESS

Another vitally important issue about PN modeling involves the liveness of the resulting PN. Of course, when the plant to be modeled is inherently nonlive, one cannot obtain a live model. The question is how the model reveals the property of nonliveness.

We have seen that in modeling the part production processes by POPN, the non-liveness of the model is described by the ill-behaved siphon (IBS). We have shown that, essentially, the nonliveness of a POPN results from the existence of resource circuits. Consider the two IBSs: $S_1 = \{p_{m1}, p_{m2}, p_{12}, p_{22}\}$ and $S_2 = \{p_{m1}, p_{m2}, p_{13}\}$ in the POPN models shown in Figures 4.4 and 4.5; we know that they result from $RC_1 = \{p_{m1}, t_{22}, p_{m2}, t_{22}\}$ and $RC_2 = \{p_{m1}, t_{13}, p_{m2}, t_{12}, p_{m1}\}$, respectively. By comparing RC_1 and RC_2 with the ROPN models in Figures 5.8 and 5.9, we find that these two RCs just correspond to a single PPC $\{p_1, t_{12}, p_2, t_{21}, p_1\}$. Thus, we have the following result:

Theorem 6.3: In modeling the part production processes by POPN and CROPN, when a PPC in the ROPN is full, the corresponding IBS in the POPN is emptied.

Proof: It follows from the modeling process of the ROPN that a PPC in an ROPN forms a resource circuit in the part production process. This resource circuit must correspond to one RC in the POPN, for they model the same process. When the PPC is full in the CROPN, it implies that the resources in the PPC are all occupied. It is known from the proof of Property 4.1 that all the resource places in the RC are emptied, so the IBS is emptied. ∎

Theorem 6.3 points out that in modeling part production processes, both POPN and ROPN can describe the potential deadlocks in a system, but in different ways. It should be noticed that PPCs in ROPN are structurally evident, while the IBSs are not so obvious to identify by the structure of a PN. Thus, it is more difficult to identify IBSs in POPN than PPCs in ROPN. We should also note that S_1 and S_2 are different in POPN. However, they result from the same RC, and thus they correspond to the same PPC in ROPN. This shows that RC and IBS reveal the same property.

Now consider the POPN and ROPN in Figures 4.6 and 5.10 for the same example, considering material handling. There is an IBS $S_3 = \{p_r, p_b, p_{m2}, p_5\}$ in the model shown in Figure 4.6, implying that there is potential deadlock in the system. However, after examining the model in Figure 5.10, there is no PPC, implying there is no deadlock at all. Why?

In the model shown in Figure 4.6, when p_r represents that availability of the robot is marked, t_2 can always fire, no matter whether p_b representing the space availability in the buffer is marked. If p_b is not marked and t_2 fires, it must be a deadlock. However, for the model shown in Figure 5.10, if p_2 representing the buffer is fully marked, i.e., reaching its token capacity, then t_2 can never fire, no matter whether p_r is marked. In this way, the deadlock is avoided just by ROPN modeling, not by any control law. The key point is that in ROPN, to fire t_2, the model enforces the system to check if there are enough spaces in the buffer, or it implies that the enabling of t_2 guarantees that there are enough spaces in the buffer. However, for POPN, the enabling of t_2 in Figure 4.6 just implies that the robot is available no matter whether there are spaces in the buffer or not. Hence, such deadlock can be completely avoided just by ROPN modeling.

Theorem 6.4: Potential deadlocks due to material handling devices in POPN are avoided in its corresponding ROPN.

6.4   EXAMPLE

To illustrate the result, we present the example adapted from Ezpeleta et al. (1995).

Example 6.1

The AMS adapted from Ezpeleta et al., 1995 and shown in Figure 6.1 is composed of four machines m_{1-4} and three robots r_{1-3}. Each machine can process two parts at a time.

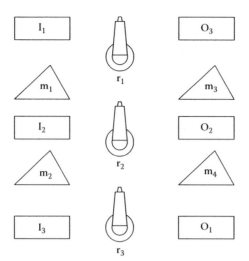

FIGURE 6.1 A manufacturing cell.

Three types of parts are to be processed, and their routes are $m_1 \rightarrow m_2$, $m_3 \rightarrow m_4$; m_2, and $m_4 \rightarrow m_3$, respectively. Robot r_1 serves for I_1, O_3, m_1, m_3; r_2 for I_2, O_2, $m_1 - m_4$; and r_3 for I_3, O_1, m_2, m_4.

The POPN model for this AMS is shown in Figure 6.2. It uses twenty-six places and twenty transitions. The production processes are modeled as

Part A: $p_{10} \rightarrow t_{10} \rightarrow p_{11} \rightarrow t_{11} \rightarrow p_{12} \rightarrow t_{13} \rightarrow p_{14} \rightarrow t_{15} \rightarrow p_{16} \rightarrow t_{17} \rightarrow p_{18} \rightarrow t_{19} \rightarrow p_{10}$, and $p_{10} \rightarrow t_{10} \rightarrow p_{11} \rightarrow t_{12} \rightarrow p_{13} \rightarrow t_{14} \rightarrow p_{15} \rightarrow t_{16} \rightarrow p_{17} \rightarrow t_{18} \rightarrow p_{18} \rightarrow t_{19} \rightarrow p_{10}$

Part B: $p_{20} \rightarrow t_{21} \rightarrow p_{21} \rightarrow t_{22} \rightarrow p_{22} \rightarrow t_{23} \rightarrow p_{23} \rightarrow t_{24} \rightarrow p_{20}$

Part C: $p_{30} \rightarrow t_{31} \rightarrow p_{31} \rightarrow t_{32} \rightarrow p_{32} \rightarrow t_{33} \rightarrow p_{33} \rightarrow t_{34} \rightarrow p_{34} \rightarrow t_{35} \rightarrow p_{35} \rightarrow t_{36} \rightarrow p_{30}$

The resource sharing is modeled as

m_1: $m_1 \rightarrow t_{11} \rightarrow p_{12} \rightarrow t_{13} \rightarrow m_1$

m_2: $m_2 \rightarrow t_{15} \rightarrow p_{16} \rightarrow t_{17} \rightarrow m_2$, $m_2 \rightarrow t_{22} \rightarrow p_{22} \rightarrow t_{23} \rightarrow m_2$

m_3: $m_3 \rightarrow t_{12} \rightarrow p_{13} \rightarrow t_{14} \rightarrow m_3$, $m_3 \rightarrow t_{34} \rightarrow p_{34} \rightarrow t_{35} \rightarrow m_3$

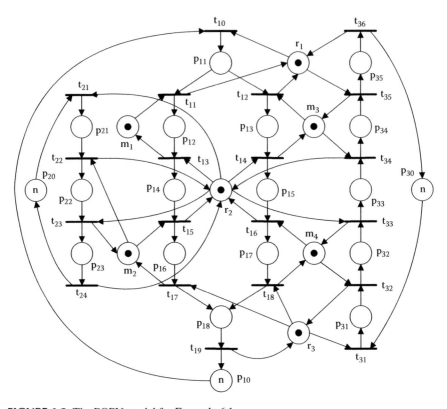

FIGURE 6.2 The POPN model for Example 6.1.

$m_4: m_4 \rightarrow t_{16} \rightarrow p_{17} \rightarrow t_{18} \rightarrow m_4, m_4 \rightarrow t_{32} \rightarrow p_{32} \rightarrow t_{33} \rightarrow m_4$

$r_1: r_1 \rightarrow t_{10} \rightarrow p_{11} \rightarrow t_{11} \rightarrow r_1, r_1 \rightarrow t_{10} \rightarrow p_{11} \rightarrow t_{12} \rightarrow r_1, r_1 \rightarrow t_{35} \rightarrow p_{35} \rightarrow t_{36} \rightarrow r_1$

$r_2: r_2 \rightarrow t_{13} \rightarrow p_{14} \rightarrow t_{15} \rightarrow r_2, r_2 \rightarrow t_{14} \rightarrow p_{15} \rightarrow t_{16} \rightarrow r_2, r_2 \rightarrow t_{21} \rightarrow p_{21} \rightarrow t_{22} \rightarrow r_2,$
$r_2 \rightarrow t_{23} \rightarrow p_{23} \rightarrow t_{24} \rightarrow r_2, r_2 \rightarrow t_{33} \rightarrow p_{33} \rightarrow t_{34} \rightarrow r_2$

$r_3: r_3 \rightarrow t_{17} \rightarrow p_{18} \rightarrow t_{19} \rightarrow r_3, r_3 \rightarrow t_{18} \rightarrow p_{18} \rightarrow t_{19} \rightarrow r_3, r_3 \rightarrow t_{31} \rightarrow p_{31} \rightarrow t_{32} \rightarrow r_3$

It can be seen that there are PMEs and SMEs. $(m_2, (t_{22},), (t_{15}, t_{17}))$, $(m_3, (t_{12}, t_{14})$, $(t_{34}, t_{35}))$, $(m_4, (t_{16}, t_{18}), (t_{32}, t_{33}))$, $(r_1, (t_{10}, t_{11}), (t_{10}, t_{12}), (t_{35}, t_{36}))$, $(r_2, (t_{21}, t_{22}), (t_{23}, t_{24}), (t_{13},$ $t_{15}), (t_{14}, t_{16}), (t_{33}, t_{34}))$, and $(r_3, (t_{17}, t_{19}), (t_{18}, t_{19}), (t_{31}, t_{32}))$ are all PME, and $(r_2, (t_{21}, t_{22}),$ $(t_{23}, t_{24}))$ is SME.

In Ezpeleta et al. (1995), eighteen IBSs are identified:

$S_1 = \{p_{32}, p_{18}, m_4, r_3\}$

$S_2 = \{p_{21}, p_{23}, p_{14}, p_{15}, p_{34}, m_3, r_2\}$

$S_3 = \{p_{21}, p_{23}, p_{14}, p_{15}, p_{35}, m_1, m_3, r_1, r_2\}$

$S_4 = \{p_{23}, p_{16}, p_{15}, p_{33}, m_2, r_2\}$

$S_5 = \{p_{23}, p_{16}, p_{15}, p_{34}, m_2, m_3, r_2\}$

$S_6 = \{p_{23}, p_{16}, p_{15}, p_{35}, m_1, m_2, m_3, r_1, r_2\}$

$S_7 = \{p_{21}, p_{23}, p_{14}, p_{17}, p_{31}, m_4, r_2\}$

$S_8 = \{p_{21}, p_{23}, p_{14}, p_{17}, p_{34}, m_3, m_4, r_2\}$

$S_9 = \{p_{21}, p_{23}, p_{14}, p_{17}, p_{35}, m_1, m_3, m_4, r_1, r_2\}$

$S_{10} = \{p_{23}, p_{16}, p_{17}, p_{33}, m_2, m_4, r_2\}$

$S_{11} = \{p_{23}, p_{16}, p_{17}, p_{34}, m_2, m_3, m_4, r_2\}$

$S_{12} = \{p_{21}, p_{16}, p_{17}, p_{35}, m_1, m_2, m_3, m_4, r_1, r_2\}$

$S_{13} = \{p_{21}, p_{23}, p_{14}, p_{18}, p_{33}, r_3, r_2\}$

$S_{14} = \{p_{21}, p_{23}, p_{14}, p_{18}, p_{34}, m_3, m_4, r_3, r_2\}$

$S_{15} = \{p_{21}, p_{23}, p_{14}, p_{18}, p_{35}, m_1, m_3, m_4, r_1, r_3, r_2\}$

$S_{16} = \{p_{23}, p_{18}, p_{33}, m_2, m_4, r_3, r_2\}$

$S_{17} = \{p_{23}, p_{18}, p_{34}, m_3, m_2, m_4, r_3, r_2\}$

$S_{18} = \{p_{23}, p_{18}, p_{31}, m_1, m_3, m_2, m_4, r_1, r_3, r_2\}$

These IBSs correspond to the following RCs:

$RC_1 = \{m_4, t_{32}, r_3, t_{18}, m_4\}$

$RC_2 = \{m_3, t_{34}, r_2, t_{14}, m_3\}$

$RC_3 = \{m_1, t_{11}, r_1, t_{35}, m_3, t_{34}, r_2, t_{13}, m_1\}$

$RC_4 = \{m_2, t_{15}, r_2, t_{23}, m_2\}$

$RC_5 = \{m_2, t_{15}, r_2, t_{14}, m_3, t_{34}, r_2, t_{23}, m_2\}$

$RC_6 = \{m_1, t_{11}, r_1, t_{35}, m_3, t_{34}, r_2, t_{23}, m_2, t_{15}, r_2, t_{13}, m_1\}$

$RC_7 = \{r_2, t_{33}, m_4, t_{16}, r_2\}$

$RC_8 = \{r_2, t_{14}, m_3, t_{34}, r_2, t_{33}, m_4, t_{16}, r_2\}$

$RC_9 = \{r_2, t_{13}, m_1, t_{11}, r_1, t_{35}, m_3, t_{34}, r_2, t_{33}, m_4, t_{16}, r_2\}$

$RC_{10} = \{r_2, t_{33}, m_4, t_{16}, r_2, t_{23}, m_2, t_{15}, r_2\}$

$RC_{11} = \{r_2, t_{23}, m_2, t_{15}, r_2, t_{14}, m_3, t_{34}, r_2, t_{33}, m_4, t_{16}, r_2\}$

$RC_{12} = \{r_2, t_{13}, m_1, t_{11}, r_1, t_{35}, m_3, t_{34}, r_2, t_{23}, m_2, t_{15}, r_2, t_{33}, m_4, t_{16}, r_2\}$

$RC_{13} = \{r_2, t_{33}, m_4, t_{32}, r_3, t_{18}, m_4, t_{16}, r_2\}$

$RC_{14} = \{r_2, t_{14}, m_3, t_{34}, r_2, t_{33}, m_4, t_{32}, r_3, t_{18}, m_4, t_{16}, r_2\}$

$RC_{15} = \{r_2, t_{13}, m_1, t_{11}, r_1, t_{35}, m_3, t_{34}, r_{2}, t_{33}, m_4, t_{32}, r_3, t_{18}, m_4, t_{16}, r_2\}$

$RC_{16} = \{r_2, t_{23}, m_2, t_{15}, r_2, t_{33}, m_4, t_{32}, r_3, t_{18}, m_4, t_{16}, r_2\}$

$RC_{17} = \{r_2, t_{23}, m_2, t_{15}, r_2, t_{14}, m_3, t_{34}, r_2, t_{33}, m_4, t_{32}, r_3, t_{18}, m_4, t_{16}, r_2\}$

$RC_{18} = \{r_2, t_{13}, m_1, t_{11}, r_1, t_{35}, m_3, t_{34}, r_2, t_{23}, m_2, t_{15}, r_2, t_{33}, m_4, t_{32}, r_3, t_{18}, m_4, t_{16}, r_2\}$

This demonstrates the conclusion presented in Chapter 4, that in POPN, IBSs result from RCs. It should be pointed out that some of these RCs are not elementary circuits. In fact, only $RC_1 - RC_4$ and RC_7 are elementary circuits.

An ROPN model is built as shown in Figure 6.3 and Table 6.1. Because input and output storages I_1, I_2, I_3, O_1, O_2, and O_3 do not make contributions to IBS, we merge them into p_0. The capacity of each machine is two, i.e., $K(p_i) = 2$, $i = 1, 2, 3$, and 4. It contains eight places and ten transitions. The part flow of part A is modeled: $t_{01} \rightarrow p_1$

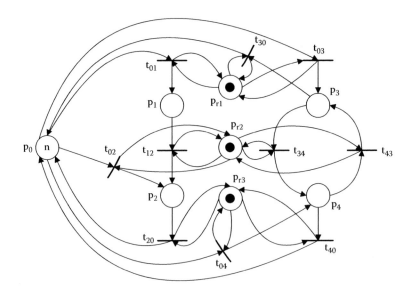

FIGURE 6.3 ROPN model for Example 6.1.

TABLE 6.1
Interpretation of Places and Transitions for Figure 6.3

Places		Transitions	
p_0	Part waiting in storage	t_{01}	Load a part to m_1 by r_1
p_1	Part being processed by m_1	t_{02}	Load a part to m_2 by r_2
p_2	Part being processed by m_2	t_{03}	Load a part to m_3 by r_1
p_3	Part being processed by m_3	t_{04}	Load a part to m_4 by r_3
p_4	Part being processed by m_4	t_{12}	Deliver a part from m_1 to m_2 by r_2
p_{r1}	r_1 available	t_{34}	Deliver a part from m_3 to m_4 by r_2
p_{r2}	r_2 available	t_{43}	Deliver a part from m_4 to m_3 by r_2
p_{r3}	r_3 available	t_{40}	Move a finished part from m_4 to storage by r_3
		t_{20}	Move a finished part from m_2 to storage by r_3
		t_{30}	Move a finished part from m_3 to storage by r_1

$\rightarrow t_{12} \rightarrow p_2 \rightarrow t_{20}$ or $t_{03} \rightarrow p_3 \rightarrow t_{34} \rightarrow p_4 \rightarrow t_{40}$; part B, $t_{02} \rightarrow p_2 \rightarrow t_{20}$; and part C, $t_{04} \rightarrow p_4 \rightarrow t_{43} \rightarrow p_3 \rightarrow t_{30}$. The material handling processes are modeled by self-loops: (p_{r1}, t_{01}), (p_{r1}, t_{03}), (p_{r1}, t_{30}), (p_{r2}, t_{02}), (p_{r2}, t_{12}), (p_{r2}, t_{34}), (p_{r2}, t_{43}), (p_{r3}, t_{04}), (p_{r3}, t_{40}), and (p_{r3}, t_{20}).

Consider that the places representing the material handling devices in the model make no contribution to IBS; they can be removed without affecting system analysis. The ROPN model with such places and its input and output arcs removed is shown in Figure 6.4. It can be seen that in the ROPN model, there is only one PPC $\{p_3, t_{34}, p_4,$

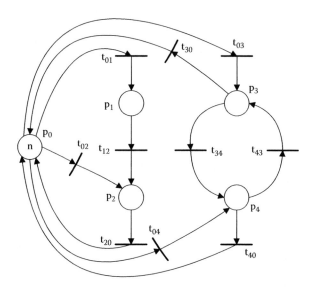

FIGURE 6.4 The ROPN model for Example 6.1, with places for material handling devices removed.

TABLE 6.2

Comparison between POPN and ROPN via Examples

	Example	Example 4.1	Example 4.2	Example 4.3	Example 6.1
	$|P|$	8	6	10	26
POPN	$|T|$	6	4	6	20
	Number of IBS	1	1	1	18
	$|P|$	3	3	5	8
ROPN	$|T|$	6	4	4	10
	Number of PPC	1	1	0	1

TABLE 6.3

Comparison between POPN and ROPN Modeling in Terms of Properties

	POPN	ROPN
Modeling power	High	Low
Model size	Large	Small
Conservativeness	Yes	Yes
Complexity for analysis and control	High	Low

t_{43}, p_3} that corresponds to only one IBS. This PPC corresponds to RC_8 in the POPN model. Thus, the other IBSs identified in Ezpeleta et al. (1995) are avoided. Based on this model, it is very easy to control the system without leading to any deadlocks since one needs to guarantee that only this PPC is never full. This shows the advantage of the ROPN in AMS control.

The results of comparing POPN and ROPN via four examples are summarized and given in Tables 6.2 and 6.3.

6.5   SUMMARY

Automated manufacturing systems (AMS) exhibit complex discrete event system characteristics that are difficult to analyze and control. Because Petri nets can describe the concurrency, choice, mutual exclusion, and synchronization in the system, they are widely used to model AMS.

In modeling AMS by Petri nets, one mainly uses a process-oriented method. Recently, in studying the deadlock avoidance problem, another method, called resource-oriented modeling, was proposed. In this chapter, we compare their resulting models, i.e., POPN and ROPN, for modeling AMS in which each operation requires a single unit of a single resource. POPN models the part production processes straightforwardly and is powerful for modeling the system in detail. ROPN has a very compact structure, and the part production processes can be described by introducing colors. In modeling the production processes, a production process circuit in ROPN corresponds to an ill-behaved siphon in POPN. However, in modeling the material handling processes, the ill-behaved siphons in POPN can be avoided by ROPN modeling. In general, ROPN can significantly decrease the complexity for analyzing and controlling the system. In this book, we discuss how the ROPN will be used for the analysis and control of AMS.

REFERENCES

Ezpeleta, J., J. M. Colom, and J. Martinez. 1995. A Petri net based deadlock prevention policy for flexible manufacturing systems. *IEEE Transactions on Robotics and Automation* 11:171–84.

Wu, N. Q., M. C. Zhou, and G. Hu. 2007. On the Petri net modeling of automated manufacturing systems. In *Proceedings of 2007 IEEE International Conference on Networking, Sensing and Control*, London, 228–33.

7 Control of Flexible and Reconfigurable Manufacturing Systems

7.1 INTRODUCTION

Flexible manufacturing systems (FMS) and reconfigurable manufacturing systems (RMS) are characterized by the ability to process multiple part types simultaneously. The parts are manufactured by routing raw material through the machines according to a prescribed sequence of operations for each part type. The multiple parts move from buffers to machines, machines to buffers, or buffers to buffers concurrently, competing for a finite set of resources in the system, such as machines, material handling devices, tools, and buffers. Thus, resources in the system are shared by multiple processes. It is a great challenge to operate such systems effectively. It is important that the flexibility of the FMS (RMS) should be exploited sufficiently. This requires effective application of control and management techniques and theories to model and analyze the behavior of the systems (Jensen, 1986).

One of the important issues to be addressed in operating an RMS (FMS) is deadlock resolution, since deadlock disables the operation of the whole system and makes automation impossible. Thus, in recent years, great attention has been paid to deadlock resolution for RMS (FMS). One way to control deadlock is to synthesize a live and bounded PN (Zhou and DiCesare, 1991, 1993; Zhou et al., 1992; Ezpeleta et al., 1995; Li and Zhou, 2004, 2008; Li et al., 2008; Viswanadham et al., 1990) because a live PN is deadlock-free. This is a static way to control deadlock, called deadlock prevention. The second way to control deadlock is so-called deadlock detection and recovery. By this technique, deadlock is detected based on a model, and then a recovery strategy is applied (Wysk et al., 1991; Cho et al., 1995). The third way is deadlock avoidance. This is a dynamic way to control deadlock, and deadlock is avoided by allocating the limited resources based on the state of the system. Different techniques are used for deadlock avoidance in RMS (FMS). Based on the graph theoretic approach description, various deadlock avoidance policies are proposed (Fanti et al., 1997a, 1997b, 2004; Kim and Kim, 1997; Yim et al., 1997; Lawley, 1999). A supervisory controller is synthesized to avoid deadlock in RMS by using automata theory (Lawley et al., 1998a, 1998b; Reveliotis and Ferreira, 1996). Petri nets are widely used as a tool to develop deadlock avoidance control policies in RMS (Banaszak and Krogh, 1990; Hsieh and Chang, 1994; Xing et al., 1996; Wu, 1997, 1999; Wu and Zhou, 2001; Chu and Xie, 1997; Viswanadham et al., 1990; Abdallah and Elmaraghy, 1998). For a survey of deadlock avoidance in FMS, see Fanti and Zhou (2004).

Chapter 5 presented the resource-oriented Petri net (ROPN) modeling method for RMS. It is shown that the resulting model is compact and has the advantage to describe deadlock situations. Based on the CROPN presented in Chapter 5, this chapter presents deadlock avoidance policies. By using CROPN, necessary and sufficient conditions are obtained for a kind of FMS. A maximal permissive control policy for deadlock-free operation for a class of RMS is then proposed.

7.2   DEADLOCK IN FMS

The phenomenon of deadlock exists in computer systems and other concurrent systems and has been studied extensively (Coffman et al., 1971; Gligor and Shuttuck, 1980; Silberschatz and Galvin, 1994). In concurrent systems, several processes may compete for a finite number of resources. If a process requests resources that are not available at that time, the process enters a wait state. It is then a set of processes trapped in a wait state, for the requested resources are held by other waiting processes. Four conditions are identified for deadlock to occur (Coffman et al., 1971):

1. Mutual exclusion: Processes claim exclusive control of the resources they require, and only one process at a time can use the resource.
2. Hold and wait: Processes hold resources already allocated to them while waiting for additional resources.
3. No preemption: Resources cannot be removed from the processes holding them until they are released by the processes themselves.
4. Circular wait: Each process holds one or more resources that are being requested by the next process in a chain. There exists a set $\{p_0, p_1, \ldots, p_k\}$ of waiting processes such that p_0 is waiting for a resource that is held by p_1, p_1 is waiting for a resource that is held by $p_2, \ldots, p_{k-1}$ is waiting for a resource that is held by p_k, and p_k is waiting for a resource that is held by p_0.

In manufacturing systems, the first three conditions are already present. Thus, deadlock occurs when two or more jobs enter a circular wait state, and to avoid deadlock in manufacturing systems is to guarantee that there is no circular wait. To illustrate deadlock situations in FMS, consider a simple FMS with four machines and the robot shown in Figure 7.1. Assume that there is no buffer and the robot transfers parts from machine to machine. Two part types, A-part and B-part, are to be processed with routes A-part, $m_1 \rightarrow m_2 \rightarrow m_4$ and B-part, $m_4 \rightarrow m_1 \rightarrow m_3$. If m_1 is processing an A-part for operation 1, m_2 is processing an A-part for operation 2, and m_4 is processing a B-part, then a circular wait occurs, or there is a deadlock, as shown in Figure 7.2.

For the circular wait condition shown in Figure 7.2, it is assumed that there are no flexible routes to the process parts. However, because of the flexibility of FMS, there are alternative machines for some operations. Let us assume that three part types are to be processed in the system shown in Figure 7.1 with the following routes: A-part, $m_1 \rightarrow \{m_2, m_3\} \rightarrow m_4$; B-part, $m_4 \rightarrow m_1 \rightarrow \{m_2, m_3\}$; and C-part, $m_2 \rightarrow \{m_1, m_3\} \rightarrow m_4$. If machines m_1 and m_3 are processing an A-part for operations 1 and 2, respectively, and m_4 is processing a B-part for operation 1 (Figure 7.3a), no deadlock

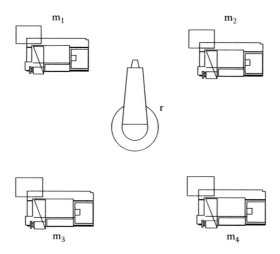

FIGURE 7.1 An illustrative FMS.

occurs. In this situation, although there is a circuit $\{m_1, m_3, m_4\}$ like that shown in Figure 7.2, it does not form a circular wait since the next operation of the A-part on m_1 can be performed by m_2. However, at the same time, if m_2 is processing a C-part, a circular wait occurs and the situation is shown in Figure 7.3b. It is seen that with routing flexibility, a deadlock situation can be more complicated.

In this chapter, routing flexibility is taken into account for deadlock avoidance. We first discuss deadlock avoidance in part production processes. A maximal permissive control policy for a class of FMS is proposed. Based on the control policy in part production processes, the deadlock avoidance problem is discussed with the material handling process taken into consideration. These results are mostly from Wu (1997, 1999) and Wu and Zhou, (2000, 2001).

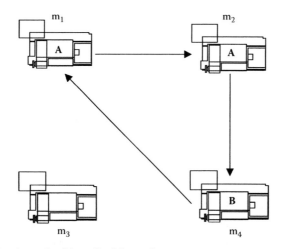

FIGURE 7.2 Circular wait without flexible routing.

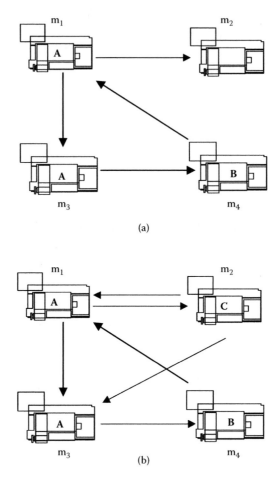

FIGURE 7.3 Circular wait condition with routing flexibility: (a) no circular wait and (b) a circular wait.

There are many kinds of resources in an FMS, such as machines, buffers, AGVs, pallets, tools, and fixtures. In general, before a part is moved to a machine for processing, the tools needed to process the part are already in the magazine of that machine. Therefore, tools do not contribute to deadlocks. Similarly, pallets and fixtures are irrelevant to deadlocks. Machines, buffers, and AGVs are the main resources in the production process in FMS and impact the occurrence of deadlocks as shown in Banaszak and Krogh (1990) and Zhou and colleagues (1991, 1992, 1993). We first deal with deadlocks in part production processes. Hence, we need to take only machines and buffers into consideration. Based on the result obtained, the deadlock avoidance problem is discussed with a material handling process taken into consideration.

7.3   SYSTEM MODELING BY CROPN

Buffers are often associated with machines (Banaszak and Krogh, 1990). Buffers can be associated with a machine in different ways, as shown in Figure 7.4. Let B_I and B_O denote the number of spaces of an input buffer and an output buffer, respectively. A machine can process one part at a time. In case a, a part in the input buffer can be loaded onto the machine immediately. Thus, if the machine is idle or there is a space in the input buffer, the resource can be used immediately. If a space in the output buffer is available, it can be used when a part is completed by the machine. In this way, in some time, the number of parts that can be loaded into the machine and its associated buffers = $B_I + B_O + 1$. This implies that the machine together with its associated buffers can hold $B_I + B_O + 1$ parts at a time, and if there is space available, it can be utilized sooner or later. For case b, if there is only one space in the buffer, the machine, together with its associated buffer, can hold only one part at a time; otherwise, it will be deadlocked. Hence, this case is equivalent to case e, where no buffer is associated with a machine. If there are two or more spaces in the buffer, we can assign one for input, another for output, and the others can be used as both input and output. In this situation, the case is

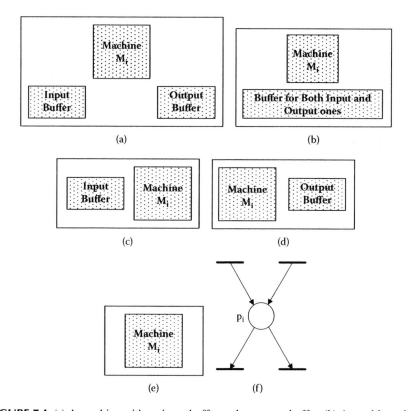

FIGURE 7.4 (a) A machine with an input buffer and an output buffer. (b) A machine with a buffer for both input and output buffers. (c) A machine with an input buffer. (d) A machine with an output buffer. (e) A machine without a buffer. (f) PN model for a machine and its associated buffer.

equivalent to case a and the number of parts that can be held at a time is $B + 1$, where B is the number of spaces in the buffer. Similarly, for cases c and d, the number of parts that can be held at a time is $B_I + 1$ and $B_O + 1$, respectively.

In the sense of deadlock avoidance, we are concerned only with the number of parts that can be held by a resource. Thus, based on the analysis above, a machine and its associated buffers can be seen as a single resource, with the number of parts that can be held most being the capacity. It should be pointed out that if a buffer is shared by two or more different machines, this buffer should be seen as a different resource, because the spaces in the buffer are dynamically assigned to the different machines.

Therefore, in the ROPN model for an FMS, machine i and its associated buffer are modeled by an H-place p_i, as shown in Figure 7.4f. The capacity of p_i represents the number of parts that can be held by the buffer and machine as discussed above. The multiple input transitions of place p_i in Figure 7.4f represent the sharing of the machine and its associated buffer. The multiple output transitions of place p_i represent the selective output. Thereafter, when we mention m_i, we means m_i and its associated buffer. Without loss of generality, in this chapter it is assumed that each machine is associated with an input buffer. By this assumption, if there is only one token in p_i that represents machine i, this token represents a part that is being processed. If there are two or more tokens in p_i, one of them must represent a part that is being processed. Furthermore, in p_i at marking M, only the token representing a part that is being processed is removed from p_i. Thus, it is necessary to identify the token representing a part that is being processed from the other tokens in p_i at any marking M. Hence, thereafter, we use θ_i to denote the token representing a part that is being processed in p_i at a marking M. With the model for machines the part production processes can be modeled by CROPN according to the method presented in Chapter 5.

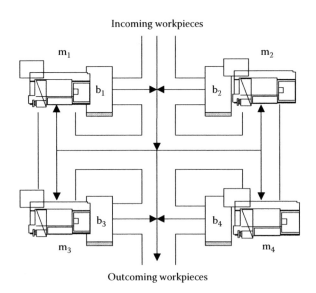

Incoming workpieces

FIGURE 7.5 An FMS.

Example 7.1

Consider the FMS shown in Figure 7.5 that contains four machines and three types of parts. The operation sequences of the three types of parts are A-part, $m_1 \rightarrow m_3 \rightarrow \{m_4, m_1\}$; B-part, $m_3 \rightarrow m_1 \rightarrow \{(m_3, m_4), (m_2, m_1)\}$; and C-part, $m_3 \rightarrow m_4 \rightarrow m_3 \rightarrow m_1$.

By following the ROPN modeling method, the A-, B-, and C-subnets are obtained (Figure 7.6). Then three subnets are merged to obtain the ROPN for the whole system. It is shown in Figure 7.7 that A-parts can be released into the system from p_0 to p_1 by going through t_{01}. The third operation can be processed by m_1 or m_4, or the parts can go from p_3 to p_1, or p_4 through t_{31} or t_{34}. When a part is completed by m_1 (m_4), it goes back to p_0 through t_{10} (t_{40}). The flow of the other part types is modeled similarly by the ROPN in Figure 7.7. It should be pointed out that according to the operation sequences, there are four operations for a B-part. Its third operation can be processed by m_3 or m_2. If it is processed by m_3, the fourth operation should be processed by m_4. If it is processed by m_2, the fourth operation should be processed by m_1.

The color of transition t_i is c_i. For example, t_{01}'s color is c_{01} and t_3's color is c_3. Colors of tokens in different places for different part types and operations can be defined as follows. Let $C(p_i, A_k)$ denote the set of colors for a token representing an A-part at its kth operation in place p_i, with 0th operation being its raw part. For other part types, it denotes the colors in a similar way. Then the colors for tokens in different places are as follows:

Place p_0: $C(p_0, A_0) = \{c_{01}\}$, $C(p_0, B_0) = C(p_0, C_0) = \{c_{03}\}$

Place p_1: $C(p_1, A_1) = \{c_{13}\}$, $C(p_1, A_3) = \{c_{10}\}$, $C(p_1, B_2) = \{c_{13}, c_{12}\}$, $C(p_1, B_4) = C(p_1, C_4) = \{c_{10}\}$

Place p_2: $C(p_2, B_3) = \{c_{21}\}$

Place p_3: $C(p_3, A_2) = \{c_{31}, c_{34}\}$, $C(p_3, B_1) = C(p_3, C_3) = \{c_{31}\}$, $C(p_3, B_3) = C(p_3, C_1) = \{c_{34}\}$

Place p_4: $C(p_4, A_3) = C(p_4, B_4) = \{c_{40}\}$, $C(p_4, C_2) = \{c_{43}\}$

Thus, the system is accurately modeled. Because CROPN is a finite capacity PN, the enabling and firing rules given by Expressions (2.3), (2.4), and (2.5) in Chapter 2 should be applied. When condition 2.3 is met, this indicates that there are jobs waiting for processing, and thus it is called process-enabled. When condition 2.4 is met, there are enough resources for the firing, and thus it is called resource-enabled. Hence, a transition is enabled if it is both process and resource-enabled. Because of the exclusive use of a resource in manufacturing systems, the weight of each arc in an ROPN is 1.

7.4   EXISTENCE OF DEADLOCK

Because of strong connectedness ROPN contains a number of circuits. The production process circuit (PPC) in an ROPN is a special class of circuits that plays an important role for the liveness of the ROPN. We use v to denote a PPC. Due to the routing complexity in FMS, there may be many PPCs in an ROPN, but only some PPCs in an ROPN need to be identified.

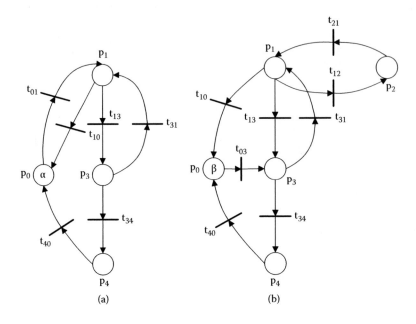

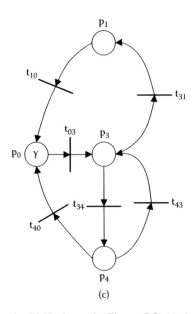

FIGURE 7.6 Subnets for the FMS shown in Figure 7.5: (a) A-subnet, (b) B-subnet, and (c) C-subnet.

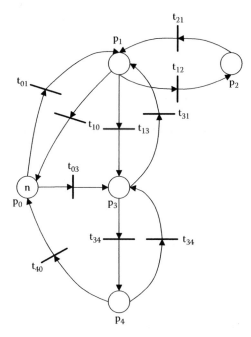

FIGURE 7.7 The ROPN for the FMS shown in Figure 7.5.

Definition 7.1: A PPC v in an ROPN is said to be an elementary PPC if it goes from one node, through a series of nodes, back to this node such that no node is repeated.

If the condition given by Definition 7.1 is not satisfied for a PPC, then the PPC is a nonelementary PPC.

Property 7.1: A nonelementary PPC in an ROPN is composed of several elementary PPCs.

Proof: Consider a sequence of places on a nonelementary PPC along the direction of the circuit. Assume that the sequence is $\sigma = \{p_1, p_2, \ldots, p_i, p_{i+1}, \ldots, p_n\}$, such that $p_1 = p_{i+1}$, and p_{i+1} is the first repeated place. This sequence can be divided into two subsequences: $\{p_1, \ldots, p_i\}$ and $\{p_{i+1}, \ldots, p_n\}$. Obviously, the first subsequence of places forms an elementary PPC. The second subsequence forms another PPC, and it may not be an elementary one. If it is not an elementary PPC, change the starting place such that the sequence has the same characteristic as σ. Then do the same as before. By repeating this procedure, a number of elementary PPCs can be obtained. This proves the property. ∎

In Figure 7.7, $v_1 = \{p_1, t_{13}, p_3, t_{31}, p_1\}$, $v_2 = \{p_1, t_{12}, p_2, t_{21}, p_1\}$, $v_3 = \{p_3, t_{34}, p_4, t_{43}, p_3\}$, $v_4 = \{p_1, t_{13}, p_3, t_{31}, p_1, t_{12}, p_2, t_{21}, p_1\}$, and $v_5 = \{p_1, t_{13}, p_3, t_{34}, p_4, t_{43}, p_3, t_{31}, p_1\}$ are PPCs, but only v_1, v_2, and v_3 are elementary. Property 7.1 implies that only the elementary PPCs in an ROPN need to be considered. Thereafter, when we mention a PPC, we refer to an elementary PPC.

In an ROPN, the number of places on a PPC v must be equal to the number of transitions on v, and the input place of a transition on v must also be on v. Let $P(v) = \{p_1, \ldots, p_k\}$ and $T(v) = \{t_1, \ldots, t_k\}$ be the sets of places and transitions on v with $p_i \in {}^{\bullet}t_i$, respectively. We use $M(p_i, v)$, to denote the number of tokens in place p_i that enable only transition t_i in v. Because these tokens enable only t_i, after the firing of t_i, these tokens remain in v; we call them *cycling tokens* of v. The other tokens in p_i can leave PPC v after firing the transitions other than t_i, and we call such tokens possible *leaving tokens* of PPC v. For example, in Figure 7.7, the token representing A-part for the processing of the first operation in place p_1 is a cycling token of v_1, for it enables only t_{13}, which is on v_1. The token for the same part type for the processing of the second operation in place p_3 enables both t_{31} and t_{34}, and it can leave v_1 by firing t_{34}. Thus it is not a cycling token of v_1, but a possible leaving token of v_1.

Definition 7.2: In a CROPN, define $M(v) = \Sigma M(p_i, v)$, $p_i \in P(v)$ as the number of cycling tokens of PPC v at marking M.

We will see the important role played by expression $M(v)$ in our discussion. In CROPN, the interaction of PPCs complicates the problem of liveness of the net. We discuss interactive PPC subnet as follows:

Definition 7.3: In ROPN a subnet formed by a number of PPCs is said to be an interactive PPC subnet, if every PPC in the subnet has shared places and transitions with at least one other PPC in the subnet and the subnet is strongly connected.

In Figure 7.7, v_1 and v_2 share p_1, and v_1 and v_3 share p_3. Hence, v_1, v_2, and v_3 form an interactive PPC subnet. We will briefly call the interactive PPC subnet the interactive subnet. Because place p_0 is not in any PPC, it is not in any interactive subnet either. We denote an interactive subnet formed by n PPCs by v^n, where n stands for the number of PPCs that form the subnet. For example, in Figure 7.7, the interactive subnet formed by v_1, v_2, and v_3 is a v^3. Let $P(v^n)$ be the set of places and $T(v^n)$ the set of all transitions in v^n. For $p_i \in P(v^n)$, let $T_i = \{t \in p_i{}^{\bullet} \cap T(v^n)\}$.

If the transitions enabled by a token in place $p_i \in P(v^n)$ are all in T_i, then after firing any transition enabled by the token, the token remains in v^n; such a token is called a cycling token of subnet v^n. For example, the token representing an A-part for its processing of the second operation in p_3 is a cycling one, for it enables both t_{31} and t_{34}, which are both in $T(v^3)$. If the transitions enabled by a token in place $p_i \in P(v^n)$ contain a transition that is not in T_i, then this token is not a cycling token of subnet v^n, because this token leaves the subnet v^n when one of its output transitions not in v^n fires. For example, the token representing an A-part for its processing of the third operation in p_4 enables t_{40}. Hence, it is not a cycling one. Letting $M(p_i, v^n)$ be the number of cycling tokens of subnet v^n in place p_i, we define the number of cycling tokens in v^n as follows:

Definition 7.4: In a CROPN, define $M(v^n) = \Sigma M(p_i, v^n)$, $p_i \in P(v^n)$ as the number of cycling tokens in subnet v^n at marking M.

Based on the discussion for CROPN model and the deadlock condition given in Section 7.2, we immediately have the following property:

Property 7.2: Assume that a CROPN is marked. Deadlock occurs in a CROPN if there exists a transition set T_d in marking $M \in R(M_0)$ such that every transition in T_d is process-enabled and none is resource-enabled.

In a PN, if it is possible for deadlock to occur, we say that there exist potential deadlocks. With CROPN the existence of potential deadlocks in an FMS can easily be analyzed, because a PPC in a CROPN models a possible circular wait situation. Deadlocks in an FMS are strongly related to nonliveness of the PN model of the system. It is known that an infinite capacity state machine with initial marking M_0 is live if and only if the net is strongly connected and M_0 has at least one token. We assume that there are always tokens in an CROPN, or there are parts in an AMS to be processed to make the discussions meaningful. Then due to the fact that a CROPN is a strongly connected state machine, we can conclude that it is the finiteness of capacity in a CROPN that causes the nonliveness. This implies that the buffer space is the key resource in deadlock control.

Let $K(v_i) = \sum K(p_j)$, $p_j \in P(v_i)$, denote the capacity of PPC v_i. To see the deadlock situation, let us observe Figure 7.7. If a marking M is reached such that PPC v_1 is full of tokens, the token θ_1 in p_1 is A-part for the first operation that enables only t_{13}, and the token θ_3 in p_3 is C-part with the third operation just completed that enables only t_{31}, then no transitions on v_1 can fire any more and a deadlock in v_1 occurs. At this time, $M(v_1) = K(v_1)$. In fact, at this time the overall system is disabled, and no transitions can fire after some time. Lemmas 5.1 and 5.2 and Theorem 5.1 present the properties of the system modeled by CROPN. In the following discussion, we assume that the potential deadlocks exist in the system considered, or there are enough active parts in the system to cause deadlocks.

7.5   DEADLOCK AVOIDANCE POLICY

Before discussing the deadlock-free conditions and control policy, we first present some definitions and notation needed for the discussion.

Definition 7.5: A transition t in a CROPN is said to be controlled if the firing of t is determined by a control policy when t is both process and resource-enabled according to the enabling rule of CROPN.

Therefore, a control policy for the CROPN is restrictive. It determines if each controlled transition can fire by observing the state of the net even if it is both process and resource-enabled. When a controlled transition t can fire according to a control policy, we say t is control-enabled under this policy.

Definition 7.6: A CROPN is said to be controlled if at least one transition in the net is controlled.

For example, to avoid deadlock, transitions t_{01} and t_{03} in Figure 7.7 should be controlled. Hence, the CROPN is controlled. According to Definitions 7.5 and 7.6, it is known that a controlled transition in a controlled CROPN is enabled only when it is process, resource, and control-enabled. The CROPN considered here is not live. The goal is to control the CROPN by a restrictive policy such that the

CROPN becomes live. The set of reachable markings of a CROPN under a control policy is changed. We use $R_c(M_0)$ to denote the set of reachable markings of the CROPN under control.

Definition 7.7: A PPC v in a CROPN is said to be live if every transition on v is live.

We know that it is the PPCs in a CROPN that contribute to the occurrence of deadlocks in FMS (RMS). Therefore, to avoid deadlock, the control policy must control the number of tokens in all PPCs in the CROPN. Often, PPCs are interactive and a number of interactive subnets are formed.

Definition 7.8: A transition t is said to be an output transition of an interactive subnet v^n in a CROPN if t is not in v^n and ${}^\bullet t$ is in v^n, i.e., $t \notin T(v^n)$ and ${}^\bullet t \in P(v^n)$. It is an input transition of v^n if $t \notin T(v^n)$ and $t^\bullet \in P(v^n)$.

It is the firing of the input and output transitions of v^n that changes the number of tokens in v^n. An interactive subnet v^n may have more than one output (input) transition. Let $T_O(v^n)$ $(T_I(v^n))$ denote the set of the output (input) transitions of interactive subnet v^n. For example, in Figure 7.7, there is a v^3, and $T_O(v^3) = \{t_{10}, t_{40}\}$ and $T_I(v^3) = \{t_{01}, t_{03}\}$. When we discuss the deadlock avoidance problem for an interactive subnet, we assume that every $t \in T_O(v^n)$ is resource-enabled, and later we will show that every such transition is live when all the interactive subnets are live under control. Similarly, we use $T_O(v)$ and $T_I(v)$ to denote the set of the output and input transitions of the subnet formed by a single PPC v, respectively.

Definition 7.9: Define $S(v_i) = \sum_{p_j \in P(v_i)} (K(p_j) - M(p_j))$ and $S'(v_i) = K(v_i) - M(v_i)$ as the number of free spaces and the number of current potential spaces available in PPC v_i in a CROPN at marking M.

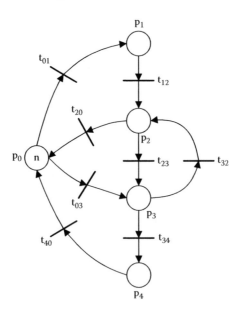

FIGURE 7.8 CROPN containing one PPC.

Below we discuss the control problem in different interactive subnet situations. In this section we assume that there are tokens in a subnet considered and let $L(v, M)$ denote the number of possible leaving tokens of PPC v in marking M.

7.5.1 CASE 1: SUBNET FORMED BY ONE PPC

This is the special case of an interactive subnet formed by only one PPC. The CROPN shown in Figure 7.8 describes the concurrent contention for resources by two types of parts with sequences $m_1 \rightarrow m_2 \rightarrow m_3 \rightarrow \{m_4, m_2\}$ and $m_3 \rightarrow m_2$. In this net, there is a PPC $v = \{p_2, t_{23}, p_3, t_{32}\}$. We use v to denote the subnet formed by PPC v.

It is clear that the number of places and transitions is finite in any PPC in a CROPN, because the number of resources in an FMS is finite. We call such PPC a finite PPC. For the subnet v we have the following result:

Theorem 7.1: A subnet formed by a PPC v in a CROPN is live if and only if for any marking $M \in R_c(M_0)$,

$$S'(v) \geq 1 \tag{7.1}$$

Proof: It is clear and is omitted. ∎

Similar to the concept of enabling a transition in a CROPN, we can discuss the enabling of a PPC v. A PPC v in a CROPN in a marking M is said to be potentially process-enabled if for every $t_i \in T(v)$, once $M(p_i) = K(p_i)$, then either $M(p_i, v) \geq 1$ or the token θ_i in p_i enables transitions in $T_O(v^n)$, where p_i is the input place of t_i. It should be noticed that when p_i is full of tokens, or $M(p_i) = K(p_i)$, the condition guarantees that t_i is process-enabled, and token θ_i in p_i enabling a transition $t \in T_O(v^n)$ can be seen as a potential space. If $S(v) > 0$ or there are tokens in $P(v)$ that enable transitions in $T_O(v^n)$, then PPC v is potentially resource-enabled.

Definition 7.10: A PPC v in a CROPN is said to be enabled if PPC v is both potentially process-enabled and potentially resource-enabled.

From the discussion above, we know that there are spaces in an enabled PPC v, or some spaces will move into v after some possible leaving tokens leave v. Once there are spaces in the enabled PPC, the transitions on the PPC can fire one by one. Thus, tokens and spaces can move into the PPC. In a subnet formed by a PPC v, all the possible leaving tokens must enable the transitions in $T_O(v)$. Thus we have the following corollary:

Corollary 7.1: If the condition in Theorem 7.1 is satisfied for a subnet formed by a PPC v, then PPC v is an enabled PPC.

7.5.2 CASE 2: INTERACTIVE SUBNET FORMED BY TWO PPCS

In this subsection we discuss the deadlock control problem for the interactive subnet formed by two PPCs, a subnet more complex than the one discussed in the last subsection. In accordance with the symbol of interactive subnet v^n, v^2 denotes an interactive subnet formed by two PPCs. A CROPN containing an interactive subnet formed by two PPCs is shown in Figure 7.9.

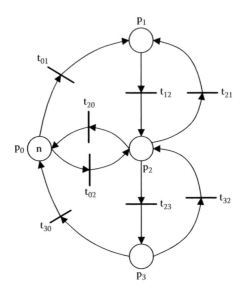

FIGURE 7.9 A CROPN containing a subnet formed by two PPCs.

From Figure 7.9 we find that v^2 may not be live, although the condition given by Theorem 7.1 is satisfied for both PPCs in v^2. The CROPN shown in Figure 7.9 contains two PPCs: $v_1 = \{p_1, t_{12}, p_2, t_{21}\}$ and $v_2 = \{p_2, t_{23}, p_3, t_{32}\}$. When all the places in the subnet are full of tokens, and the token θ_1 in place p_1 enables t_{12}, the token θ_2 in p_2 enables both t_{23} and t_{21}, and the token θ_3 in p_3 enables t_{32}, then $S(v_1) \geq 1$ and $S(v_2) \geq 1$. But no transitions in the subnet can fire, or it is not live. In this state, there is no space in the subnet and no token in the subnet enables t_{40} or t_{20}, or no PPC in the subnet is potential resource-enabled. This implies that there is no enabled PPC in the subnet. This concludes that the condition given in Theorem 7.1 is not adequate to make v^2 deadlock-free. We must add some more conditions to it to make it live.

By the definition of an interactive subnet, we know that the two PPCs in v^2 have shared places (the shared resources by two PPCs) and transitions. In v^2, these shared places are connected by the shared transitions and form a shared direct place path (SDPP). Note that a single place shared by multiple PPCs is also an SDPP. It can be observed that the first (last) place on the SDPP has two input (output) transitions belonging to different PPCs. We call these input (output) transitions intercircuit input (output) transitions (IITs and IOTs). Let t_{iik} denote the IIT, which is external to PPC v_k, and t_{iok} denote the IOT, which is on v_k. For example, in the CROPN shown in Figure 7.9, SDPP = $\{p_2\}$, t_{32} and t_{12} are two IITs that are external to PPCs v_1 and v_2, respectively, and t_{21} and t_{23} are two IOTs that are on v_1 and v_2, respectively.

Let the two PPCs in an interactive subnet v^2 be v_1 and v_2, and P_{12} be the set of shared places of these two PPCs, with $p_r \in P_{12}$ being the last place of SDPP. Let $\eta(v^n, M)$ denote the number of enabled PPCs in v^n at marking M.

Lemma 7.1: If $\eta(v^2, M) \geq 1$ in an interactive subnet v^2 at marking M, then $S'(v_i) \geq 1$, for both $i = 1$ and 2.

Proof: It is sufficient to show that the lemma holds when $\eta(v^2, M) = 1$. We assume that v_1 is enabled, then t_{io1}, the intercircuit output transition on v_1, must be process-enabled. This implies (1) there is a free space in p_r, (2) the token θ_r in p_r enables t_{io1} and a transition in $T_O(v^2)$, (3) the token θ_r enables both t_{io1} and t_{io2}, (4) the token θ_r enables only a transition in $T_O(v^2)$, or (5) the token θ_r in p_r enables only t_{io1}. In the first four cases, $S'(v_i) \geq 1$ holds for both $i = 1$ and 2, or the lemma holds. In the last case, $S'(v_2) \geq 1$ holds, and to make PPC v_1 enabled, there must be at least one space in v_1 or there is at least one token in v_1 that enables a transition in $T_O(v^2)$. This implies that $S'(v_1) \geq 1$ also holds. ■

Theorem 7.2: An interactive subnet v^2 in a CROPN is live if and only if $\eta(v^2, M) \geq 1$ at any marking $M \in R_c(M_0)$.

Proof: *Necessity:* We show that if the condition given in the theorem does not hold, or $\eta(v^2, M) = 0$, then v^2 is not live. By assumption there are tokens in the subnet, so when $\eta(v^2, M) = 0$, place p_r must be full (or the place has tokens); otherwise, both PPCs are enabled. If the token θ_r in p_r enables both t_{io1} and t_{io2}, then both PPCs in v^2 are potentially process-enabled. This time $\eta(v^2, M) = 0$ implies that all the places in v^2 are full and no token enables a transition in $T_O(v^2)$. Hence, in any firing sequence in the CROPN, no space can move into the subnet. Therefore, all the transitions in v^2 are process-enabled, but no transition in v^2 is resource-enabled, and the subnet is not live. If the token θ_r in p_r enables only one of the IOTs, say t_{io1}, then PPC v_1 is potentially process-enabled. But $\eta(v^2, M) = 0$, v_1 must not be potentially resource-enabled, or there is no space in v_1 and no token in v_1 enables a transition in $T_O(v^2)$. Therefore, PPC v_1 is deadlocked, and the subnet is not live.

Sufficiency: By assumption there are tokens in v^2. If there is only one space in v^2 and this space is in a place on SDPP, then no token in the subnet enables a transition in $T_O(v^2)$. This space can move into the first place of SDPP by firing the transitions on SDPP one by one. If the token θ_r in p_r enables only one of the IOTs, say t_{io1} on v_1, then v_1 is enabled, the transitions on v_1 can fire one by one. The first transition to fire is t_{ii2}, the IIT on v_1. After the firing of t_{ii2}, the space moves into the input place of t_{ii2}, which is on v_1. The space then moves into the output place of t_{io1} in some time so that t_{io1} is process and resource-enabled and can fire. The firing of t_{io1} moves the space into SDPP again. If the space is in the first place of SDPP and the token θ_r in p_r enables both t_{io1} and t_{io2}, then both PPCs are enabled, or $\eta(v^2, M) = 2$, so the transitions on one PPC, v_1 or v_2, can fire one by one. Thus, one of the IITs, t_{ii1} or t_{ii2}, can be selected to fire. If t_{ii1} is selected, then the space moves into v_2. In some time the space can move into the output place of t_{io2} and then into SDPP with the firing of t_{io2}. Therefore, in both cases the subnet is live.

Now we assume that the only space is not in a place on SDPP. If the token θ_r in p_r enables only one IOT, say t_{io1}, then by $\eta(v^2, M) \geq 1$ the space must be in v_1. From the above discussion, we know that the subnet is live. If the token θ_r in p_r enables both t_{io1} and t_{io2}, then the space can be in any PPC and the subnet is still live. Clearly, if there are two or more spaces in the subnet and the condition in the theorem holds, the subnet is live.

When there is no space available in the subnet, there must be some tokens that enable transitions in $T_O(v^2)$. If the token θ_r in p_r enables only one IOT, say t_{io1}, then to make $\eta(v^2, M) \geq 1$, there must a token in v_1 that enables transition $t \in T_O(v^2)$. By assumption t is resource-enabled, and after the firing of t, a space moves into v_1. Hence, the subnet is live. If the token θ_r in p_r enables both t_{io1} and t_{io2}, there must be a token in the subnet that enables $t \in T_O(v^2)$, but the token can be in any PPC, and the subnet is still live. ∎

It is known from the above discussion that in an interactive subnet, only the transitions on the enabled PPCs can fire immediately. Therefore, the firing order of the IITs is very important to the liveness of an interactive subnet. The IITs on the enabled PPCs should fire first, though other IITs may be process and resource-enabled at the same time. After the firing of the transitions on the enabled PPCs, the IOT on another PPC may become process-enabled and the PPC becomes enabled. In other words, the PPCs in the subnet become enabled one by one. For example, we assume that the capacity of all places other than p_0 in v^2 in Figure 7.9 is 2, and only p_2 has a free space; all other places are full, and θ_2 in p_2 enables t_{21} and θ_3 in p_3 enables t_{32}. This time, v_1 is enabled and v_2 is not. Then if t_{12} fires, the subnet is live. However, if t_{32} fires, the subnet is dead. This is because the IIT t_{12} is on v_1, but t_{32} is not.

7.5.3   CASE 3: INTERACTIVE SUBNET FORMED BY MULTIPLE PPCs

An interactive subnet formed by multiple PPCs is much more complex than v^2. In v^n, the places can be shared by PPCs in any complex way, and a number of SDPPs can be formed. PPCs in v^n can be interactive in any complex way. There may be many IOTs on a PPC. The condition $\eta(v^n, M) \geq 1$ does not guarantee that $S'(v_i) \geq 1$ holds for every PPC v_i in the subnet. This can be seen in Figure 7.7. If all the places except p_3 in the subnet are full of tokens, the token θ_1 in p_1 enables only t_{12}, and the token θ_3 in p_3 enables both t_{31} and t_{34}. Then, both $v_2 = \{p_1, t_{13}, p_3, t_{31}\}$ and $v_3 = \{p_3, t_{34}, p_4, t_{43}\}$ are potentially resource-enabled and v_3 is potentially process-enabled. Thus, v_3 is enabled, or $\eta(v^n, M) \geq 1$ holds. But it is easy to verify that $S'(v_1) = 0$ for $v_1 = \{p_1, t_{12}, p_2, t_{21}\}$.

Because of the complexity of the PPC interaction in v^n, we define two particular classes of interactive subnets.

Definition 7.11: Let $v_1 - v_k$ be k PPCs in v^n. If v_i and v_{i+1} have shared places with SDPP P_i, $\forall i \in N_{k-1}$; $P_i \cap P_j = \emptyset$, $\forall i \neq j$, $i, j \in N_{k-1}$. If v_1 and v_k have no shared place, $v_1 - v_k$ form a PPC chain; otherwise, $v_1 - v_k$ form a PPC ring.

Let $L(v, M)$ denote the number of possible leaving tokens of PPC V at marking M. We have the following lemma:

Lemma 7.2: In an interactive subnet v^n, there exists at least one PPC that is potentially process-enabled at any reachable marking M.

Proof: First, we show that the lemma holds when there is no PPC ring in v^n. To show that, it is sufficient to show that there exists at least one PPC v in the subnet such that the IOTs on that PPC are all process-enabled. By assumption there are tokens in the subnet, and if every t_i in the IOTs on v is process-enabled when $M(p_i) = K(p_i)$, with p_i being the input place of t_i, then the lemma holds. First, we show this by assuming that a token in a place enables only one of the output transitions of that place.

If all the n PPCs in the subnet have shared places with the same SDPP, then there is only one IOT on each PPC and the token θ_r in p_r enables one transition, which is the only IOT on a PPC. Therefore, the lemma holds.

If the n PPCs form a PPC chain, we can check these PPCs beginning with v_1 of the chain. If the only IOT on v_1 is process-enabled, then the lemma is true. If it is not, then we check v_2. If the IOT on v_2 is the output transition of P_2 and is process-enabled, then the lemma is true. Otherwise, we check v_3, and so on. If the IOT on v_{n-1}, which is the output transition of P_{n-2}, and the IOT on v_{n-1}, which is the output transition of P_{n-1}, are both process-enabled, then the lemma holds. Otherwise, the only IOT on v_n is process-enabled.

If a PPC v has shared places with several PPCs with different SDPPs, then the net must be a star form, or a PPC that has shared places with v is the end PPC of a PPC chain. If all IOTs on v are process-enabled, then the lemma holds. If one of the IOTs on v, say t, is not process-enabled, then t is the output transition of an SDPP of v and another PPC, and we assume that the place chain is P_1, the SDPP of v and v_1. Let CH be the PPC chain, with v_1 being the end PPC. Surely there must be at least one PPC v_2 that is on the PPC chain CH such that all IOTs on v_2 are process-enabled.

Any interactive subnet without a PPC ring can be composed of the subnets discussed above. Therefore, the lemma holds for the case that a token enables only one transition. When a token can enable more than one transition, more transitions are process-enabled. Hence, the lemma still holds.

Now we show that the lemma holds when there is a PPC ring in v^n. We need only show that there exists at least one PPC on which the IOTs are all process-enabled when each token in the subnet enables only one transition. Assume that k PPCs v_1, $v_2, \ldots, v_k$ in the subnet form a PPC ring, and all the places in this ring are full, and for every PPC v_i, $i = 1, 2, \ldots, k$, $L(v_i, M) = 1$; otherwise, there exists at least one PPC v in these k PPCs such that the IOTs on v are all process-enabled. We assume also that p_{rh} is the last place of SDPP P_h of v_h and v_{h+1}, p_{rk} is the last place of P_k of v_k and v_1, and the token θ_{r1} in p_{r1} enables transition t_1, which is on v_1. To satisfy the assumption that $L(v_i, M) = 1$, $i = 1, 2, \ldots, k$, the token θ_{r2} in p_{r2} must enable t_2 on v_2, and θ_{rk} in p_{rk} enables t_k on v_k. In this case, there must be a path from t_1 to P_k, and thus to t_k; a path from t_k to $t_{k-1}, \ldots$; and a path from t_2 to t_1. Transitions t_1, t_2, and t_k are the IOTs on the PPC v composed of these paths. We know that all the IOTs on v are process-enabled. Obviously, when a token can enable more than one transition, this still holds. Therefore, the proof is completed. ∎

Figure 7.10 shows a subnet containing a PPC ring; the three PPCs, $v_1 = \{p_1, t_{12}, p_2, t_{23}, p_3, t_{34}, p_4, t_{41}\}$, $v_2 = \{p_3, t_{35}, p_5, t_{56}, p_6, t_{67}, p_7, t_{73}\}$, and $v_3 = \{p_6, t_{69}, p_9, t_{91}, p_1, t_{18}, p_8, t_{86}\}$, form a PPC ring. If t_{34}, t_{18}, and t_{67} are process-enabled, then all the IOTs on PPC $v = \{t_{34}, p_4, t_{41}, p_1, t_{18}, p_8, t_{86}, p_6, t_{67}, p_7, t_{73}, p_3\}$ are process-enabled.

A potential process-enabled PPC becomes enabled if it is also potential resource-enabled. It follows from Corollary 7.2 that there can be enabled PPCs in an interactive subnet v^n in a CROPN if there are enough spaces in the interactive subnet. Based on this observation, now we can present the necessary and sufficient conditions of deadlock-free operation for the general interactive subnet v^n.

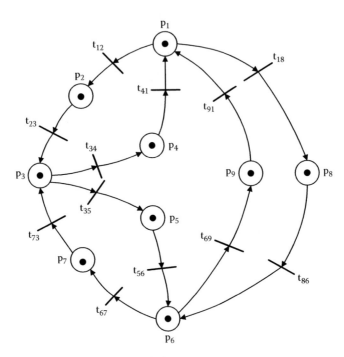

FIGURE 7.10 A subnet containing a PPC ring.

Theorem 7.3: An interactive subnet in a CROPN is live if and only if at any marking $M \in R_c(M_0)$ reachable from M_0,

$$S'(v_i) \geq 1, \text{ for every } v_i \tag{7.2}$$

and

$$\eta(v^n, M) \geq 1 \tag{7.3}$$

Proof: *Necessity:* If there exists a PPC v in the subnet such that $S'(v) = S(v) + L(v, M) = 0$ at marking M, then there is no space available in v and there is no possible leaving token of v. This implies that every transition, including IOTs on v, is process-enabled, but no transition on v is resource-enabled. Furthermore, in this situation, every transition on v can become resource-enabled only when a space occupied by another token on the same PPC v is released. This is impossible and PPC v is deadlocked, or the subnet is not live.

If $\eta(v^n, M) = 0$, then there is no enabled PPC in the subnet. By assumption that there are tokens in the subnet, there exists at least one potential process-enabled PPC in the subnet. By $\eta(v^n, M) = 0$, all the potential process-enabled PPCs in the subnet are not potentially resource-enabled. Therefore, there is no space available in any of these PPCs, and no token in any of these PPCs enables a transition in $T_0(v^n)$. This implies that no space can move into any of these PPCs, and the transitions on these

PPCs cannot fire. Obviously, the transitions on the PPCs that are not potentially process-enabled cannot fire either. Thus, the subnet is not live.

Sufficiency: It is sufficient to show that the theorem is true when $S'(v_i) = 1$ for every v_i in the subnet and $\eta(v^n, M) = 1$. When $\eta(v^n, M) = 1$, there is a PPC in the subnet, say v_1, that is enabled. The other $n - 1$ PPCs in the subnet are not enabled. For a nonenabled PPC v_i in the subnet, when $S'(v_i) = 1$ there is no space and no token that enables a transition in $T_O(v^n)$, but there is a possible leaving token. By the definition of an enabled PPC, every IOT on v_1 is process-enabled, or the token in the input place of the IOT enables a transition in $T_O(v^n)$, or the input place is empty. We assume, without loss of generality, that all the IOTs on v_1 are process-enabled. Therefore, the transitions on v_1 can fire sequentially. Finally, the space moves into an SDPP shared by v_1 with some other PPCs, and the token θ_r in p_r does not enable the IOT on v_1, but instead enables an IOT on another PPC, say v_2. Because the space is in the SDPP, v_2 becomes enabled. This time the transitions on v_2 can fire, and then another PPC becomes enabled. For any nonenabled PPC v_i, assume that the possible leaving token is in place p_h. Place p_h must be the last place of an SDPP, and the IOT t_h on v_i is the output transition of place p_h. When one IOT enabled by the token θ_h in place p_h fires, a token in the place representing the associated buffer of p_h becomes the new θ_h in p_h, and t_h on v_i can be process-enabled. Meanwhile, the space is in the SDPP shared by v_i with other PPCs, and v_i is enabled. In other words, all the PPCs in the subnet can become enabled sequentially, and all the transitions can fire. This means that the subnet is live. ■

From the proof of Theorem 7.3, we know that to make an interactive subnet live, there must be some spaces in the subnet. The problem is how many spaces are necessary and where these spaces should be in a marking M. Theorem 7.3 provides such conditions.

It can be observed that, at a marking M, if the token θ_r in every p_r, the last place of an SDPP, enables all output transitions of p_r, then all IOTs in the subnet are process-enabled. Thus, all the PPCs in the subnet are potentially process-enabled and $S'(v_i) \geq 1$ for every v_i. Similar to $S'(v)$, we use $S'(v^n) = K(v^n) - M(v^n)$ to denote the potential spaces available in v^n at marking M. From Theorem 7.3, we have the following corollary:

Corollary 7.3: In an interactive subnet v^n, if all the IOTs are process-enabled in marking M, then the subnet is live if and only if

$$S'(v^n) \geq 1 \tag{7.4}$$

Corollary 7.3 indicates that in this situation only one space is required to make the subnet live, and this space can be in any PPC. It is the extreme case that a token in a place enables all the output transitions of the place, or an operation in the system can be processed by any machine; this time the CROPN becomes a real free choice. This indicates that the routing flexibility can reduce the possibility of deadlock occurrence in some way. In fact, if any operation can be processed on any machine in the system, then we can schedule the system such that all the parts go in the same direction.

Because the original CROPN is not live, to make it live, some transitions in the net cannot fire freely, and a control policy is necessary to restrict the firing of some transitions according to the marking reached. From the proof of Theorem 7.3, we know that to make the subnet live, the firing order of some transitions in the subnet must be carefully controlled such that no matter what marking is reached, the conditions given in Theorems 7.2 and 7.3 are always satisfied.

Definition 7.12: A PPC in a CROPN is said to be an entering PPC of transition t, if t is not on the PPC, but the output place of t is on the PPC.

We call t the input transition of the PPC. It may have more than one entering PPC. Let $V_{en}(t)$ denote the set of its entering PPCs. We have the following result for the deadlock control law for interactive subnet v^n:

Theorem 7.4: An interactive subnet v^n in a CROPN is deadlock-free if and only if all of the following conditions are satisfied: (1) any transition $t \in T_I(v^n)$ and any IIT in the subnet are controlled; (2) at marking M, before a controlled t fires, for every $v_i \in V_{en}(t)$, $S'(v_i) \geq 2$; and (3) after t fires, the CROPN's marking is changed from M to M' such that $\eta(v^n, M') \geq 1$.

Proof: *Necessity:* From Theorem 7.3 it is obvious.

Sufficiency: If the control law given in the theorem is applied, then firing any $t \in T_I(v^n)$ or any IIT guarantees that Theorems 7.2 and 7.3 are satisfied. Thus, from Theorem 7.3, such firing guarantees the liveness of the subnet. Notice that no firing of other transitions can move tokens and spaces from or into a PPC in the subnet. Hence, the firing of other transitions does not impact the liveness of the subnet. Therefore, the proof is completed. ∎

The control law given by Theorem 7.4 is a restrictive policy; by controlling the firing order of transitions, it restricts the firing of $T_I(v^n)$ and IITs in the subnet such that these transitions can fire only in some markings. By this control law, only $T_I(v^n)$ and the IITs should be controlled, and the other transitions can fire freely. Notice that this control law is necessary and sufficient for deadlock-free operation in FMS with each machine having an associated buffer. Thus, it permits as many active parts as possible in the system, while deadlock is totally avoided. In scheduling an FMS, two things should to be avoided. One is deadlock and the other is starvation. To avoid starvation requires that there are as many parts as possible in the system. Therefore, when this control law is embedded into a real-time scheduler, it provides a good opportunity to avoid starvation or to improve resource utilization while avoiding deadlocks.

7.6 LIVENESS OF OVERALL SYSTEM

The control law presented above is for subnets in a CROPN. We now show that if every subnet in the CROPN is controlled by this law, then the overall controlled CROPN is live.

When we discuss the deadlock avoidance problem for interactive subnets, we assume that the transitions in $T_O(v^n)$ are resource-enabled. It can be observed that a transition t in $T_O(v^n)$ is either an input transition of another interactive subnet in the CROPN or an input transition of p_0. An input transition of p_0 is always resource-enabled, so to show that the CROPN is live is to show that all the input transitions of all the interactive subnets in the CROPN are live. We assume that there are k interactive subnets in the CROPN considered: they are $v_1^{n_i} - v_k^{n_k}$. First, we present the following lemma:

Lemma 7.3: Assume that a CROPN is formed by k subnets $v_1^{n_i} - v_k^{n_k}$, and is under control by the law given in Theorem 7.4. If a subnet $v_i^{n_i}$ is replaced by a place p_i^n, with $K(p_i^n) = K(v_i^{n_i})$ and $T_I(v_i^{n_i})$ and $T_O(v_i^{n_i})$ being the set of input and output transitions, respectively, then the liveness of the net does not change.

Proof: The control law given in Theorem 7.4 guarantees the liveness of the subnet v^n by assuming that the transitions in $T_O(v^n)$ are resource-enabled. Thus to show the lemma is to show that the liveness of the transitions that are not in v^n is not changed. These transitions are in $T_I(v^n)$ and $T_O(v^n)$. From Theorem 7.4 we know that for each v^n the controlled transitions are in $T_I(v^n)$ and IITs. By controlling the IITs, we control the token distribution within a subnet. This does not impact the liveness of transitions not in the subnet. By controlling transitions in $T_I(v^n)$, we restrict the number of tokens in the subnet. With $K(p_i^n) = K(v_i^{n_i})$, when $M(p_i^n) = K(p_i^n)$ no transition in $T_I(p_i^n)$ can fire; this represents the controlling process. Each part in a subnet can be completed and leaves the subnet by firing one of the transitions in $T_O(v^n)$. A token in $v_i^{n_i}$ enabling a transition in $T_O(v_i^{n_i})$ represents this process. Therefore, when a subnet $v_i^{n_i}$ is replaced by a place p_i^n, the overall net structure is not changed and the enabling of transitions in $T_I(v_i^{n_i})$ and $T_O(v_i^{n_i})$ is not changed either. Thus, the proof is completed. ■

The resulting net from Lemma 7.3 is called a simplified CROPN. In a simplified CROPN, the key point is that no transition in $T_O(v_i^{n_i})$ that is not resource-enabled in the original CROPN is made resource-enabled. Therefore, if the simplified CROPN is live, then any transition in $T_O(v^n)$ in the original CROPN will be resource-enabled in some time, or the assumption that the transitions in $T_O(v^n)$ are resource-enabled is justified. A simplified CROPN containing six subnets is shown in Figure 7.11.

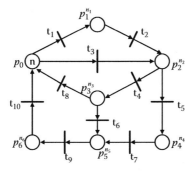

FIGURE 7.11 A simplified CROPN.

Observing the simplified CROPN, we immediately have the following lemma:

Lemma 7.4: In a simplified CROPN, place p_0 for the load/unload station is a common place of all circuits in the net.

It follows from Lemma 7.4 that for the liveness of the simplified CROPN, we have the following lemma:

Lemma 7.5: The simplified CROPN is live.

From Lemma 7.5, the following result is obvious:

Theorem 7.5: A CROPN is live if and only if it is controlled by the control law given in Theorem 7.4.

It is seen that if each subnet is live, then the tokens in the system can flow from subnet to subnet, then to place p_0. Thus, the key is how to make each subnet live—this is what we have.

7.7 ILLUSTRATIVE EXAMPLE

Consider the CROPN in Figure 7.12. It contains three PPCs: $v_1 = \{p_1, t_{12}, p_2, t_{21}\}$, $v_2 = \{p_1, t_{13}, p_3, t_{31}\}$, and $v_3 = \{p_3, t_{34}, p_4, t_{43}\}$. These three PPCs form an interactive subnet v^3, and the input transition set of this interactive subnet is $T_I(v^3) = \{t_{01}, t_{03}\}$.

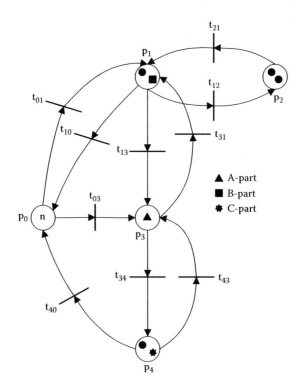

FIGURE 7.12 The CROPN for the illustrative example.

At the initial marking M_0 there are n (> 4) tokens (parts) in place p_0, and the other places in the net are all empty. This implies that there are enough parts to be processed.

We assume, without loss of generality, that all the places other than p_0 have a capacity of two with a machine and a buffer space. According to the control law given in Theorem 7.4, at the initial marking M_0, both transitions t_{01} and t_{03} are pro-cess, resource, and control-enabled and can fire. According to the control law, the following events can be executed:

1. $B_1: p_0 \to p_3$ (the first B-part is loaded from p_0 to p_3), $C_1: p_0 \to p_3$, and $A_1: p_0 \to p_1$
2. $B_1: p_3 \to p_1$, $C_1: p_3 \to p_4$, $A_1: p_1 \to p_3$, and $B_2: p_0 \to p_3$
3. $A_1: p_3 \to p_4$, $B_1: p_1 \to p_2$, $B_2: p_3 \to p_1$, $B_3: p_0 \to p_3$, and $A_2: p_0 \to p_1$
4. $B_2: p_1 \to p_2$, $A_2: p_1 \to p_3$, and $B_3: p_3 \to p_1$
5. $A_3: p_0 \to p_1$

After these events are executed (or transitions are fired), only the buffer space in place p_3 is empty, and the other places in the subnet are full, with the token in p_3 representing A-part, the token θ_1 in p_1 representing B-part (operation 2 is just completed) and enabling both t_{12} and t_{13}, and the token θ_4 in p_4 representing C-part. The marking is shown in Figure 7.12, where ● denotes a part regardless of its type. At this marking, t_{01} is not resource and control-enabled, and t_{03} is not control-enabled, though it is process and resource-enabled according to the con-trol policy. Thus, at this marking, these two transitions cannot fire, and deadlock can be avoided. In this marking, both PPCs v_2 and v_3 are enabled, t_{13} or t_{43} can be selected to fire. If, however, the token θ_4 in p_4 represents A-part and enables t_{40}, an output transition of the subnet, then transition t_{03} is still control-enabled and can fire. Hence, sometime later, the parts in the subnet can be completed and leave the subnet; then t_{01} and t_{03} can fire again.

From the example, we can see that deadlock in a CROPN can be totally avoided if the control law given in Theorem 7.4 is used in real time. By observing the state of the system, the control law controls the holding order of resources by the jobs that compete for the resources.

7.8 IMPLEMENTATION

The control policy is to be applied in real time. Hence, its computational com-plexity is very important. It seems that to implement the control policy we need to check each PPC to see if the conditions given in Theorem 7.4 are satisfied. However, the number of PPCs may be exponential with the number of machines. Thus, it is computationally inefficient. Notice that if a transition t is not fireable according to the control policy and t fires, then the CROPN must be deadlocked. Therefore, to determine whether a controlled transition t is fireable or not, we can just fire t to see whether the CROPN is deadlocked. If it is, it implies that the conditions given by the control policy are violated; if not, t can fire. The problem remaining is to find an efficient approach to see whether a PPC subnet is deadlocked at a given marking.

For a given CROPN let $P_{full} \subseteq P(v^n)$ be a set of places in v^n that are full of tokens at marking M, $T_1 \subseteq T(v^n)$ be a set of transitions in v^n that are process-enabled at marking M, and $T_2 \subseteq T_O(v^n)$ be a set of transitions that are process-enabled at marking M, and let $T_{en} = T_1 \cup T_2$. Clearly, we have $P_{full} \subseteq P$ and $T_{en} \subseteq T$. Further, let $G = (P_{full} \cup T_{en}, \alpha)$ be a digraph such that $(p, t) \in \alpha$, if $p \in P_{full}$, $t \in T_{en}$, and (p, t) is an arc in the CROPN, and $(t, p) \in \alpha$, if $p \in P_{full}$, $t \in T_{en}$, and (t, p) is an arc in the CROPN.

Definition 7.13: A strongly connected component V in G is said to be a knot if there is no arc that goes out of V.

With Definition 7.13 we have the following result:

Theorem 7.6: Subnet v^n in the CROPN at marking M is deadlocked if and only if there exists a knot V in $G = (P_{full} \cup T_{en}, \alpha)$.

Proof: Subnet v^n is deadlocked $\Leftrightarrow$ some circuits in v^n are deadlocked $\Leftrightarrow$ all the places on these circuits are full, and all transitions on these circuits are process-enabled and no transition that is not on these circuits is process-enabled. Because circuits are strongly connected $\Leftrightarrow$ there exists a knot V. ∎

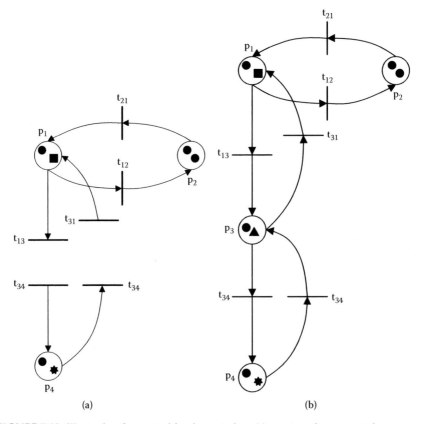

(a)   (b)

FIGURE 7.13 Illustration for control implementation: (a) no strongly connected component and (b) a strongly connected component.

Therefore, if we want to determine whether a transition $t \in T_I(v^n)$ or $t \in T(v^n)$ is fireable, we fire it and change the marking. Then we remove the places in v^n that are not full and the transitions in v^n and $T_O(v^n)$ that are not process-enabled. In this way, $G = (P_{full} \cup T_{en}, \alpha)$ is formed. Then, a test is carried out to see whether there exists a knot. To form G, we need to check all places in $P(v^n)$ and all transitions in $T(v^n)$ and $T_O(v^n)$. Thus, the computational complexity is $O(|P(v^n)|+| T(v^n) \cup T_O(v^n)|)$. To determine if knots exist, the following computation needs to be performed: (1) find the strongly connected components in G, $V = \{V_1, \ldots, V_k\}$; (2) remove such components with one node from V; and (3) for each V_i in V determine if there is an arc that goes out of V_i. There is an efficient algorithm to find strongly connected components in a digraph (Cormen, 1990) with computational complexity $O(|P_{full} \cup T_{en}|+|\alpha|)$. For step 2, we need to check each node in G to see if it forms a strongly connected component itself, and the computational complexity is $O(|P_{full} \cup T_{en}|)$. For step 3, we need to check every arc in G, and the computational complexity is $O(|\alpha|)$. Hence, in the worst case, the computational complexity for the control policy is $O(3|P \cup T| + 2|\alpha|)$, where P is the set of places, T is the set of transitions, and α is the number of arcs in the CROPN.

To illustrate the implementation of the control policy, let us consider the state shown in Figure 7.12. At this marking, p_3 is not full and thus removed. Transitions t_{10} and t_{40} are not process-enabled, and are also removed. Thus, G is obtained as shown in Figure 7.13a. Because t_{13} is a sink node, though there is a circuit $\{p_1, t_{12}, p_2, t_{21}, p_1\}$, there is an arc to t_{13} that is not on the circuit. Thus, there is no deadlock. However, if at that marking, t_{03} fires, then G can be obtained as shown in Figure 7.13b. This time, there exists a strongly connected component without an arc going out of the component, or the CROPN is deadlocked. This means t_{03} cannot fire according to the control policy.

7.9 DEADLOCK AVOIDANCE WITH SHARED MATERIAL HANDLING SYSTEM

We have discussed the deadlock avoidance policy in manufacturing processes without considering the material handling system (MHS). However, if an MHS is a shared resource in the manufacturing process, deadlock may occur due to the competition for the MHS. Chapters 5 and 6 point out that the deadlock caused by MHS sharing can be avoided by ROPN modeling. In this section, we discuss the problem of deadlock avoidance by taking the MHS into consideration and show that the conclusions given in Chapters 5 and 6 are true. Here, we assume the material handling system can do all the delivery tasks.

7.9.1 DEADLOCK SITUATIONS

First we show the deadlock situations in the manufacturing processes with the MHS considered. Consider the PN models for manufacturing processes shown in Figure 7.14, the capacity of places p_1 and p_2 that represent machines m_1 and m_2 is being two. The material handling process by an MHS in the models shown in Figure 7.14 is modeled by the process-oriented Petri net (POPN) method, where

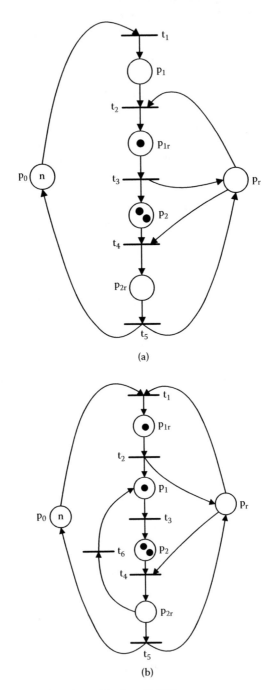

FIGURE 7.14 Deadlock situations with shared MHS.

place p_r represents a material handling device, a token in place p_{1r} or p_{2r} represents that the device is delivering a part from one place to another. We assume, without loss of generality, that places except p_0 can hold only one part at a time. In Figure 7.14a, there is no PPC. Thus, if we do not consider the material handling process, there will be no deadlock at all. At the marking shown in Figure 7.14a, the MHS is delivering a part from p_1 to p_2. At the same time, p_2 is full and the MHS can be released only when a token in p_2 is removed. However, a token in p_2 can be removed only when the MHS is released. Thus, the system is deadlocked because of the competition for the MHS.

In Figure 7.14b, there is a PPC $v = \{p_1, t_3, p_2, t_4, p_{2r}, t_6, p_1\}$, or potential deadlocks exist. In PPC v, place p_{2r} is to model the operation of the MHS; it cannot be treated as a space in v, even though p_{2r} is empty. In the marking shown in Figure 7.14b, we assume that one token in p_2 should be delivered to p_1 for further processing. Since there is a free space in p_1, PPC v is not in a deadlock state if the MHS is not taken into consideration. However, the MHS is delivering a part from p_0 to p_1. In this state, if the part in p_{1r} is loaded to p_1, PPC v will be deadlocked. If not, the MHS cannot be released. Thus, the system is deadlocked too.

By this observation, it is not sufficient to guarantee deadlock-free operation in manufacturing processes by the control policy presented in the previous sections if the MHS is considered. However, if the material handling process is modeled by the ROPN method, the manufacturing process is indeed deadlock-free if the policies presented in the previous sections are applied.

7.9.2 Deadlock Avoidance with MHS via ROPN Modeling

Consider the deadlock situation shown in Figure 7.14a. The key point is that transition t_3 is not resource-enabled. This means that to avoid deadlock, MHS cannot pick up the part in p_1. However, with the PN model shown in Figure 7.14a, only if there is a token in p_1 and the MHS is free (a token in p_r) will transition t_2 be process and resource-enabled. This implies that t_2 can fire and a deadlock occurs. The situation shown in Figure 7.14b is similar.

To avoid the deadlocks caused by MHS sharing, one may develop a control policy. Instead of giving more control to avoid such deadlocks, we attempt to improve the modeling of the material handling process by taking advantage of the deadlock avoidance policy presented in the previous sections. We use the model for the material handling process presented in Chapter 5, or the MHS is treated as a G-resource. Then the MHS modeled as G-place is added to the CROPN for the manufacturing process. To do that is to add a place p_r representing the MHS with a token in it and a number of arcs into the CROPN such that p_r and some transitions in the CROPN form a number of self-loops. In doing so, the resultant Petri net will be called augmented CROPN (ACROPN). Because there is no self-loop in the CROPN, all the self-loops in an ACROPN are associated with p_r. The ACROPNs in Figure 7.14, by changing the model of the material handling process, are obtained and are shown in Figure 7.15. We will show that an ACROPN is live if it is controlled by the policy presented in the previous sections.

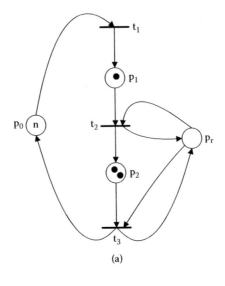

(a)

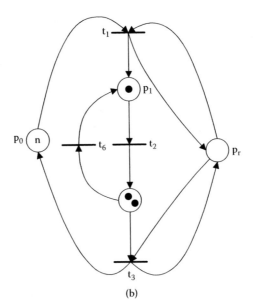

(b)

FIGURE 7.15 Augment CROPNs.

Lemma 7.6: If the original CROPN of an ACROPN is a single circuit and this circuit is not a PPC, then the augment CROPN is always live.

Proof: By assumption, the circuit in the original CROPN is not a PPC, or place p_0 is on the circuit, and p_0 is never to be full. We first show the net is live when there is only one self-loop associated with place p_r, and assume transition t is on the self-loop.

To show the liveness of the net is to show that all the transitions in the net can fire again, no matter what marking is reached. Notice that in a circuit, if one transition can fire again in a marking, then all the transitions on the circuit can fire again in some way. Now we show that transition t on the circuit can fire again in any reachable marking. We only need to show that in any marking t can fire after a sequence of firings of other transitions. Assume that places $p_1 = {}^\bullet t$ and $p_2 = t^\bullet$ are the input and output places of t in the original CROPN, respectively; then in any marking, transition t is in one of the following cases: (1) there is a token in p_1 and p_2 is empty. (2) p_1 is empty and p_2 is full. (3) both p_1 and p_2 are full, or (4) both p_1 and p_2 are empty. In case 1, t can fire immediately because there is always a token in p, unless t is firing. In case 2, we can start from p_2 and go in the direction of the circuit and check the state of each place on the path. We can certainly find some places that have space available, because p_0 will never be full. Assume the first place found with a space available is p_3; then the input transition t of p_3 can fire, and this makes the input place of t empty. Therefore, after a number of firings of transitions, p_2 will become empty. Similarly, if we start from p_1 and go in the opposite direction of the circuit, we will find a place that has a token in it. Therefore, after several firings of transitions, this token will go into p_1, then this becomes case 1. In cases 3 and 4, p_2 can be made empty and p_1 can be made full in a similar way, or case 1 can be reached from both cases. Thus, we have shown that the net is live.

Now we show that the net is live when there are multiple self-loops, or place p has several self-loops with several transitions that are on the circuit. Let T_r denote the set of these transitions. If transition $t \in T_r$ is enabled in the original CROPN, then t can fire in the ACROPN, for there is a token in p_r. If several transitions, say a transition set $T_1 \subset T_r$, are enabled in the original CROPN simultaneously, then one of these transitions, say t_1, can be chosen to fire. After the firing of t_1, a token goes back into p_r, and then another transition in T_1 can be chosen to fire. Thus, all the transitions in T_1 can fire one by one. This implies that once a transition is enabled in the original CROPN, the transition can fire sooner or later in the ACROPN. Therefore, any transition on the circuit can be made enabled in the original CROPN in any marking M after a sequence of firings of transitions in the augmented net. This means that the net is live. ∎

It is known that if there is no PPC in a CROPN, a transition can fire when it is both process and resource-enabled. By Lemma 7.6, no other condition is needed to avoid deadlock in an ACROPN.

Theorem 7.7: If there is no PPC in the original CROPN of an augment CROPN, then the augment CROPN is always live.

Proof: By assumption there is no PPC in the original CROPN of the augment CROPN, and the original CROPN must be formed by several parallel circuits with place p_0 as their common place. If the self-loops of place p_r connect p_r with only one of the circuits, then it follows from Lemma 7.8 that the net is live. Now we show that the net is live when the self-loops of place p_r connect p_r with k $(k > 1)$ circuits. Let T_r denote the set of transitions that are on the self-loops with $T_1 \subset T_r$, $T_2 \subset T_r$, ..., $T_k \subset T_r$ on circuits $v_1, v_2, \ldots, v_k$, respectively. If a number of transitions in T_r, say two transitions $t_1 \in T_1$ and $t_2 \in T_2$, are enabled in the original

CROPN, then t_1 on v_1 can be chosen to fire first, for there is a token in p_r. After the firing of t_1, the token that was in p_r before the firing of t_1 comes back into p_r, and this time t_2 on v_2 can fire. This means that once the transitions in T_r are enabled in the original CROPN, all of them can fire one by one, no matter which circuit they are on. Because p_0 is on all the circuits, in any marking M every transition can be made enabled in the original CROPN after a sequence of firings of transitions. Hence, the net is live. ■

Theorem 7.7 shows that an ACROPN is live if the original CROPN is free of PPCs. We will show that an ACROPN with PPCs in the original CROPN is live if it is under control by the policy presented in the previous sections.

Theorem 7.8: If the original CROPN of an ACROPN is an interactive subnet, then the ACROPN is live if ACROPN is controlled by the deadlock avoidance policy presented in the previous sections.

Proof: It follows from the discussion in the previous sections that if the original CROPN of an ACROPN is an interactive subnet and is controlled by the deadlock avoidance policy presented there, then each transition in the subnet can fire again in any reachable marking. We say a transition is enabled in the original subnet if the transition can fire under the control by the policy applied. When a transition that is not on any self-loops of p_r is enabled in the sense of control policy in the original subnet and needs to fire, this transition can fire immediately in the ACROPN. If a transition that is on a self-loop of p_r is enabled in the original subnet and needs to fire, and at the same time the token in p_r is not being used for firing of another transition, the transition can be chosen to fire immediately by allocating the token in p_r to this transition first. When a transition is enabled in the original subnet and the token in p_r is being used by another transition, the transition can fire when the token comes back into p_r and becomes available. Such firings have no effect on the distribution of tokens in the original CROPN, or do not cause additional deadlocks. Therefore, the ACROPN is live. ■

We know that if every interactive subnet under control is live, then the overall system is live. Thus, from Theorem 7.8, an ACROPN is live if it is controlled by the deadlock avoidance policy presented in the previous sections.

7.10 SUMMARY

With deadlock avoidance in FMS modeled by an ROPN method, this chapter presents a deadlock avoidance policy. It is a maximal permissive control policy. With this control policy, transitions to be controlled to avoid deadlock are specified. It is also shown that it is computationally efficient, since its computation is polynomial. Furthermore, when shared MHS is considered, deadlock can be avoided with material handling processes modeled by the ROPN method, and no additional control effort is needed.

However, a maximal permissive control policy is not necessarily optimal in the sense of productivity. Chapter 8 will discuss how to increase productivity in avoiding deadlocks in FMS, and a new control policy will be proposed.

In this chapter, each machine is considered to have an input buffer, or each resource has multiple capacity. It may not be applicable to systems with single capacity. For such systems, a new control policy is needed. It will be discussed in the following chapters.

REFERENCES

Abdallah, I., and A. Elmaraghy. 1998. Deadlock prevention and avoidance in FMS: A Petri net based approach. *International Journal of Advanced Manufacturing Technology* 14:704–15.

Banaszak, Z. A., and B. H. Krogh. 1990. Deadlock avoidance in flexible manufacturing systems with concurrently competing process flows. *IEEE Transactions on Robotics and Automation* 6:724–34.

Cho, H., T. K. Kumaran, and R. A. Wysk. 1995. Graph-theoretic deadlock detection and resolution for flexible manufacturing systems. *IEEE Transactions on Robotics and Automation* 11:413–21.

Chu, F., and X. L. Xie. 1997. Deadlock analysis of Petri nets using siphons and mathematical programming. *IEEE Transactions on Robotics and Automation* 13:793–804.

Coffman, E. G., Jr., M. J. Elphick, and A. Shoshani. 1971. System deadlocks. *ACM Computing Surveys* 3:67–78.

Ezpeleta, J., J. M. Colom, and J. Martinez. 1995. A Petri net based deadlock prevention policy for flexible manufacturing systems. *IEEE Transactions on Robotics and Automation* 11:171–84.

Fanti, M. P., B. Maione, S. Mascolo, and B. Turchiano. 1997a. Event-based feedback control for deadlock avoidance in flexible production systems. *IEEE Transactions on Robotics and Automation* 13:347–63.

Fanti, M. P., B. Maione, and B. Turchiano. 1997b. Comparing digraph and Petri net approaches to deadlock avoidance in FMS modeling and performance analysis. *IEEE Transactions on Systems, Man, and Cybernetics A* 30:783–98.

Fanti, M. P., and M. C. Zhou. 2004. Deadlock control methods in automated manufacturing systems. *IEEE Transactions on Systems, Man, and Cybernetics A* 34:5–21.

Gligor, V. D., and S. H. Shuttuck. 1980. On deadlock detection in distributed systems. *IEEE Transactions on Software Engineering* 6(5).

Hsieh, F., and S. Chang. 1994. Dispatching-driven deadlock avoidance controller synthesis for flexible manufacturing systems. *IEEE Transactions on Robotics and Automation* 10:196–209.

Jensen, K. (1986). *Colored Petri nets*, 248–99. Lecture Notes in Computer Science, vol. 254, part 1. Berlin: Springer-Verlag.

Kim, C. O., and S. S. Kim. 1997. An efficient real-time deadlock-free control algorithm for automated manufacturing systems. *International Journal of Production Research* 35:1545–60.

Lawley, M. A. 1999. Deadlock avoidance for production systems with flexible routing. *IEEE Transactions on Robotics and Automation* 15:1–13.

Lawley, M. A., S. Reveliotis, and P. Ferreira. 1998a. The application and evaluation of banker's algorithm for deadlock-free buffer space allocation in flexible manufacturing systems. *International Journal of Flexible Manufacturing Systems* 10:73–100.

Lawley, M. A., S. Reveliotis, and P. Ferreira. 1998b. A correct and scalable deadlock avoidance policy for flexible manufacturing policy. *IEEE Transactions on Robotics and Automation* 14:796–809.

Li, Z. W., and M. C. Zhou. 2004. Elementary siphons of Petri nets and their application to deadlock prevention in flexible manufacturing systems. *IEEE Transactions on Systems, Man, and Cybernetics A* 34:38–51.

Li, Z. W., and M. C. Zhou. 2008. Control of elementary and dependent siphons in Petri nets and their applications. *IEEE Transactions on Systems, Man, and Cybernetics A* 38:133–48.

Li, Z. W., M. C. Zhou, and N. Q. Wu. 2008. A survey and comparison of Petri Net-based deadlock prevention policy for flexible manufacturing systems. *IEEE Transactions on Systems, Man, and Cybernetics C* 38:173–88.

Reveliotis, S. A., and P. M. Ferreira. 1996. Deadlock avoidance policies for automated manufacturing cells. *IEEE Transactions on Robotics and Automation* 12:845–57.

Silberschatz, A., and P. G. Galvin. 1994. *Operating system concepts*. 4th ed. Reading, MA: Addision-Wesley.

Viswanadham, N., Y. Narahari, and T. L. Johnson. 1990. Deadlock prevention and deadlock avoidance in flexible manufacturing systems using Petri net models. *IEEE Transactions on Robotics and Automation* 6:713–23.

Wu, N. Q. 1997. Avoiding deadlocks in automated manufacturing systems with shared material handling system. In *Proceedings of 1997 IEEE International Conference on Robotics and Automation*, 2427–33.

Wu, N. Q. 1999. Necessary and sufficient conditions for deadlock-free operation in flexible manufacturing systems using a colored Petri net model. *IEEE Transactions on Systems, Man, and Cybernetics C* 29:192–204.

Wu, N. Q., and M. C. Zhou. 2000. Resource-oriented Petri nets for deadlock avoidance in automated manufacturing. In *Proceedings of 2000 IEEE International Conference on Robotics and Automation*, 3377–82.

Wu, N. Q., and M. C. Zhou. 2001. Avoiding deadlock and reducing starvation and blocking in automated manufacturing systems. *IEEE Transactions on Robotics and Automation* 17:657–68.

Wysk, R. A., N. S. Yang, and S. Joshi. 1991. Detection of deadlocks in flexible manufacturing cells. *IEEE Transactions on Robotics and Automation* 7:853–59.

Xing, K. Y., B. S. Hu, and H. X. Chen. 1996. Deadlock avoidance policy for Petri net modeling of flexible manufacturing systems with shared resources. *IEEE Transactions on Automatic Control* 41:289–95.

Yim, D.-S., J.-I. Kim, and H.-S. Woo. 1997. Avoidance of deadlocks in flexible manufacturing systems using a capacity-designated directed graph. *International Journal of Production Research* 35:2459–75.

Zhou, M. C., and F. DiCesare. 1991. Parallel and sequential mutual exclusions for Petri net modeling of manufacturing systems with shared resources. *IEEE Transactions on Robotics and Automation* 7:515–27.

Zhou, M. C., and F. DiCesare. 1993. *Petri net synthesis for discrete event control of manufacturing systems*. Boston: Kluwer Academic Publications.

Zhou, M., F. DiCesare, and A. Desrochers. 1992. A hybrid methodology for synthesis of Petri nets for manufacturing systems. *IEEE Transactions on Robotics and Automation* 18:350–61.

8 Avoiding Deadlock and Reducing Starvation and Blocking

8.1 INTRODUCTION

A good policy for avoiding deadlock should enforce the least restrictions on flexible manufacturing systems (FMS) to increase resource utilization and productivity, or it should be maximally permissive. Toward this, much outstanding work based on various discrete event models was reported (Kim and Kim, 1997; Reveliotis and Ferreira, 1996; Fanti et al., 1997; Xing et al., 1996; Lawley et al., 1998; Wu, 1999). A necessary and sufficient condition for deadlock-free operation and a maximally permissive control policy for a class of FMS are derived using the resource-oriented Petri net (ROPN) model (Wu, 1999). Its significance lies in the identification of a boundary for deadlock to occur in automated manufacturing systems (AMS). However, a new question arises regarding whether such a maximally permissive policy is optimal in terms of production rate in an environment where dispatching rules dominate and optimal scheduling is impossible due to either the unaffordable computation required or changing operational parameters and structures.

It is known that two factors affect the production rate in an FMS and reconfigurable manufacturing systems (RMS): starvation and blocking. When a machine completes a part and there is no part to be fed to it, it is in starvation. Release of too few parts to an AMS may starve some machines, lowering their production rate. This is why we pursue the maximally permissive control policy in deadlock avoidance for releasing as many jobs as possible into the system. When a machine completes a part that cannot be unloaded because of the lack of buffer spaces, it is blocked. Blocking is caused by the excessive job releases and limited buffer spaces. Blocked machines are forced idle, thereby losing productivity. The more parts in the system, the more likely it is deadlocked and the machines are blocked.

To operate an AMS effectively, the system should be well scheduled and deadlocks should be completely avoided such that starvation and blocking are reduced as much as possible. One can do the scheduling and control concurrently, as in Ramaswamy and Joshi (1996), Lee and DiCesare (1994), and Xiong and Zhou (1999), to obtain an optimal solution. Their approaches may not require any deadlock control policies. The maximally permissive deadlock control policy is best if an optimal schedule can be first determined under it, and then executed without interruption and changes in operational parameters and system structure. To our knowledge, not much work has been done to find an optimal schedule under such a maximally permissive policy. In addition, this strategy faces two tremendous challenges in practice. First, extensive computation likely prevents one from determining of an optimal schedule. Second, its optimality is no longer guaranteed when an FMS (RMS) is subject to changes of operation times, job routes, machine status, and addition/cancellation of rush jobs. That is why rule-based scheduling/sequencing policies dominate in reality, and the researchers have been seeking good heuristics and rules, as done in Chen and Lin (1999), Byeon et al. (1998), and Park et al. (1997). However, these scheduling heuristics and rules do not always prevent or avoid deadlocks. Thus, another way to operate an AMS is to do scheduling and deadlock control hierarchically. Under such a context, two approaches emerge. One is to derive good scheduling algorithms and rules under the maximally permissive control policy. The other is to first derive good deadlock control policies to reduce starvation (when compared with conservative policies) and blocking (when compared with the maximally permissive policy). Then, one can just adopt commonly used rule-based scheduling, or other good scheduling algorithms if needed.

If an FMS (RMS) operates at the deadlock boundary, i.e., under the maximally permissive control policy, it will not be deadlocked, but blocking is more likely. Consider an analogy between the liveness of PNs and the stability of continuous systems and between the deadlock boundary and stable boundary of continuous systems. To make the response of a continuous system fast enough, its eigenvalues should be assigned in the stable area with an appropriate distance to the stable boundary. Analogously, one may operate an FMS (RMS) near but not at the deadlock boundary in order to gain the highest productivity. We present such a policy and show, via examples, that without being too conservative, it can effectively reduce or even eliminate the blocking possibility that exists under a maximally permissive control policy. The presented material is based on Wu and Zhou (2001).

8.2   A SIMPLE EXAMPLE

A simple example is taken to show that a maximally permissive deadlock avoidance policy can lead to undesired blocking situations in FMS (RMS). This motivates the study of a new control policy for efficient operation of FMS (RMS).

Example 8.1

The FMS shown in Figure 8.1 contains three machines (m_{1-3}) and a robot as its material handling system. Machines m_{1-3} have buffers b_{1-3} with capacity 1, 2, and 2, respectively. The capacity of a load/unload station can be considered unlimited. The raw parts are delivered to the machines for processing from the load/unload station, and come back to the load/unload station after being completed. Consider two types of jobs/parts, C and D, either of which has three operations. The sequences and processing times of these operations are shown in Table 8.1.

According to the processing times in Table 8.1, if the number of parts for either type is the same (assume that each part type has ten parts to be processed), then the workload for each machine is also the same. In other words, this is a reasonable production plan. Now consider how the processing is carried out. Denote L/U as the load/unload station. According to the control law presented in Chapter 7, the

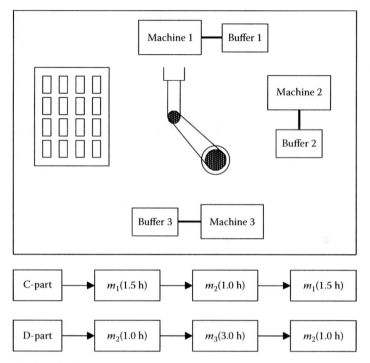

FIGURE 8.1 An AMS example.

TABLE 8.1
Operation Sequences and Processing Time

Operations	Operation 1		Operation 2		Operation 3	
Part Types	Machine	Time (h)	Machine	Time (h)	Machine	Time (h)
C	m_1	1.5	m_2	1	m_1	1.5
D	m_2	1	m_3	3	m_2	1

events that occur at times 0 to 5 are shown as follows, where "processed" means that a machine completes an operation. Notice that the subscript j in C_j and D_j is used to indicate the jth job of C and D, respectively.

1. Time 0: C_1 in L/U $\rightarrow m_1$ and D_1 in L/U $\rightarrow m_2$
2. Time 1: D_1 on m_2 (processed) $\rightarrow b_2 \rightarrow m_3$ and D_2 in L/U $\rightarrow m_2$
3. Time 1.5: C_1 on m_1 (processed) $\rightarrow b_1$ and C_2 in L/U $\rightarrow m_1$
4. Time 2: D_2 on m_2 (processed) $\rightarrow b_2$ and C_1 in $b_1 \rightarrow m_2$
5. Time 3: C_2 on m_1 (processed) $\rightarrow b_1$, C_3 in L/U $\rightarrow m_1$, C_1 on m_2 (processed) $\rightarrow b_2$, and D_3 in L/U $\rightarrow m_2$

TABLE 8.2

The State of Resource Assignment (□ and ◆ denote jobs C and D)

Resources	Capacity	T0	T1	T1.5	T2	T3	T4	T4.5	T5
m_1	1	□	□	□	□	□	□	□	□
b_1	1			□		□		□	□
m_2	1	◆	◆	◆	□	◆	□	□	□
b_2	2				◆	◆□	◆□	◆□	◆□
m_3	1	◆	◆	◆	◆	◆	◆	◆	◆
b_3	2						◆	◆	◆

6. Time 4: D_1 on m_3 (processed) → b_3, D_2 in b_2 → m_3, D_3 on m_2 (completed) → b_2, and C_2 in b_1 → m_2
7. Time 4.5: C_3 on m_1 (processed) → b_1 and C_4 in L/U → m_1
8. Time 5: C_2 on m_2 processed but cannot be unloaded

Table 8.2 shows the state of resource assignment at different times, where symbols □ and ◆ denote jobs C and D, respectively. It should be noted that a machine can process only one part at a time, or the capacity of each machine is one. From Table 8.2 we know that at time 5 only buffer b_3 has an open space; all the other resources in the system are occupied by jobs. According to the necessary and sufficient conditions presented in Chapter 7, at this time the system does not get deadlocked. But the job just completed on m_2 cannot be unloaded and delivered to b_2, i.e., the job is blocked. It is easy to verify that at time 6, the job completed on m_1 is also blocked. In fact, the job on m_2 can be unloaded into b_2 only after the job on m_3 is completed at time 7. This means that the blocking lasts for 2 h.

It is known that if the deadlock avoidance policy is too conservative, the number of active parts allowed to be in the system may be too small, leading some machines into starvation, which results in lower resource utilization and productivity. This motivates one to find a maximally permissive control policy for deadlock avoidance to reduce starvation. But, from the above example, we know that when active parts in the system are too many (although there is no deadlock), blocking may occur, affecting the resource utilization and productivity. Therefore, a good deadlock avoidance policy should take not only starvation but also blocking into consideration, and make a trade-off between them. Fortunately, the deadlock boundary identified by the control policy presented in Chapter 7 can be used to derive a control policy that keeps the state of the system near to the boundary, but not at the boundary. Thus, it controls an FMS (RMS) to achieve high productivity by reducing unnecessary starvation and blocking.

8.3   RELAXED CONTROL POLICY

Consider the control policy derived from a necessary and sufficient condition for deadlock avoidance. It requires that there is at least one leaving token in every production process circuit (PPC). A leaving token in a PPC is not necessary to leave the PPC immediately, even though the operation is completed. It may stay there and take a space in the PPC for some time. Although the system is not deadlocked, it is blocked for some time, since there is no space to move the tokens in the PPC. Hence, if every PPC in a CROPN keeps a space at any marking M reached from the initial marking M_0, less blocking may happen, and at the same time it is not too conservative. In other words, by doing so, we keep a liveness margin for the system. The problem is whether such a control policy makes the interactive subnets in the CROPN live. We next show that this is true.

From Chapter 7, we need to discuss control policies only for individual interactive subnets. We use the same notation used in Chapter 7. Two interactive subnets are shown in Figure 8.2. Denote $K(v_i)$ as the capacity of PPC v_i. Let $S(v_i) = \sum_{p_j \in P(v_i)} (K(p_j) - M(p_j))$ be the number of free spaces available in v_i, and $S(p_i) = K(p_i) - M(p_i)$ the free spaces in p_i in marking M. Similarly, we use $S(v^n)$ to denote the free spaces in an interactive subnet v^n. In the following discussion, we assume that there are tokens in the interactive subnets, implying that there are parts to be processed.

Because of the existence of shared places in v^n, when a free space is in one of the shared places, this space is shared by both PPCs. To make v^n deadlock-free and reduce the blocking possibility, it may be better to keep a free space for each PPC in the subnet. A free space in a shared direct place path (SDPP) can be allocated to any PPC that shares the SDPP, but one and only one PPC. Once it is allocated, it is called a non-shared space. For example, in the subnet shown in Figure 8.2a, if there is a free space in p_2 and other places are full, this space can be allocated to $v_1 = \{p_1, t_{12}, p_2, t_{21}\}$ or $v_2 = \{p_2, t_{23}, p_3, t_{32}\}$, but not to both. This implies that only one of the PPCs has a free space.

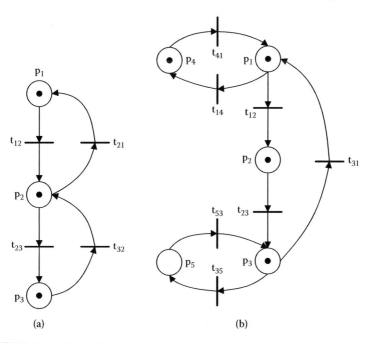

(a)   (b)

FIGURE 8.2 Interactive subnets.

However, if there is another free space in p_1, then this space can be allocated to v_1 and the space in p_2 can be allocated to v_2. This time each PPC has a nonshared free space.

Definition 8.1: The L-condition holds for PPC v and marking M if and only if there is at least one nonshared free space available in v at M.

Theorem 8.1: An interactive subnet v^n composed of n PPCs in a CROPN is live if L-condition holds for any marking $M \in R_c(M_0)$ and any PPC in v^n.

Proof: First we consider v^1 (or v), the case with $n = 1$. If $T_I(v) = \varnothing$ and $T_O(v) = \varnothing$, v is called an isolated PPC. Due to a finite number of nodes in a CROPN, v has a finite number of nodes. Obviously, an isolated PPC is live if the L-condition holds.

If $T_I(v) \neq \varnothing$ and $T_O(v) \neq \varnothing$, assume that $S(v) = 1$, or the L-condition, holds. At a marking M, if $T_O(v)$ is not enabled, then we can treat it as an isolated PPC, or v is not deadlocked. When $T_O(v)$ is enabled in a marking M, we can fire $T_O(v)$ and one token is removed from v, or the L-condition still holds. Noticing the nature of manufacturing systems, all the tokens can be removed from v in this way. When $S(v) > 1$ after some tokens are removed from v, we can fire $T_I(v)$ to move some tokens into v, but control the times of firing $T_I(v)$ such that the L-condition is not violated. Thus, if the L-condition holds, v^1 is live.

Now we show v^n is live, with n being an arbitrary finite positive integer if the L-condition holds. Assume v_i is a PPC in v^n, and let $P_{ns}(v_i) \subset P(v_i)$ be the set of places on v_i that are not shared with other PPCs, and $P_s(v_i) \subset P(v_i)$ be the set of places on v_i that are shared with other PPCs. Further, we let p_h (p_r) denote the first (last) place on the SDPP. Assume that there are n free spaces in v^n or $S(v^n) = n$ and the L-condition holds. If there are tokens in an SDPP and $t_{iol} \subset T_{iok}$ on v_1 is enabled in marking M, then we can treat v_1 as an isolated PPC, and firing any transition on v_1 will keep the L-condition satisfied. If the free space allocated to v_1 is in $P_s(v_1)$ and t_{iol} is not enabled, this time the input transition of p_h on v_1 is able to fire and the L-condition still holds. If there is no token in an SDPP, then let v_D be the set of PPCs whose free spaces are in the SDPP. We only need to forbid the firing of transitions t_{iik} with $k \in D$ that is associated to the SDPP, and the other t_{iik} associated to the SDPP can fire. Thus, some tokens can be moved to the SDPP. If there are more than n free spaces in v^n, then we only need to keep each PPC with a free space, and the remaining free paces can be used by any PPC, or we can fire some transitions $t \in T_I(v^n)$ and move some tokens into v^n. In this way, v^n is live.

When some transitions in $T_O(v^n)$ are enabled, we can fire them and some tokens are removed from v^n, this time $S(v^n) > n$. Sometimes, all the tokens in v^n can be removed from v^n and the transitions in $T_I(v^n)$ can fire. Therefore, in all the cases, v^n is live. ∎

From Theorem 8.1, we know that to make an interactive subnet v^n in a CROPN live, we have to control the number of tokens in the subnet and the firing order of its intercircuit input transitions (IITs). In the following we define the control policy called *L-policy*, where v^n is composed of n PPCs in a CROPN.

Definition 8.2: The L-policy for v^n is a control policy under which (1) the L-condition holds for any v in v^n after t's firing if $t \in T_I(v^n)$ is selected to fire, and (2) the L-condition holds for any v in subnet v^{n-1} of v^n after an IIT's firing if it is external to v^{n-1} and selected to fire.

Theorem 8.2: An interactive subnet v^n composed of n PPCs in a CROPN is live under the L-policy for v^n.

According to Theorem 8.2, for subnet v^n to be live, the transitions that need to be controlled are those in $T_j(v^n)$ and IITs. In fact, controlling the transitions in $T_j(v^n)$ controls the number of jobs released to v^n. Controlling the IITs decides the holding order of a free resource when multiple jobs compete for one resource in the sense of deadlock avoidance.

8.4   DEPENDENT PPCs IN INTERACTIVE SUBNETS

From Theorem 8.1, each PPC in an interactive subnet requires a nonshared space. Then, how many PPCs can a subnet formed by h machine primitives have, and how many free spaces are required? It can be shown that the number of possible PPCs is $\alpha(h) = \sum_{k=2}^{h} h(h-1) \cdots (h-k+1) / k$ in an interactive subnet formed by h machines with or without associated buffers. It is independent of the jobs and how they route. Let $P(h)$ denote the set of places in such a subnet, and $|P(h)|$ be the number of places in $P(h)$. Then $\alpha(h) > |P(h)|$, when $h > 3$. Thus, the number of free spaces required to make the subnet live may be greater than the capacity of the subnet if $K(p) = 1$ for every $p \in P$.

By Theorem 8.1, when a new PPC is added into v^n, one more space is required to guarantee the liveness of the resulting subnet. However, it may be the case that the resulting subnet is live even if no more free space is added. The concept of independent and dependent PPCs in an interactive subnet is proposed to characterize these two cases. In fact, only an independent PPC in a subnet needs a nonshared free space.

Definition 8.3: Assume that v^k is composed of k PPCs in v^n with $n > k$. Let V_b be the set of the k PPCs in v^k and V_d the set of the other $n - k$ PPCs in v^n. Suppose that v^n is live when the L-policy for v^k applies. Then, all PPCs in V_d are said to be dependent on the PPCs in V_b.

The dependence of PPCs is closely related to the connection of places. If a place in a CROPN has two or more input transitions, we say that a *merger* is formed. If a place has two or more output transitions, a *choice* is formed. The interaction of PPCs in an interactive subnet forms the mergers and choices. The mergers and choices divide the connected places into a number of segments that are called place path segments, denoted by CH. Given a PPC and the places $p_1, p_2, \ldots, p_i, p_{i+1}, \ldots, p_k$ on it, suppose that $p_{i+1}, \ldots, p_k$ are the only places shared by another PPC. Then two place path segments are $p_1 - p_i$ and $p_{i+1} - p_k$. Clearly, the shared places of two PPCs form one place path segment, e.g., p_2 in Figure 8.2a. If $p_1, p_2, \ldots, p_i, p_{i+1}, \ldots, p_k$ are the shared places of PPCs v_1 and v_2, and $p_{i+1}, \ldots, p_k$ are the shared places of PPCs v_1, v_2 and v_3, then we have two different place path segments: $CH_1 = p_1 - p_i$ and $CH_2 = p_{i+1} - p_k$. We have the following theorem for an interactive subnet with dependent PPCs:

Theorem 8.3: In a subnet v^{n+r}, r (>0) PPCs are dependent on the other n (>0) PPCs that form v^n without r dependent PPCs if there are $n + 1$ place path segments in the subnet v^{n+r} and each PPC contains at least two place path segments.

Proof: It is sufficient to show that subnet v^{n+r} is live when there are only n spaces in it. Distribute these n spaces such that each place path segment has one space for arbitrary n place path segments. Because each PPC contains at least two place path segments and only one place path segment has no space, we assert that each PPC has one nonshared space in every subnet v^n that contains all the places. Since each PPC contains at least two place path segments, no place path segment in the subnet forms a circuit. Thus, each transition of the output transitions of the last place p_r on a place path segment is an input transition of the first place p_h on another place path segment. The transitions on the place path segments are not controlled, or they can fire freely. Therefore, a space in a place path segment can move into the first place p_h on the place path segment automatically. Consider the place path segment that has no space and denote it by CH_1. Assume that there are transitions of IITs in the output transitions of the last place p_r on CH_1, and one of them, say t, is process-enabled. It is sure that the output place p (the first place on another place path segment) of t has a space, or it is resource-enabled. Meanwhile, all PPCs that contain p have at least two free spaces because there are at least two spaces in every PPC that does not contain place path segment CH_1. Assume that this place is on CH_2. Then, fire t and a space moves from CH_2 to CH_1. If there is no transition of IITs in the output transitions of the last place on CH_1, or no such transition is process-enabled, then there must be a transition that is not an IIT and is process-enabled. This transition can fire spontaneously, or a space moves to place path segment CH_1 from the next place path segment. It is clear that the condition guaranteeing a live subnet is unchanged after the firing of a transition under the L-policy for v^n. In this way, every transition can fire no matter what marking $M \in R_c(M_0)$ is reached, or v^{n+r} is live. ∎

To make v^{n+r} live, each PPC in v^n must have at least one nonshared free space, and meanwhile, each dependent PPC must share at least one space with those n PPCs in v^n. This implies that to control an interactive subnet with dependent PPCs, the L-policy should be applied to the subnet composed of the independent ones such that every dependent PPC shares at least one space with the independent PPCs. Definition 8.3 and Theorem 8.3 present a criterion to identify dependent PPCs in an interactive subnet.

Since a v^2 always contains three place path segments, two PPCs in it are always independent of each other. For example, in Figure 8.2b the three place path segments are $CH_1 = \{p_1\}$, $CH_2 = \{p_2\}$, and $CH_3 = \{p_3\}$. A single PPC interactive subnet has only one place path segment.

Figure 8.3 shows three examples of interactive subnets with a dependent PPC. In Figure 8.3a, let $V_b = \{v_1, v_2\}$, $v_1 = \{p_1, t_{12}, p_2, t_{24}, p_4, t_{41}\}$, $v_2 = \{p_2, t_{23}, p_3, t_{32}\}$, and $v = \{p_1, t_{13}, p_3, t_{32}, p_2, t_{24}, p_4, t_{41}\}$, and three place path segments are $CH_1 = \{p_4, t_{41}, p_1\}$, $CH_2 = \{p_2\}$, and $CH_3 = \{p_3\}$. For Figure 8.3b, let $V_b = \{v_1, v_2\}$, $v_1 = \{p_1, t_{12}, p_2, t_{21}\}$, $v_2 = \{p_2, t_{23}, p_3, t_{32}\}$, and $v = \{p_1, t_{12}, p_2, t_{23}, p_3, t_{31}\}$, and three place path segments are $CH_1 = \{p_1\}$, $CH_2 = \{p_2\}$, and $CH_3 = \{p_3\}$. In Figure 8.3c, let $V_b = \{v_1, v_2, v_3\}$, $v_1 = \{p_1, t_{12}, p_2, t_{23}, p_3, t_{34}, p_4, t_{41}\}$, $v_2 = \{p_1, t_{13}, p_3, t_{31}\}$, $v_3 = \{p_1, t_{12}, p_2, t_{23}, p_3, t_{31}\}$, and $v = \{p_1, t_{13}, p_3, t_{34}, p_4, t_{41}\}$, and four place path segments are $CH_1 = \{p_1\}$, $CH_2 = \{p_2\}$, $CH_3 = \{p_3\}$, and $CH_4 = \{p_4\}$. In these three subnets every PPC contains at least two place path segments. Therefore, for all three cases, the conditions in Theorem 8.3 are satisfied, and thus PPC v is dependent on the others.

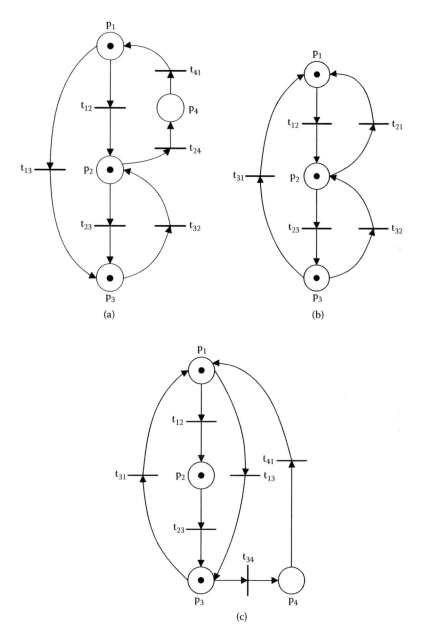

FIGURE 8.3 Subnets with a dependent PPC.

Now consider the three subnets shown in Figure 8.4. In Figure 8.4a, there are three PPCs, $v_1 = \{p_1, t_{12}, p_2, t_{24}, p_4, t_{41}, p_1\}$, $v_2 = \{p_1, t_{12}, p_2, t_{23}, p_3, t_{31}, p_1\}$, and $v_3 = \{p_2, t_{23}, p_3, t_{32}, p_2\}$, and four place path segments, $CH_1 = \{p_1\}$, $CH_2 = \{p_2\}$, $CH_3 = \{p_3\}$, and $CH_4 = \{p_4\}$. In Figure 8.4b there are also three PPCs, $v_1 = \{p_1, t_{12}, p_2, t_{21}, p_1\}$, $v_2 = \{p_2, t_{23}, p_3, t_{34}, p_4, t_{42}, p_2\}$, and $v_3 = \{p_1, t_{14}, p_4, t_{42}, p_2, t_{21}, p_1\}$, and four place path

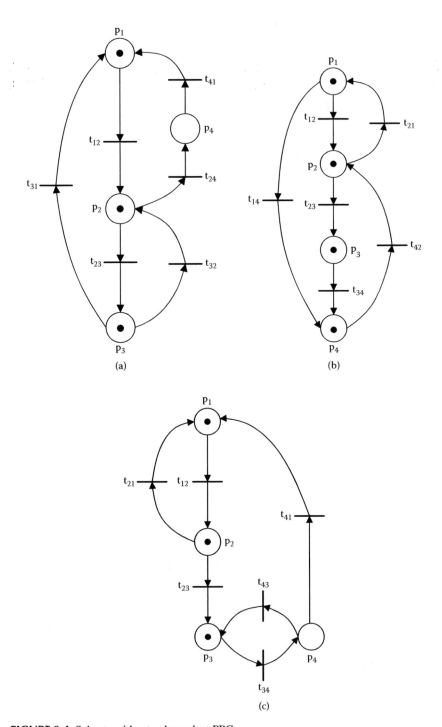

FIGURE 8.4 Subnets without a dependent PPC.

segments, $CH_1 = \{p_1\}$, $CH_2 = \{p_2\}$, $CH_3 = \{p_3\}$, and $CH_4 = \{p_4\}$. It can be seen that in both Figures 8.4a and b, every PPC contains at least two place path segments. Thus, there is no dependent PPC. In Figure 8.4c there are three PPCs, $v_1 = \{p_1, t_{12}, p_2, t_{21}, p_1\}$, $v_2 = \{p_3, t_{34}, p_4, t_{43}, p_3\}$, and $v_3 = \{p_1, t_{12}, p_2, t_{23}, p_3, t_{34}, p_4, t_{41}, p_1\}$, and two place path segments, $CH_1 = \{p_1, t_{12}, p_2\}$ and $CH_2 = \{p_3, t_{34}, p_4\}$. However, each of the PPCs v_1 and v_2 contains only one place path segment. Hence, there is no dependent PPC. In fact, in Figure 8.4c, v_1 and v_2 cannot form an interactive subnet v^2, although all three form v^3.

Note that nonelementary PPCs must be dependent on the elementary PPCs in an interactive subnet. Besides, an elementary PPC may be a dependent PPC, e.g., v in Figure 8.3.

Up to now we have presented the dependency of the PPCs in an interactive subnet. The question is how many independent PPCs an interactive subnet formed by h primitives can have. The following study of completely connected v^n answers this and provides the absolute maximum number of free spaces needed for a general interactive subnet to stay live.

Definition 8.4: An interactive subnet formed by h machines in a CROPN is completely connected if any two places p_i and p_j, representing machines m_i and m_j, are connected by two transitions, t_{ij} from p_i to p_j, and t_{ji} from p_j to p_i.

Let $V(v^n)$ be the set of all the PPCs in subnet v^n. We define a set of basic PPCs in an interactive subnet as follows:

Definition 8.5: A set V_b of PPCs is the set of basic PPCs in an interactive subnet v^n if any PPC $v \in V(v^n) - V_b$ is dependent on V_b, and no PPC in V_b is dependent on the others in V_b.

Denote G_h as a completely connected interactive subnet formed by h machines in a CROPN. G_3 is shown in Figure 8.5. Its five elementary PPCs are $v_1 = \{p_1, t_{12}, p_2, t_{21}\}$,

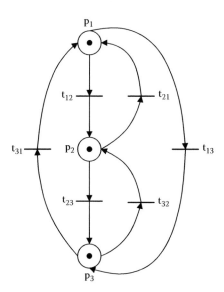

FIGURE 8.5 The completely connected interactive subnet G_3.

$v_2 = \{p_2, t_{23}, p_3, t_{32}\}$, $v_3 = \{p_1, t_{13}, p_3, t_{31}\}$, $v_4 = \{p_1, t_{13}, p_3, t_{32}, p_2, t_{21}\}$, and $v_5 = \{p_1, t_{12}, p_2, t_{23}, p_3, t_{31}\}$. Let $V_b = \{v_1, v_2\}$ and $|V_b| = 2$ for G_3.

In G_h, let $v_1 = \{p_1, t_{12}, p_2, t_{21}\}$, $v_2 = \{p_2, t_{23}, p_3, t_{32}\}$, . . . , $v_{h-1} = \{p_{h-1}, t_{h-1, h}, p_h, t_{h, h-1}\}$, and $V_b = \{v_1, v_2, \ldots, v_{h-1}\}$. Let $H(D) = 1$, if $S(D) > 0$, and otherwise, $H(D) = 0$, where D denotes a set of places or a place path segment, and $S(D)$ denotes the number of free spaces in D. Let γ denote an arbitrary subnet of a PN and $P(\gamma)$ the set of places in γ. Define $H(\gamma) = 1$ if $S(P(\gamma)) \geq 1$ at a marking M, and $H(\gamma) = 0$ if $S(P(\gamma)) = 0$. The following theorem implies that all other PPCs in G_h are dependent on V_b.

Theorem 8.4: V_b is a basic set of PPCs in a completely connected subnet G_h in a CROPN.

Proof: It is clear that PPCs in V_b are not dependent on each other. Let $F(G_h) = H(p_1) + H(p_2) + \cdots + H(p_h)$, and v^{h-1} be composed of the PPCs in V_b. If $F(G_h) = h - 1$, then the condition in Theorem 8.1 is satisfied for v^{h-1}. Assume that $H(p_h) = 0$ and $H(p_i) = 1$, $i = 1, 2, \ldots, h - 1$; then according to Theorem 8.2, one of the output transitions t_{hi} ($i = 1, 2, \ldots, n - 1$) of p_h can fire, for there are two spaces in every two-place PPC that does not contain p_h. Assume that t_{hk} is fired. Then, there is at least one space in each place except p_k, and thus every output transition of p_k is enabled by the control policy (or control-enabled). In this way, at any time, there are at least $h - 1$ spaces distributed in at least $h - 1$ places. Thus, there exists at least one place whose output transitions are enabled by the control policy. Any one of the enabled transitions can be chosen to fire. By choosing a different transition in the enabled transitions to fire, we choose a different place whose output transitions are to be enabled in the next step. Therefore, every transition can fire in finite steps no matter what marking $M \in R_c(M_0)$ is reached, and this process can repeat endlessly. This means that G_h is live. Therefore, V_b is a basic set of PPCs in G_h. ∎

G_h has the most PPCs among all interactive subnets in a CROPN formed by h places, and $|V_b| = h - 1$. By Theorem 8.4, it needs $h - 1$ free spaces to make subnet G_h live. Therefore, it needs no more than $h - 1$ free spaces to make a live interactive subnet involving h places (machines). If each machine has an associated buffer, it contains at least two spaces. Thus, an interactive subnet formed by h machines has at least $2h$ spaces, i.e., its capacity is at least $2h$. Therefore, we can have at least $2h - (h - 1) = h + 1$ tokens representing parts in it. This shows that the L-policy causes no deadlock and can allow sufficient parts in AMS for high productivity. G_h contains h place path segments, and each place path segment is formed by a machine and its associated buffer. Then, we have the following corollary:

Corollary 8.1: For an AMS composed of h places (machines), if at any time there are $h - 1$ free spaces that are distributed to $h - 1$ different machines, the system is deadlock-free.

In the existing techniques for deadlock avoidance (e.g., Fanti et al., 1997), the part types to be processed and their routes (with routing flexibility or not) should be known in advance, and thus a deterministic model can be built. By Corollary 8.1, to make the system live, it is not necessary to have such information, and the part types and their routes can be changed on-line. Furthermore, using Corollary 8.1, no circuits need to be identified.

It should be pointed out that although we discuss the deadlock avoidance policy by assuming that there is an associated buffer for each machine, or the resources have multiple capacity, the control policy is a correct deadlock avoidance control policy for the systems where the resources are single-capacity ones. The following example demonstrates this.

Example 8.2

Consider the PNs shown in Figure 8.6, where every place except p_0 has a capacity of one, or it is a single-capacity system. In the PNs, the token in p_2 needs to move to p_5 and the token in p_6 needs to move to p_4.

In Figure 8.6a, the PN is not deadlocked, since t_5 and t_9 are both enabled. However, it will be deadlocked no matter which transition fires. This situation is called second-level deadlock in Fanti et al. (1997). It is easy to check that at this marking, the control policy presented here is violated. In Figure 8.6b, at this marking the control policy presented here is satisfied. According to the control policy, we can fire t_9, then t_6, and finally t_5, resulting in no deadlock.

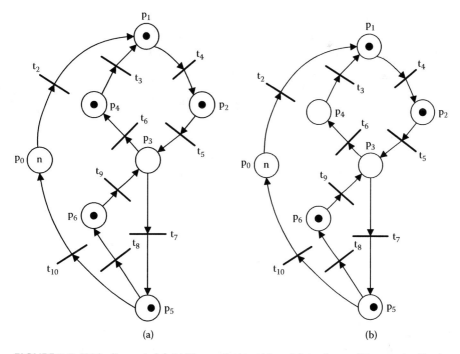

(a)   (b)

FIGURE 8.6 PN for Example 8.2. (a) The reached marking violates the condition required by the control policy, though it is not a deadlock state, but it will reach a deadlock state. (b) The condition required by the L-policy is met and the PN will not be deadlocked under the L-policy.

8.5 COMPLEXITY IN APPLYING THE CONTROL LAW

The presented policy tells the controller what states should be avoided by disabling some transitions. To implement it, we need to build the model and identify the PPCs if the jobs and their routes are known (if these are not known, we can use the control policy presented by Corollary 8.1, and no PPCs need to be identified). Although this is not trivial, it can be done off-line. After the PPCs are identified, for the real-time control we need only check the free spaces in the PPCs. The next question is: How many PPCs does one need to check in real time for a subnet formed by h places?

We analyze the complexity by considering the completely connected interactive subnet G_h. In G_h, any two places (machines) p_i and p_j, $i, j = 1, \ldots, h$ with $i \neq j$, form a PPC, denoted by v_{ij}. Let $V_H = \{v_{ij}\}$ be the set of all such PPCs. For example, in G_3 shown in Figure 8.5 there are five elementary PPCs, $v_{12} = \{p_1, t_{12}, p_2, t_{21}\}$, $v_{23} = \{p_2, t_{23}, p_3, t_{32}\}$, $v_{13} = \{p_1, t_{13}, p_3, t_{31}\}$, $v_4 = \{p_1, t_{13}, p_3, t_{32}, p_2, t_{21}\}$, and $v_5 = \{p_1, t_{12}, p_2, t_{23}, p_3, t_{31}\}$, and $V_H = \{v_{12}, v_{23}, v_{13}\}$. It is easy to observe that $P(v_{12}) \cup P(v_{23}) \cup \ldots \cup P(v_{h1})$ = $P(G_h)$ and $T(v_{12}) \cup T(v_{23}) \cup \ldots \cup T(v_{h1}) = T(G_h)$. This means that the set of places (transitions) in V_H contains all the places (transitions) in G_h. By applying the control law presented above, if we check all the PPCs in V_H, then we guarantee that all the transitions in V_H are live, or G_h is live. Thus, we only need to check the PPCs in V_H instead of all the elementary PPCs in G_h when applying the control law. In fact, in G_h, if every PPC in V_b has a nonshared space and every PPC in $V_H - V_b$ has a shared space, then all other PPCs in G_h have a shared space. This is why we need to check only the PPCs in V_H. The number of PPCs in V_H is $|V_H| = \frac{h(h-1)}{2}$, or the computational complexity is $O(h^2)$ in the real-time control. Notice that G_h has the most PPCs in an interactive subnet formed by h places, and the complexity of implementing this control law is from the number of PPCs to be checked. Hence, we only need to trace V_b and V_H, and at most, $|V_H|$ PPCs need to be checked in real time in any subnet formed by h places. Therefore, the complexity is $O(h^2)$ in applying the proposed control law, although the number of PPCs in a subnet formed by h places is exponential with respect to h.

8.6 PERFORMANCE IMPROVEMENT THROUGH EXAMPLES

Up to now, we have shown deadlock-free operation of an AMS under the L-policy, but not its performance improvement. In this section we show the performance improvement by the proposed control policy through two examples.

Example 8.3 (continued from Example 8.1)

The CROPN for the example is shown in Figure 8.7, and it contains a subnet v^2. The two PPCs in the figure are $v_1 = \{p_1, t_{12}, p_2, t_{21}\}$ and $v_2 = \{p_2, t_{23}, p_3, t_{32}\}$. From Example 8.1, we know that if the maximally permissive control policy presented in Wu (1999) is applied, then at time 5, m_2 is blocked. But, by applying the proposed control policy in this chapter, at time 4.5 we do not deliver a C-part from the load/unload station to m_1, but deliver the part in buffer b_2 to m_1 by firing transition t_{21} instead of t_1. The progress is as follows, and Table 8.3 shows the states of

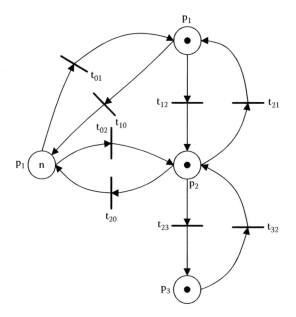

FIGURE 8.7 The CROPN for Example 8.3.

TABLE 8.3

The States of Resource Allocation in Progress

Resources	M_1	B_1	M_2	B_2	M_3	B_3
Capacity	1	1	1	2	1	2
T4.5	□	□	□	◆	◆	◆
T5	□		□	□◆	◆	◆
T6	□		◆	□◆	◆	◆
T7	□		◆	□◆	◆	◆
T7.5	□		◆	◆	◆	◆
T8	□		◆	◆	◆	
T9	□		◆	◆	◆	
T10	□		◆	◆	◆	
T10.5	□	□	◆	◆	◆	
T11	□		□	◆	◆	

resource allocation. Note that "completed" means that all operations for that part are completed.

1. Time 4.5: C on m_1 (processed) $\rightarrow b_1$ and C in $b_2 \rightarrow m_1$
2. Time 5: C on m_2 (processed) $\rightarrow b_2$ and C in $b_1 \rightarrow m_2$
3. Time 6: C on m_1 (completed) $\rightarrow b_1 \rightarrow$ L/U, C in $b_2 \rightarrow m_1$ and D in L/U $\rightarrow m_2$
4. Time 7: D on m_3 (processed) $\rightarrow b_3$, D in $b_2 \rightarrow m_3$, D on m_2 (processed) $\rightarrow b_2$, and D in $b_3 \rightarrow m_2$
5. Time 7.5: C on m_1 (completed) $\rightarrow b_1 \rightarrow$ L/U and C in $b_2 \rightarrow m_1$
6. Time 8: D on m_2 (completed) $\rightarrow b_2 \rightarrow$ L/U and D in $b_3 \rightarrow m_2$

7. Time 9: C on m_1 (completed) → m_1 → L/U, C in L/U → m_1, D on m_2 (completed) → b_2 → L/U, and D in L/U → m_2
8. Time 10: D on m_3 (processed) → b_3, D in b_2 → m_3, D on m_2 (processed) → b_2, and D in b_3 → m_2
9. Time 10.5: C on m_1 (processed) → b_1, C in L/U → m_1
10. Time 11: D on m_2 (completed) → b_2 → L/U and C in b_1 → m_2

Up to time 11, three parts for each part type have been completed, and there is no blocking or starvation. It can also be seen that at time 11 the state is the same as at time 2. Hence, the process can repeat, and blocking and starvation will never occur. This shows that the L-policy completely eliminates the blocking, ending with the 11 h cycle time. Under the maximally permissive policy, blocking can occur if a C-part is loaded at Time 4.5, leading to a 13 h cycle time. The throughput improvement using the L-policy is over 15%.

Example 8.4

The system has four machines and needs to deal with flexible routes. Products, routes, and average operation hours in the parentheses are as follows:

Product A:
  Route 1: m_1 (8) → m_3 (10) → m_4 (6)
  or
  Route 2: m_1 (8) → m_3 (10) → m_1 (8)

Product B:
  Route 1: m_3 (5) → m_1 (12) → m_3 (6) → m_4 (15)
  or
  Route 2: m_3 (5) → m_1 (12) → m_2 (18) → m_1 (3)

Product C: m_3 (5) → m_4 (10) → m_3 (5)

The uniform distribution (0.9x, 1.1x) is assumed, where x is the average operation time shown above. We tested both random and shortest processing time first dispatching rules under the maximally permissive policy and L-policy. A number of

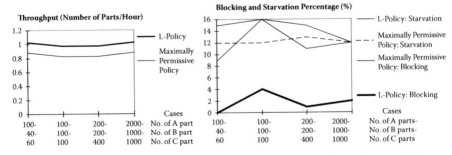

FIGURE 8.8 System throughput. Machine 3's blocking and starvation under two policies.

randomly selected job mixes were tested with the maximum set of two thousand jobs of A, one thousand of B, and one thousand of C. The results of the system throughput, blocking, and starvation of machine 3 are shown in Figure 8.8 for the L-policy and maximally permissive policy. All the cases show that on average, the L-policy outperforms the maximally permissive policy by over 10% in throughput. Similar throughput gains were obtained when we randomized all the operation times and sequences. With the L-policy, all machines are significantly less likely to be blocked (around 10%), but have a small increase in starvation (3%).

8.7 SUMMARY

Based on the maximal deadlock avoidance control policy presented in Chapter 7, this chapter presents a control policy from a sufficiency condition. It is shown that this policy can reduce blocking such that productivity is increased. At the same time, this policy is applicable to single-capacity systems and can deal with the so-called second-level deadlocks (Fanti et al., 1997). Because of this property, this control policy can be applied to avoid deadlocks in automated guided vehicle (AGV) systems.

The theoretical question of whether the proposed L-policy is the optimal one under some fixed scheduling rule remains open. In other words, is there an even more relaxed policy that allows the system to perform at the highest productivity, i.e., neither blocking nor starvation? Clearly, both theoretical and experimental work must be performed to answer such questions.

REFERENCES

Byeon, E.-S., S. D. Wu, and R. H. Storer. 1998. Decomposition heuristics for robust job-shop scheduling. *IEEE Transactions on Robotics and Automation* 14:303–13.

Chen, S.-J., and L. Lin. 1999. Reducing total tardiness cost in manufacturing cell scheduling by a multi-factor priority rule. *International Journal of Production Research* 37:2939–56.

Fanti, M. P., B. Maione, S. Mascolo, and B. Turchiano. 1997. Event-based feedback control for deadlock avoidance in flexible production systems. *IEEE Transactions on Robotics and Automation* 13:347–63.

Kim, C. O., and S. S. Kim. 1997. An efficient real-time deadlock-free control algorithm for automated manufacturing systems. *International Journal of Production Research* 35:1545–560.

Lawley, M. A., S. Reveliotis, and P. Ferreira. 1998. A correct and scalable deadlock avoidance policy for flexible manufacturing policy. *IEEE Transactions on Robotics and Automation* 14:796–809.

Lee, D. Y., and F. DiCesare. 1994. Scheduling FMS using Petri nets and heuristic search. *IEEE Transactions on Robotics and Automation* 10:123–32.

Park, S. C., N. Raman, and M. J. Shaw. 1997. Adaptive scheduling in dynamic flexible manufacturing systems: A dynamic rule selection approach. *IEEE Transactions on Robotics and Automation* 13:486–502.

Ramaswamy, S. E., and S. B. Joshi. 1996. Deadlock-free schedules for automated manufacturing workstations. *IEEE Transactions on Robotics and Automation* 12:391–400.

Reveliotis, S. A., and P. M. Ferreira. 1996. Deadlock avoidance policies for automated manufacturing cells. *IEEE Transactions on Robotics and Automation* 12:845–57.

Wu, N. Q. 1999. Necessary and sufficient conditions for deadlock-free operation in flexible manufacturing systems using a colored Petri net model. *IEEE Transactions on Systems, Man, and Cybernetics C* 29:192–204.

Wu, N. Q., and M. C. Zhou. 2001. Avoiding deadlock and reducing starvation and blocking in automated manufacturing systems. *IEEE Transactions on Robotics and Automation* 17:657–68.

Xing, K. Y., B. S. Hu, and H. X. Chen. 1996. Deadlock avoidance policy for Petri net modeling of flexible manufacturing systems with shared resources. *IEEE Transactions on Automatic Control* 41:289–95.

Xiong, H. H., and M. C. Zhou. 1999. A Petri net method for deadlock-free scheduling of flexible manufacturing systems. *International Journal of Intelligent Control and Systems* 3:277–95.

9 Control and Routing of Automated Guided Vehicle Systems

9.1 INTRODUCTION

It is well known that deadlock may occur due to limited resources in flexible manufacturing systems (FMS), leading to a system-wide standstill. One type of limited resources is caused by the competition by parts for the manufacturing resources, such as machines, buffers, and the material handling system (MHS). It occurs during the part processing. The other type is due to the competition for nodes and lanes by automated guided vehicles (AGVs) when multiple AGVs are used in MHS. The deadlock issue in part production processes has been discussed in the previous chapters. This chapter discusses a deadlock avoidance problem in AGV systems.

In an AGV system, when there are several AGVs, some serious problems may arise in managing them, e.g., blocking, conflict, deadlock, and collision (Koff, 1987; Malmbog, 1990; Kim and Tanchoco, 1991; Zeng et al., 1991; Reveliotis, 2000). Differences can be observed between these two types of deadlocks. To improve the productivity and resource utilization, it is desired to have as many parts as possible in a system. However, the more parts in FMS or reconfigurable manufacturing systems (RMS), the more likely a system is to be deadlocked. The parts need to be delivered from one station to another by MHS. Often, only a few AGVs are available in an AGV system, mainly due to their high cost and easily satisfied transportation demand. The total number of AGVs likely remains constant. On the other hand, the following four differences lead to more challenging deadlock control problems for AGVs:

1. All AGVs always stay in the system. Jobs in FMS (RMS) are assumed to enter and leave the system after its completion. Thus, some deadlock avoidance policies for part production processes may not be applicable to AGV systems. For example, applying the policies proposed in Fanti et al. (1997) to some AGV systems may allow only one AGV in such a system, making them infeasible.

2. While the route of each part type in FMS (RMS) is often known in advance, the routes of an AGV frequently change based on real-time transportation requests. Hence, the policies obtained for handling the first type of deadlock requiring known routes become difficult to apply, primarily because of their computational requirements (e.g., Banaszak and Krogh, 1990; Wu, 1999; Wu and Zhou, 2001a; Xing et al., 1996).

3. In AGV systems, a lane and a node may hold one and only one AGV. This leads to a single-capacity resource system, invalidating many elegant deadlock policies that require multiple capacity resources (e.g., Banaszak and Krogh, 1990; Lawley, 1999; Wu, 1999).
4. In the part production process, when a part is completed it leaves the system and occupies no resource. When an AGV fulfills a mission it holds a node and may block other AGVs to complete their missions.

The most widely used technique for vehicle management of AGV systems is zone control. Guided paths are divided into several disjoint zones. A zone can accommodate only one AGV at a time. The guided paths may be unidirectional or bidirectional. The deadlock problem for an AGV system with unidirectional guided paths is studied using zone control (Lee and Lin, 1995; Yeh and Yeh, 1998). Their algorithms predict deadlock based on the current routings of the AGVs, and then make decisions to avoid deadlocks. However, in an AGV system with bidirectional paths, AGVs need to compete for not only zones, like in a unidirectional system, but also lanes, increasing the control complexity. Note that this chapter calls such zones nodes to avoid any confusion.

One way to solve this problem is to carefully design the configuration of an AGV system such that its management is simplified. For example, tandem configuration (Bozer and Srinivasan, 1989, 1991, 1992) partitions all the stations into nonoverlapping, single-vehicle, closed loops with additional *pickup* and *deposit* locations provided as an interface between adjacent loops. Each station is assigned to only one loop, and each loop is served by exactly one AGV. Instead of partitioning the paths into nonoverlapping loops, a segmented flow approach (Sinriech and Tanchoco, 1995) partitions the paths into nonoverlapping segments. Each segment is comprised of one or more nodes and is served by a single AGV. Transfer buffers are located at both ends of each segment and serve as interface devices between the segments. The buffers are designed to be able to serve both sides of the segments simultaneously. Therefore, the only possible conflicts are the use of the interface buffers. As pointed out in Zeng et al. (1991), the drawback of tandem and segment flow AGV systems is that a load might have to be handled by two or more AGVs before reaching its destination. In addition, extra pickup and deposit stations are required to interface with each other, increasing the system scale and cost.

The problem of conflict detection is studied for bidirectional systems using colored PN (Zeng et al., 1991). An approach is presented to detect the competition for a lane by AGVs in the adjacent nodes. However, this is not enough to control the system. A conflict may occur when two AGVs in nonadjacent nodes compete for some lanes. Furthermore, when such a conflict occurs, it in fact leads to deadlock. In Reveliotis (2000), the deadlock avoidance problem for an AGV system with bidirectional paths is discussed and an approach is proposed. It assumes that the AGV system contains docking stations so that when an AGV reaches its destination, it can go to a docking station to unblock other AGVs. However, a system may have no such docking station. This calls for a more comprehensive approach. Based on results presented in Wu and his colleagues (2002, 2004, 2007), new approaches for deadlock resolution in AGV systems are presented in this chapter.

9.2 CONTROL OF AGV SYSTEMS WITH UNIDIRECTIONAL PATHS

9.2.1 MODELING AGV SYSTEMS WITH UNIDIRECTIONAL PATHS BY CROPN

First we discuss the deadlock avoidance problem for AGV systems with unidirectional paths. Then, based on the results obtained for AGV systems with unidirectional paths, we discuss the deadlock avoidance problem for AGV systems with bidirectional paths. The most widely used technique for vehicle management of AGV systems is zone control. Guided paths are divided into several disjoint zones. A zone can accommodate only one AGV at a time. Thus, an AGV system is a single-capacity system. Each zone (or node) can be treated as an H-resource with capacity 1, and can be modeled by the resource-oriented Petri net (ROPN) model presented in Chapter 5.

An AGV system is composed of AGVs and built paths. Often, a zone contains a workstation or an intersection. Thus, we may refer to a zone as a node. There are lanes between workstations (intersections), and they form the paths. The configuration of an AGV system is determined by the configuration of the paths. Because the lanes are built in advance and cannot be changed often, the configuration of an AGV system is relatively static. In an AGV system with unidirectional paths, an AGV can go in the given direction on any lane; in other words, for any lane the direction is determined in advance. An AGV system with unidirectional paths adapted from Yeh and Yeh (1998) is shown in Figure 9.1. This AGV system contains six workstations, four intersections, and four AGVs. It is divided into ten zones, each of them containing a workstation or an intersection.

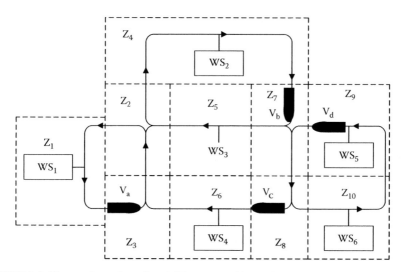

FIGURE 9.1 The configuration of an AGV system with unidirectional paths.

In modeling the manufacturing processes of parts, the colored ROPN (CROPN) is obtained by modeling the part flows based on the part routings, where the part routings are known. In an AGV system, the routings of AGVs are changed dynamically, but the configuration of the system cannot be changed. Thus, we can obtain the CROPN for the AGV system with unidirectional paths by modeling the zones (nodes) as places and the directed lane as transitions. In this way, the CROPN for the AGV system shown in Figure 9.1 can be obtained and is shown in Figure 9.2.

With the structure of the CROPN obtained, it is easy to model the routings of AGVs in the system. When an AGV is assigned a mission to fulfill, it starts from a zone, goes through a number of zones, and finally reaches its destination according to the paths in the CROPN. Thus, by the colors for the CROPN, the routings of AGVs are well modeled. For example, if the next mission for AGV V_a in zone 3 is to pick up a part at WS_1 and deliver it to WS_5, its route is $3 \rightarrow 2 \rightarrow 1 \rightarrow 3 \rightarrow 2 \rightarrow 4 \rightarrow 7 \rightarrow 8 \rightarrow 10 \rightarrow 9$. Then this process can be modeled in Figure 9.2 by $p_3 \rightarrow t_3 \rightarrow p_2 \rightarrow t_1 \rightarrow p_1 \rightarrow t_2 \rightarrow p_3 \rightarrow t_3 \rightarrow p_2 \rightarrow t_4 \rightarrow p_4 \rightarrow t_5 \rightarrow p_7 \rightarrow t_{10} \rightarrow p_8 \rightarrow t_{12} \rightarrow p_{10} \rightarrow t_{13} \rightarrow p_9$. Let $C(t_i) = \{c_i\}$, $V_{a(k-1)}$ denote that V_a is at the beginning of its kth leg, and $C(p_j(V_{a(k-1)}))$ denote the color when V_a is in p_j at the beginning of its kth leg. Then $C(p_3(V_{a0})) = \{c_3\}$, $C(p_2(V_{a1})) = \{c_1\}$, $C(p_1(V_{a2})) = \{c_2\}$, $C(p_3(V_{a3})) = \{c_3\}$, $C(p_2(V_{a4})) = \{c_4\}$, $C(p_4(V_{a5})) = \{c_5\}$, $C(p_7(V_{a6})) = \{c_{10}\}$, $C(p_8(V_{a7})) = \{c_{12}\}$, and $C(p_{10}(V_{a9})) = \{c_{13}\}$. Note that when V_a reaches's p_9, its mission is fulfilled, or each transition on the route is required to fire only once. Thus, we require the system to be L1-live.

9.2.2 DEADLOCK AVOIDANCE POLICY

In manufacturing processes, when a part is completed and removed from the system, it does not affect the processing of other parts in the system. However, when an AGV reaches its destination and a mission is completed, the AGV may stay here and block

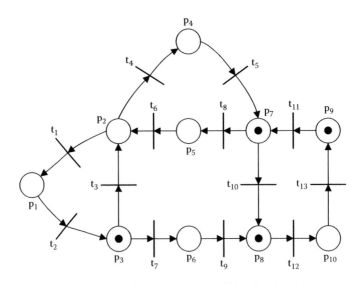

FIGURE 9.2 The CROPN for an AGV system with unidirectional paths.

other AGVs. To avoid this situation, we assume that when an AGV reaches its destination, it can be asked to travel to any zone if necessary.

In manufacturing processes, it is reasonable to assume that at the beginning (or at the initial marking) there is no part (no tokens) in the system. In an AGV system we cannot take the AGVs out of the system. Thus, it is important that the system is initially live. In an AGV system, the location and routing for an AGV are dynamically changed. Thus, the terminal state of one mission affects the initial state of the next mission. Here, we focus on the deadlock avoidance policy, and we always assume that the system is initially live. To make the CROPN for an AGV system live, some necessary condition must be satisfied.

Lemma 9.1: Let $\gamma = |P|$ denote the number of places (the number of zones) in the CROPN model. If the CROPN model for an AGV system is initially live in marking M_0, then the number of tokens (the number of AGVs in the system) must be less than γ.

The condition given in Lemma 9.1 is necessary to make the CROPN for an AGV system live. In fact, it is always satisfied. Often, only a few AGVs are available in an AGV system mainly due to their high cost and easily satisfied transportation demand.

As discussed before, whether the system will be deadlocked, the key is the circuits in the CROPN. In the CROPN for manufacturing processes, only production process circuits (PPCs) can be deadlocked, for place p_0 has infinite capacity. In the CROPN for an AGV system, every place represents a single-capacity resource. Thus, every circuit may be a potential deadlock source.

As pointed out in Chapter 8, the deadlock avoidance policy presented can be applied in a single-capacity system. Therefore, the policy is directly applied here and is stated as follows:

Theorem 9.1: The CROPN model for an AGV system with unidirectional paths is live if every v_i in the CROPN has at least one nonshared space at any marking M.

The problem is how many circuits there are in a CROPN. This number depends on the configuration of its underlying AGV system. This is because a transition in the CROPN models a lane in the AGV system. Thus, there is a connection transition between two places in a CROPN only if a lane exists between the two corresponding nodes in the real system. However, theoretically, there may be a large number of circuits if we ignore its configuration. In this case, applying Theorem 9.1 may create a problem. For example, with a five-node system, the number of possible circuits may be well above 5. Thus, no AGV is allowed in the system according to Theorem 9.1, which makes no sense. Thus, in spite of only a limited number of circuits in most real AGV systems, we present the circuit dependency to guarantee the theoretical correctness and completeness of our approach as follows. In Chapter 8, the circuit dependency is discussed based on multicapacity systems. It is now discussed based on single-capacity systems.

First, we consider the CROPN shown in Figure 9.3a that is from a fictitious AGV system. There are three AGVs in the system with the following routes: $V_1, p_6 \rightarrow p_1 \rightarrow p_2 \rightarrow p_3 \rightarrow p_4 \rightarrow p_5 \rightarrow p_{10}$; $V_2, p_7 \rightarrow p_1 \rightarrow p_2 \rightarrow p_5 \rightarrow p_1 \rightarrow p_9$; and $V_3, p_8 \rightarrow p_2 \rightarrow p_4 \rightarrow p_1 \rightarrow p_{11}$. There are five circuits: $v_1 = \{p_1, t_1, p_2, t_2, p_3, t_3, p_4, t_4, p_5, t_8, p_1\}$, $v_2 = \{p_1, t_1, p_2, t_2, p_3, t_3, p_4, t_7, p_1\}$, $v_3 = \{p_1, t_1, p_2, t_5, p_4, t_7, p_1\}$, $v_4 = \{p_1, t_1, p_2, t_5, p_4, t_4, p_5, t_8, p_1\}$,

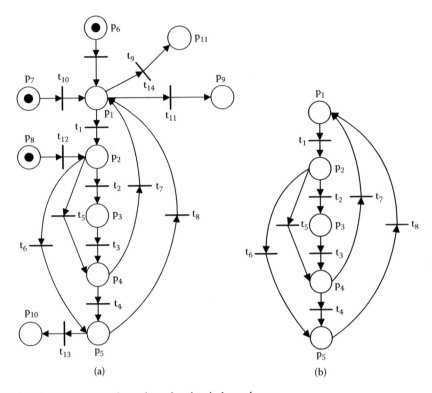

FIGURE 9.3 An example to show the circuit dependency.

and $v_5 = \{p_1, t_1, p_2, t_6, p_5, t_8, p_1\}$. These five circuits together have five places. Hence, only if no AGV enters any of the circuits can each circuit be allocated with one non-shared space to meet Theorem 9.1's condition. This implies that no AGV can reach its destination. Fortunately, we can show that it is unnecessary to require each circuit to have a nonshared space to avoid deadlock by the concept of dependent circuits.

Definition 9.1: (1) A set of places $p_1, p_2, \ldots,$ and p_k $(k > 1)$ in an interactive subnet is a place pool iff $|p_i{}^{\bullet}| = |{}^{\bullet}p_{i+1}| = 1$, $p_i{}^{\bullet} = {}^{\bullet}p_{i+1}$, for $i = 1, 2, \ldots, k-1$, $|{}^{\bullet}p_1| > 1$, and $|p_k{}^{\bullet}| > 1$; and (2) a single place is a place pool if it is not contained in another place pool.

AGVs can enter a place pool only from p_1, go through the only path, and leave the pool through p_k. For example, the subnet shown in Figure 9.3b has four place pools: one formed by p_1 and p_2 plus three single place pools, p_3, p_4, and p_5. We use pl to denote a place pool thereafter.

Definition 9.2: Assume that an interactive subnet in a CROPN is composed of $n + r$ circuits and there are $n + 1$ place pools in the subnet, with every circuit containing at least two place pools. Then any n circuits containing all $n + 1$ place pools are called the basic circuits in the subnet, and other r circuits the dependent ones.

In the subnet shown in Figure 9.3b, $n = 3$. Clearly, v_1, v_2, and v_3 contain all the place pools. Thus, they are the basic circuits and v_4 and v_5 are the dependent ones.

Theorem 9.2: In an interactive subnet composed of $n + r$ circuits, assume that there are tokens in the subnet and r circuits are dependent on n basic circuits. Then the subnet can be deadlock-free if each basic circuit has at least one nonshared space and each dependent circuit can share at least one space with a basic one.

Proof: We only need to show that when each basic circuit has one nonshared space, the subnet is live. Initially, we can control the distribution of the n free spaces such that each place pool has one free space for each of n place pools and one pool is full. Thus, by the definition of circuit dependency, every basic circuit has one nonshared space and every dependent circuit can share at least one space with a basic circuit. Without loss of generality, assume that pl_i, $1 \le i \le n$, has a free space and pl_{n+1} has no space. Then, we can move a token from pl_{n+1} to one of the other place pools, so that the condition given in the theorem still holds. By doing so, the subnet is live. Because of the strong connectedness of the subnet, a token in the subnet can go to any place in the subnet. ∎

A dependent circuit's v_1 sharing a space with a basic circuit's v_2 means that we can allocate a space to v_2 and the space is in a common place of v_1 and v_2. From Theorem 9.2, it is easy to verify that if there are two tokens in the interactive subnet shown in Figure 9.3b, the subnet is still live. Thus, in the CROPN in Figure 9.3a, two AGVs can enter the subnet at a time and the CROPN can still be live.

According to Definition 9.2, when an interactive subnet involves n places (nodes), the number of place pools is no more than n, or the number of basic circuits is no more than $n - 1$. Thus, no matter how many circuits there are in an interactive subnet, at least one AGV can be allowed to enter the subnet.

Whether there are dependent circuits in a CROPN of an AGV system depends on the number of circuits in the CROPN. In a real AGV system, because of the high investment and space requirement to build a path between nodes, the number of paths between nodes is limited. This restricts the number of possible circuits. For example, the number of circuits will be no more than four for the AGV system shown in Figure 9.1. Thus, a real AGV system likely admits no dependent circuit. It should also be pointed out that when there is no dependent circuit, the condition in Theorem 9.2 is the same as that in Theorem 9.1, implying that Theorem 9.1 is a special case of Theorem 9.2.

In most cases, the number of AGVs in the system is much less than the number of zones because of the easy satisfied transportation demand. Furthermore, because of the high cost and space required, the lanes are very limited. Thus, often there are not enough AGVs to occupy all the zones in two circuits, and there is no dependent circuit. Therefore, we can apply the deadlock avoidance policy presented in Chapter 8 in the simplest way, as stated in Theorem 9.1.

9.2.3 Computational Complexity

The AGV deadlock control problem is an NP-hard problem, as pointed out in Yeh and Yeh (1998). It is meaningful to analyze the complexity of the proposed control law.

From the discussion above, we know that we need only to observe the state of the system on-line, calculate $S(v_i)$ for every v_i, and control the transitions in $T_l(v_i)$ through one-step look-ahead.

The main difficulty with implementing the control law is the identification of the circuits in the PN model for an AGV system. However, this can be done off-line, since the configuration of an AGV system is known in advance and will not change often. Of course, theoretically the number of circuits may be exponential to the number of zones, but we do not need to identify and calculate $S(v_i)$ for all the circuits; we need to do that only for the elementary circuits. Furthermore, an AGV can travel only on the built paths. The guided paths are very limited due to the high cost and space, and thus, the number of circuits in a real AGV system is limited. Thus, it is expected that for a real AGV system, in general, the number of circuits underlying its physical configuration, and the number of circuits to be controlled, will not be too large. For example, in the PN model shown in Figure 9.2, four circuits are identified: $v_1 = \{p_1, t_2, p_3, t_3, p_2, t_1, p_1\}$, $v_2 = \{p_7, t_{10}, p_8, t_{12}, p_{10}, t_{13}, p_9, t_{11}, p_7\}$, $v_3 = \{p_2, t_4, p_4, t_5, p_7, t_8, p_5, t_6, p_2\}$, and $v_4 = \{p_2, t_4, p_4, t_5, p_7, t_{10}, p_8, t_9, p_6, t_7, p_3, t_3, p_2\}$. In fact, the problem can be simplified further. According to the control law, we have the following result:

Theorem 9.3: A deadlock can occur in circuit v_i only if the capacity of v_i, $K(v_i)$, is less than or equal to the number of AGVs in the system.

Theorem 9.3 presents the necessary condition for deadlock to occur. Thus, it is only necessary to identify the circuits that meet the condition given by Theorem 9.3 and control (calculate $S(v_i)$) only these circuits. For the example shown in Figure 9.2, these circuits are v_1, v_2, and v_3, and only transitions t_3, t_6, t_7, t_5, and t_{11} need to be controlled. In a real AGV system the number of AGVs will not be large, mainly due to their high cost and easily satisfied transportation demand. Hence, there will be a limited number of circuits that satisfy the condition given in Theorem 9.3. In conclusion, it is expected that the proposed control algorithm is implementable in real time in many practical AGV configurations.

9.3   CONTROL OF AGV SYSTEMS WITH BIDIRECTIONAL PATHS

9.3.1   Modeling AGV Systems with Bidirectional Paths by CROPN

Compared with an AGV system with unidirectional paths, the system with bidirectional paths is more efficient and cost-effective, as pointed out in Egbelu and Tanchoco (1986). However, they are more complex to manage. In unidirectional systems, the only resources the AGVs compete for are the zones. In bidirectional systems, the resources to be competed for include both zones and lanes. This makes conflict and deadlock avoidance in such a system more complicated than in unidirectional AGV systems.

In an AGV system with unidirectional paths, discussed in Section 9.2.1, we model a zone (node) by a place, and a lane by a transition. Notice that a transition models only one direction. However, in a bidirectional path system, an AGV can travel on a lane in both directions. Thus, we model a lane (a path between two adjacent nodes) by two transitions with different directions, as shown in Figure 9.4. A token representing an AGV can flow from $p_1 \to p_2$ by firing t_1, or $p_2 \to p_1$ by firing t_2, implying that the lane is assigned to an AGV from $p_1 \to p_2$ or $p_2 \to p_1$. Note that when p_1 and p_2 are both marked, neither transition is enabled according to the resource enabling

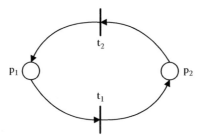

FIGURE 9.4 The PN model for a bidirectional lane between two nodes.

condition, thereby avoiding any collision on the lane. If a lane is one-way, e.g., from p_1 to p_2 only, then only t_1 is needed to model the lane.

In an AGV system with unidirectional paths, the direction in which an AGV travels on the lanes is determined in advance. Thus, we can model the system by a static CROPN. In contrast, in the AGV system with bidirectional paths, an AGV can travel on the lanes in any direction, depending on the real-time routing. Thus, we can only model the system by CROPN dynamically. Each time, when the AGVs are assigned new missions to fulfill and the routes are determined, a CROPN for the system is configured as done in Chapter 8 to describe the part flow.

In modeling the AGV system with unidirectional paths, the strong connectedness of the resulting CROPN is guaranteed by the configuration of the system. However, as pointed out above, for AGV systems with bidirectional paths, the strong connectedness of the resulting CROPN cannot be guaranteed by the configuration of the system. Thus, the following assumptions are made:

1. After an AGV reaches its destination and finishes its task, a control policy can command it to move to other nodes along the paths if such movements are feasible.
2. The configuration of the AGV system is strongly connected.

Assumption 1 allows one to change the initially requested destinations of certain AGVs, as well as the initially stationary AGVs to change their locations. The key point is that once an AGV reaches a node, one can assume that it fulfills its tasks (e.g., loading/unloading). Note that a faulty AGV or damaged lane will disable its related part of the system and degrade the performance, which is beyond the scope of our discussion. The second assumption allows one to build a strongly connected PN to contain all the traveling processes. Without such assumptions, no control law would be able to complete the above tasks physically. Yet, by appropriately routing the vehicles each time, or by optimizing the moving sequences between nodes, there will be less blocking—an important issue to be addressed in the next section.

Now with the assumptions, we are ready to develop the model for the traveling processes of AGVs. To do that, we can follow the method presented in Chapter 5: (1) model each AGV's mission for a subnet and (2) merge the subnets. However, in presenting the ROPN modeling method, only the initial state is concerned. Here, the destination marking should also be taken into consideration. Let M_{id} denote the destination

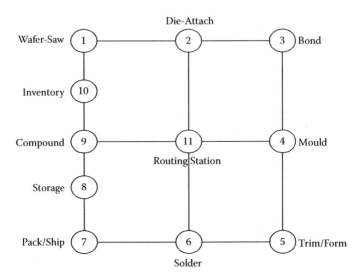

FIGURE 9.5 The configuration of an AGV system in semiconductor manufacturing.

marking for subnet PN_i and M_d denote the destination marking for the union CROPN, respectively. The following equation is needed for the union of PN_i and PN_j:

$$M_d(p) = \begin{cases} M_{id}(p), p \in P_i \\ M_{jd}(p), p \in P_j \\ M_{id}(p) + M_{jd}(p), p \in P_i \cap P_j \end{cases}$$

The configuration of an AGV system in semiconductor manufacturing, adapted from Hammond (1986), is shown in Figure 9.5. Assume that there are two AGVs, V_1 and V_2, in the system and they are initially in nodes 8 and 10, respectively. For their next missions to be fulfilled, the routes for the two AGVs are: V_1, $8 \rightarrow 7 \rightarrow 6 \rightarrow 11 \rightarrow 2 \rightarrow 1$ and V_2, $10 \rightarrow 9 \rightarrow 11 \rightarrow 4 \rightarrow 5 \rightarrow 6 \rightarrow 7$. With these routes, we can obtain the CROPN for this process, shown in Figure 9.6a, where the routes of AGVs V_1 and V_2 are modeled by $p_8 \rightarrow t_{87} \rightarrow p_7 \rightarrow t_{76} \rightarrow p_6 \rightarrow t_{6,11} \rightarrow p_{11} \rightarrow t_{11,2} \rightarrow p_2 \rightarrow t_{21} \rightarrow p_1$ and $p_{10} \rightarrow t_{10,9} \rightarrow p_9 \rightarrow t_{9,11} \rightarrow p_{11} \rightarrow t_{11,4} \rightarrow p_4 \rightarrow t_{45} \rightarrow p_5 \rightarrow t_{56} \rightarrow p_6 \rightarrow t_{67} \rightarrow p_7$, respectively. The colors are defined as $C(p_8(V_{10})) = \{c_{87}\}$, $C(p_7(V_{11})) = \{c_{76}\}$, $C(p_6(V_{12})) = \{c_{6,11}\}$, $C(p_{11}(V_{13})) = \{c_{11,2}\}$, $C(p_2(V_{14})) = \{c_{21}\}$, $C(p_{10}(V_{20})) = \{c_{10,9}\}$, $C(p_9(V_{21})) = \{c_{9,11}\}$, $C(p_{11}(V_{22})) = \{c_{11,4}\}$, $C(p_4(V_{23})) = \{c_{45}\}$, $C(p_5(V_{24})) = \{c_{56}\}$, and $C(p_6(V_{25})) = \{c_{67}\}$.

Notice that the CROPN constructed via union may not be strongly connected. Based on assumption 2 and the actual AGV system layout, one can always add a limited number of transitions and places to expand such a CROPN into a strongly connected one. For example, adding transitions t_{78} and $t_{1,10}$ (with dotted arcs) leads the CROPN in Figure 9.6a to a strongly connected one in Figure 9.6b. Observing the CROPN shown in Figure 9.6, we call $v = \{p_{11}, t_{11,4}, p_4, t_{45}, p_5, t_{56}, p_6, t_{6,11}, p_{11}\}$ a circuit and $Y = \{p_6, t_{67},$

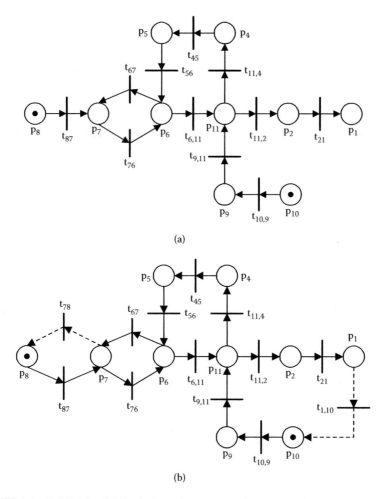

FIGURE 9.6 CROPN for AGV missions: (a) not strongly connected and (b) strongly connected after expansion.

p_7, t_{76}, p_6} a cycle. A cycle is a special circuit modeling a bidirectional lane between two nodes, as shown in Figure 9.4. Let v denote a circuit and Y a cycle.

Compared with the CROPN for AGV systems with unidirectional paths, the CROPN for AGV systems with bidirectional paths contains both circuits and cycles. This complicates the deadlock avoidance problem, which we will discuss for these two situations, respectively.

9.3.2 DEADLOCK AVOIDANCE FOR AGV SYSTEMS WITH CYCLES

If the CROPN for an AGV system with bidirectional paths is cycle-free, then the CROPN is not different from the one for the AGV system with unidirectional paths. Thus, the deadlock avoidance policy presented in Section 9.2.2 or by Theorem 9.1 can be directly applied here. Thus, here we discuss deadlock avoidance for CROPN with cycles.

In an AGV system, the conflict and deadlock caused by bidirectional paths are more difficult to avoid. When two AGVs travel on the same path and in opposite directions, cycles are formed in the CROPN of an AGV system. Cycles are the important source of conflicts and deadlocks. When two AGVs are in a cycle, they compete for both nodes and lanes in the cycle. This section discusses the problem of deadlock avoidance in the cycles in a CROPN.

A single cycle is formed by two AGVs competing for a lane between two adjacent nodes. One AGV travels in one direction, and the other in the opposite one. The CROPN shown in Figure 9.6 contains a cycle $Y = \{p_6, t_{67}, p_7, t_{76}, p_6\}$. If both AGVs take the lane, a collision (conflict) occurs. If both enter the nodes (both p_6 and p_7 have a token in the cycle Y in Figure 9.6), a deadlock occurs. Thus, we have the following result:

Lemma 9.3: A single cycle formed by routings of two AGVs in a CROPN is deadlock-free iff there is at most one token in the cycle in any marking M.

Proof: *Necessity:* If the condition in the lemma is not satisfied, then places p_1 and p_2 in the cycle shown in Figure 9.4 are all occupied by tokens, the token in p_1 is waiting to enter to p_2, and the token in p_2 is waiting to enter p_1. This is a circular wait, or a deadlock.

Sufficiency: If the condition is satisfied or there is at most one token in the cycle shown in Figure 9.4. Assume that p_1 has a token and p_2 is empty. Then t_1 can fire and the token in p_1 enters into p_2 and then leaves the cycle. After that, the token can enter into p_2 and t_2 can fire. Thus, no deadlock occurs in the cycle.   ■

Clearly, if there is no deadlock in a single cycle, two AGVs never occupy the two adjacent nodes in the cycle simultaneously. Thus, the conflict of using the lane never occurs.

A single cycle is the simplest case that causes conflict and deadlock. If there are multiple cycles distributed in the CROPN separately, then we can treat each cycle as a single cycle separately. However, the cycles may appear in a serial way.

Definition 9.3: In a CROPN of an AGV system, if p_1 and p_2 form a cycle, p_2 and p_3 form a cycle, . . . , and p_{n-1} and p_n form a cycle $(n > 2)$, then the n places form a cycle chain, which is denoted by H.

A cycle chain example is shown in Figure 9.7. It is formed by the AGV routings $V_1, p_8 \rightarrow p_2 \rightarrow p_3 \rightarrow p_4 \rightarrow p_5 \rightarrow p_{10}$ and $V_2, p_{11} \rightarrow p_5 \rightarrow p_4 \rightarrow p_3 \rightarrow p_2 \rightarrow p_1 \rightarrow p_7$. Let $Y_1 = \{p_2, t_{23}, p_3, t_{32}, p_2\}$, $Y_2 = \{p_3, t_{34}, p_4, t_{43}, p_3\}$, and $Y_3 = \{p_4, t_{45}, p_5, t_{54}, p_4\}$. The situation in a cycle chain is different from that in a single cycle. Consider the firing process in Figure 9.7. Firing t_{82} leads the token standing for V_1 in p_8 to p_2. Firing $t_{11,5}$ leads the token standing for V_2 to p_5. Now, although either cycle Y_1 or Y_3 has only one token, deadlock is inevitable. To progress, t_{23} and t_{34}, t_{54} and t_{43}, or t_{32} and t_{54} may fire, leading to two tokens in cycle Y_3, Y_1, or Y_2, resulting in deadlock.

It should be pointed out that a number of cycles form a cycle chain if there exist two AGVs that go through all the cycles in different directions. For example, in Figure 9.7, if there are three AGVs with routes $p_8 \rightarrow p_2 \rightarrow p_3 \rightarrow p_4 \rightarrow p_9, p_3 \rightarrow p_4 \rightarrow p_5 \rightarrow p_{10}$ and $p_{11} \rightarrow p_5 \rightarrow p_4 \rightarrow p_3 \rightarrow p_2 \rightarrow p_1 \rightarrow p_7$, then Y_1 and Y_2 form a cycle chain, and Y_2 and Y_3 form another cycle chain.

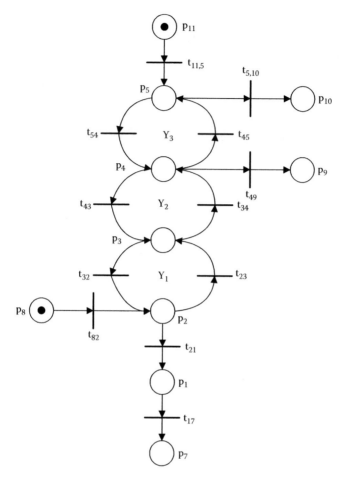

FIGURE 9.7 The CROPN containing a cycle chain: $Y_1 - Y_2 - Y_3 = \{p_2, t_{23}, p_3, t_{34}, p_4, t_{45}, p_5, t_{54}, p_4, t_{43}, p_3, t_{32}, p_2\}$.

Lemma 9.4: In a CROPN of an AGV system, if there is a cycle chain formed by the routes of two AGVs, the chain is deadlock-free iff there is at most one token in it in any marking M.

Proof: *Necessity:* If the condition is not satisfied, we assume that there are two tokens in the cycle chain. Because the two tokens go in the opposite directions, after firing some transitions they will enter a single cycle, resulting in a deadlock.

Sufficiency: If the condition is satisfied, we assume that there is one token in the cycle chain; then the token can go from one cycle to the next in the cycle chain in the prespecified direction, and finally leaves the cycle chain. After that, the other token can enter the cycle chain and goes through it. Thus, there is no deadlock. ∎

A cycle chain may be formed by routes of more than two AGVs. Assume that there is another AGV V_3 with route $p_6 \rightarrow p_1 \rightarrow p_2 \rightarrow p_3 \rightarrow p_9$. Then two cycle chains,

$H_1 = \{Y_1, Y_2, Y_3\}$ and $H_2 = \{Y_4, Y_1\}$, as shown in Figure 9.8, are formed by the routes of these three AGVs, where $Y_1 = \{p_2, t_{23}, p_3, t_{32}, p_2\}$, $Y_2 = \{p_3, t_{34}, p_4, t_{43}, p_3\}$, $Y_3 = \{p_4, t_{45}, p_5, t_{54}, p_4\}$, and $Y_4 = \{p_1, t_{12}, p_2, t_{21}, p_1\}$.

Like circuits, cycle chains can also interact with each other as a result of multiple AGV routes. The cycle chains formed by more than two AGVs may have overlaps. For example, in Figure 9.8, H_1 and H_2 have overlap Y_1.

Definition 9.4: A subnet of CROPN for an AGV system is called an interactive cycle chain if it is formed by two or more cycle chains, and each of the chains has overlap with at least one of the other chains. It is called a cycle chain subnet for short and denoted by w.

The net shown in Figure 9.8 is an interactive cycle chain subnet. Assume that n cycle chains $H_1, H_2, \ldots,$ and H_n form w that has m overlap segments $D_1, D_2, \ldots,$ and D_m. Let $F_{Di}(w) = \{V_{fDi}, V_{bDi}, V_{cDi}; D_i\}$ and $F_{Hi}(w) = \{V_{fHi}, V_{bHi}, V_{cHi}; H_i\}$, where V_f is the set of AGVs that travel forward on D_i or H_i, V_b is the set of AGVs that travel

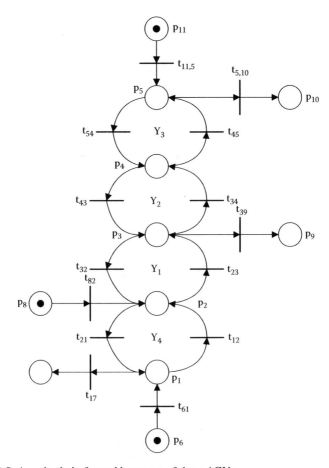

FIGURE 9.8 A cycle chain formed by routes of three AGVs.

backward on D_i or H_i, and V_c is the set of AGVs that cross D_i or H_i. Notice that all the V_f and V_b for D_i and H_i are consistent in w.

Definition 9.5: Let

$$\delta(F_\bullet(w)) = \begin{cases} 1, \ |V_f| > 0 \ \text{and} \ |V_b| > 0 \\ \\ 0, \ \text{otherwise} \end{cases} \tag{9.1}$$

where $\bullet$ stands for a cycle chain H or overlap segment D in w. If $\delta(F_\bullet(w)) = 0$ in the current marking M, H or D is said to be conflict-free.

Further, let $g(w) = \sum_{i=1}^{n} \delta(F_{H_i}(w)) + \sum_{i=1}^{m} \delta(F_{D_i}(w))$. We assume, without loss of generality, that there is no same cycle chain in w, or $H_i \neq H_j$, if $i \neq j$. Then we have the following result.

Theorem 9.4: A subnet of cycle chains formed by n cycle chains in a CROPN of an AGV system is deadlock-free iff the following condition holds in any marking M:

$$g(w) = 0 \tag{9.2}$$

Proof: The "only if" part is obvious and omitted.

Now we show the "if" part. Assume that H_1 and H_2 have overlap with D_1 as the common segment. If there is no token in D_1 in marking M, then $V_{fD1} = V_{fH1} \cup V_{fH2}$ and $V_{bD1} = V_{bH1} \cup V_{bH2}$, and Equation 9.2 requires that the tokens in H_1 and H_2 must go in the same direction, respectively. Let B_1 and B_2 be the sets of tokens belonging to H_1 and H_2, respectively. If all the tokens in B_1 and B_2 go in the same direction, it is obvious that there is no deadlock. If the tokens in B_1 go in one direction and the ones in B_2 go in the other, then we let the tokens in B_1 (or B_2) go through the subnet first, then the tokens in B_2 (or B_1) go next, and there is no deadlock either. If, however, there are tokens in D_1 in marking M, then from Equation 9.2, these tokens must go in the same direction. If all these tokens belong to one cycle chain, say H_1, then all the tokens in H_1 must go in the same direction as those in D_1 according to Equation 9.2. This time the tokens of H_1 can go through the subnet first, and the tokens in H_2 can go next, for the direction in which they go is the same. If there are tokens in D_1, and some of them belong to H_1 and the others to H_2, then because these tokens must go in the same direction, all the tokens in H_1 and H_2 must go in the same direction according to Equation 9.2, or the subnet is live. When there are two or more overlap segments, we can show similarly that the subnet is live.

If $H_1 \subset H_2$ in w, then we can treat H_1 as the overlap segment of H_1 and H_2, and the subnet is live if Equation 9.2 is met.

If there are tokens of V_c that cross the cycle chains in w in marking M, then these tokens can be made to go first, and this has no effect on the liveness of the subnet. In summary, the subnet is live if Equation 9.2 is satisfied. ∎

In a two-AGV system, only cycle chains may create deadlock. Notice that for cycle chains, Lemmas 9.3 and 9.4 are the special cases of Theorem 9.4. Thus, we can use Equation 9.2 in Theorem 9.4 to avoid deadlock and conflict in all cycle chains.

A cycle chain can also be formed by the route of a single AGV. In fact, if an AGV goes to some nodes and comes back by the same path (in the opposite direction), a cycle chain will be formed. By using the results presented in Theorem 9.4, no conflict and deadlock will occur in such a cycle chain since no other AGV enters the chain.

Definition 9.6: Transition t is said to be the input one of cycle chain H if $t \notin H$ but $t^{\bullet} \in H$.

Let $T_I(H) \subset T$ denote the set of input transitions of H. To avoid deadlock and conflict in cycle chains, control the firings of transitions in $T_I(H_i)$ such that Equation 9.2 is always satisfied. Because the condition is necessary and sufficient, it is the least restrictive control law. It is easy to calculate $g(w)$, and thus simple to implement.

Consider that a CROPN is constructed dynamically in real time. The following algorithm first finds cycles and then cycle chain subnets.

Algorithm 9.1:

1. Set CYCLES $= \varnothing$, and $T' = T$.
2. While $T' \neq \varnothing$,
   a. Select any $t \in T'$ and $T' = T' - \{t\}$.
   b. If there is $r \in T'$ such that $r^{\bullet} = {}^{\bullet}t$ and ${}^{\bullet}r = t^{\bullet}$, then $Y_{tr} = \{{}^{\bullet}t, t, t^{\bullet}, r, {}^{\bullet}t\}$, $T' = T' - \{r\}$, and CYCLES $=$ CYCLES $\cup$ Y_{tr}.
3. Let $U = UU = \varnothing$, where U and UU represent sets of single cycles and cycle chain subnets, respectively.
4. While CYCLES $\neq \varnothing$, select $Y = \{p, t, q, r\} \in$ CYCLES and let $Q = \{p, q\}$, $H = \{Y\}$, and CYCLES $=$ CYCLES $- \{Y\}$.
   a. While $Q \neq \varnothing$, select any $u \in Q$, find all $X \subset$ CYCLES such that $\forall Y' = \{p', t', q', r', p'\} \in X$, $u \in Y'$. If $X \neq \varnothing$, let $Q = Q \cup \underset{Y' \in C'}{UY' \cap P} - \{u\}$,

      CYCLES $=$ CYCLES $- X$, and $H = H \cup X$.
   b. If $H = \{Y\}$, $U = U \cup \{Y\}$; otherwise, $UU = UU \cup \{H\}$.

The outcome of the algorithm is the set of all single cycles (U) and all cycle chain subnets (UU) in the CROPN. Let $N_t = |T|$ be the number of transitions in the CROPN and $N_c = |$CYCLES$|$ be the number of cycles found. This algorithm's complexity is $O(N_t^2 + N_c^2) = O(N_t^2)$ since $N_c \leq N_t/2$ (the number of bidirectional lanes).

9.3.3 DEADLOCK AVOIDANCE IN THE CROPN

We have discussed how to avoid deadlocks and conflicts in circuits and cycle chains, respectively. In a CROPN of an AGV system, if there is no interaction between circuits and cycle chains, we can control the system by using the control laws presented in the last two sections to control circuits and cycle chains, respectively. However, the circuits and cycle chains may interact with each other. This makes the problem more complicated. Based on the results obtained in the last two sections, we present a rule to avoid deadlocks when the circuits and cycle chains interact with each other in a CROPN.

A subnet with interaction of a circuit and cycle chain is shown in Figure 9.9. This subnet contains the cycle chain made of places p_{1-5} and a circuit $v = \{p_2, t_{2,12}, p_{12}, t_{12,4}, p_4, t_{43}, p_3, t_{32}, p_2\}$. Places p_{2-4} are on both the cycle chain and v.

Let us consider Equation 9.2. By this condition, a shared path can be used only in one direction, as analogous to a time-shared system. It requires switching from one direction to another appropriately. In switching from one direction to another, we allow AGVs to go in one direction at a time.

Definition 9.7: A shared path or direction is active in a CROPN if it is selected to allow AGVs to pass. All transitions along this direction are active. A circuit containing active transitions is also called active.

In Figure 9.9, if the direction from p_5 to p_1 is selected or active (i.e., some AGV will occupy them at some time if there is no deadlock), then t_{54}, t_{43}, t_{32}, and t_{21} are active. When there exists interaction between a circuit and a cycle chain as shown in Figure 9.9,

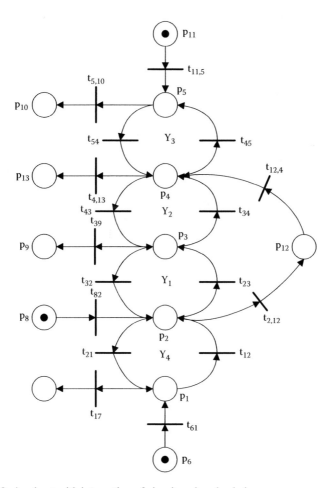

FIGURE 9.9 A subnet with interaction of circuit and cycle chain.

a part of the circuit must form a cycle chain. Because a cycle chain is composed of two directions, only the places and transitions in one direction can be on the circuit. For example, in Figure 9.9, only t_{43} and t_{32} on the cycle chain are on circuit v. Thus, only if the direction from p_5 to p_1 is active will circuit v play a role and is activated. Hence, if the direction from p_5 to p_1 is active, so is v. On the other hand, if the direction from p_1 to p_5 is active, t_{12}, t_{23}, t_{34}, and t_{45} are also, but v is not. Circular wait in a circuit may occur only when it is active. Hence, we can state the result as follows:

Theorem 9.5: A subnet in a CROPN of an AGV system with interaction of circuits and cycle chains is deadlock-free if the following conditions hold in any marking M:

1. The conditions given in Theorems 9.1 and 9.4 hold if the circuits are active.
2. Theorem 9.4 holds if the circuits are not active.

Proof: It follows from Theorems 9.1 and 9.4 directly. ■

Because a circuit will never be deadlocked in a two-AGV system, we don't need to consider the deadlocks in circuits. However, if there are multiple AGVs and we make the AGV assignments such that there are cycle chains and circuits, then the result given in Theorem 9.1 can be applied to solve the problem.

In a CROPN for an AGV system, if neither circuits nor cycles exist, transitions can fire spontaneously. If in some part there are only circuits or cycle chains, then the results presented in Theorems 9.1 and 9.4 can be applied, respectively. If the structure with interaction of circuits and cycle chains exists in some part, Theorem 9.4 needs to be applied. In this way, the overall CROPN is live.

It should be pointed out that when the number of circuits and cycles grows, the inter-action between the circuits and cycle chains may theoretically be very complex. The computation required to identify the interaction between the circuits and cycle chains may be overwhelming, and the process has, in general cases, exponential time complexity. Further studies in reducing it by restricting to certain structures are needed. Yet, for AGV systems in practice, the number of AGVs is very limited, and owing to their configurations, the number of circuits whose place count is smaller or equal to the AGV count, as well as cycle chains, is also limited. Since the proposed control law is a one-step look-ahead control policy based on all those circuits and cycle chains, it can be applied to many practical AGV systems, and all AGV systems reported so far.

9.3.4 EXAMPLES

We use some examples to show the application of the proposed approach to the AGV system shown in Figure 9.5. Two typical requests and assignments of AGVs are given below:

Example 9.1

$$V_1: 8 \rightarrow 7 \rightarrow 6 \rightarrow 11 \rightarrow 2 \rightarrow 1 \rightarrow 10$$

$$V_2: 10 \rightarrow 9 \rightarrow 11 \rightarrow 4 \rightarrow 11 \rightarrow 9$$

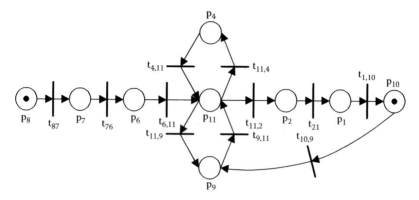

FIGURE 9.10 The CROPN for Example 9.1.

The CROPN for this example is shown in Figure 9.10. There are two cycles: $Y_1 = \{p_4, t_{4,11}, p_{11}, t_{11,4}, p_4\}$ and $Y_2 = \{p_{11}, t_{11,9}, p_9, t_{9,11}, p_{11}\}$. These two cycles form a cycle chain. However, it is easy to see that the cycle chain is due to the route of the same AGV V_2. Thus, there will be no conflict or deadlock in it. In fact, even if transition $t_{10,9}$ fires first and the token representing V_2 enters into place p_9 in the cycle chain, t_{87}, t_{76}, and $t_{6,11}$ can still fire and the token in p_8 (representing V_1) can enter p_{11}. Then we can fire $t_{11,2}$ and the token leaves p_{11} for p_2. If the token representing V_2 is in p_{11}, then $t_{6,11}$ cannot fire according to the transition enabling and firing rule. There is a circuit, but it cannot create the deadlock condition given in Theorem 9.1. Therefore, both AGVs can reach the destination with no deadlock.

Example 9.2

$$V_1: 10 \rightarrow 1 \rightarrow 2 \rightarrow 3 \rightarrow 4 \rightarrow 11 \rightarrow 9$$

$$V_2: 9 \rightarrow 11 \rightarrow 4 \rightarrow 5 \rightarrow 6 \rightarrow 7 \rightarrow 8$$

The CROPN for this example is shown in Figure 9.11. There are also two cycles, $Y_1 = \{p_4, t_{4,11}, p_{11}, t_{11,4}, p_4\}$ and $Y_2 = \{p_{11}, t_{11,9}, p_9, t_{9,11}, p_{11}\}$, and no circuit. These two cycles form a cycle chain H. H is due to the routes of V_1 and V_2. Thus, transition firings have to follow the control law specified in Theorem 9.4 to avoid deadlock. Theorem 9.4 requires that $g(H) = 0$. Since V_2 is already in H, we have to limit another token to enter H. This leads to the below transition firing order. Transitions

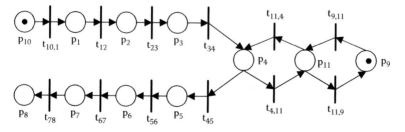

FIGURE 9.11 The CROPN for Example 9.2.

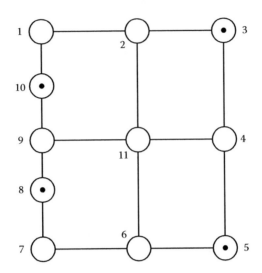

FIGURE 9.12 The initial locations of the AGVs for Example 9.3.

$t_{10,1}$, t_{12}, and t_{23} can fire any time. Transition t_{34} can fire only after $t_{9,11}$, $t_{11,4}$, and t_{45} fire. This way, neither conflict nor deadlock will occur. It is clear that it is the least restrictive control. Both cases require no CROPN expansion.

The two examples above show the cases with only two AGVs; this is because the discussed semiconductor requires only two AGVs in a manufacturing plan. To show the power of the control policy presented here, we assume four AGVs and their initial locations, as shown in Figure 9.12. It is easy to observe that only AGV V_4 has a free path to reach its destination, but when it reaches node 2, it blocks other vehicles. Thus, there is no simple way to control it.

The CROPN for this example is shown in Figure 9.13. It is possible that some AGVs may be blocked if some AGVs reach their destinations and stay there. For example, if V_3 reaches its destination first and stays there, V_2 will be blocked. Although for this example the AGV movement process exists so that such situations do not occur, the control policy presented here avoids deadlock but not such situations. However, by adding transition t_{45} (feasible according to the system configuration), the CROPN becomes strongly connected. Thus, no matter what the firing order of the transitions is, provided the control policy is obeyed, all the AGVs can reach their destinations at least once. Originally there are five circuits:

Example 9.3

$$V_1: 10 \to 1 \to 2 \to 3 \to 4 \to 11 \to 6 \to 7 \to 8$$

$$V_2: 8 \to 9 \to 10 \to 1 \to 2 \to 11 \to 6 \to 7$$

$$V_3: 3 \to 4 \to 11 \to 9 \to 10 \to 1$$

$$V_4: 5 \to 6 \to 11 \to 2$$

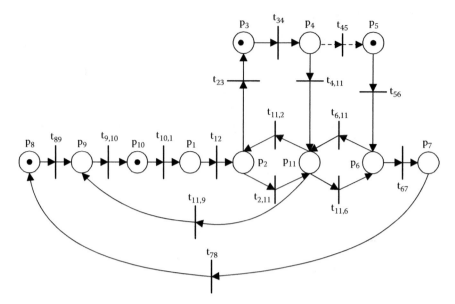

FIGURE 9.13 The CROPN for Example 9.3.

$V_1 = \{p_8, t_{89}, p_9, t_{9,10}, p_{10}, t_{10,1}, p_1, t_{12}, p_2, t_{23}, p_3, t_{34}, p_4, t_{4,11}, p_{11}, t_{11,6}, p_6, t_{67}, p_7, t_{78}, p_8\}$, $V_2 = \{p_8, t_{89}, p_9, t_{9,10}, p_{10}, t_{10,1}, p_1, t_{12}, p_2, t_{2,11}, p_{11}, t_{11,6}, p_6, t_{67}, p_7, t_{78}, p_8\}$, $V_3 = \{p_2, t_{23}, p_3, t_{34}, p_4, t_{4,11}, p_{11}, t_{11,2}, p_2\}$, $V_4 = \{p_9, t_{9,10}, p_{10}, t_{10,1}, p_1, t_{12}, p_2, t_{23}, p_3, t_{34}, p_4, t_{4,11}, p_{11}, t_{11,9}, p_9\}$, and $v_5 = \{p_9, t_{9,10}, p_{10}, t_{10,1}, p_1, t_{12}, p_2, t_{2,11}, p_{11}, t_{11,9}, p_9\}$. With the addition of transition t_{45}, three more circuits are added. They are $v_6 = \{p_2, t_{23}, p_3, t_{34}, p_4, t_{45}, p_5, t_{56}, p_6, t_{6,11}, p_{11}, t_{11,2}, p_2\}$, $v_7 = \{p_9, t_{9,10}, p_{10}, t_{10,1}, p_1, t_{12}, p_2, t_{23}, p_3, t_{34}, p_4, t_{45}, p_5, t_{56}, p_6, t_{6,11}, p_{11}, t_{11,9}, p_9\}$, and $v_8 = \{p_8, t_{89}, p_9, t_{9,10}, p_{10}, t_{10,1}, p_1, t_{12}, p_2, t_{23}, p_3, t_{34}, p_4, t_{45}, p_5, t_{56}, p_6, t_{67}, p_7, t_{78}, p_8\}$. These three circuits guarantee that when an AGV reaches its destination, it can be controlled to go to any node to allow other AGVs to finish their mission. There are also two cycles: $Y_1 = \{p_2, t_{2,11}, p_{11}, t_{11,2}, p_2\}$ and $Y_2 = \{p_{11}, t_{11,6}, p_6, t_{6,11}, p_6\}$. These two cycles form two cycle chains: $H_1 = \{Y_1, Y_2\}$ from the routes of V_2 and V_4 and $H_2 = \{Y_2\}$ from the routes of V_1 and V_4. These two cycle chains are interactive, forming a cycle chain subnet w. The cycle chains and the circuits are interactive too. Among the circuits, only v_3 meets the condition that the number of places is less than or equal to the number of AGVs. In fact, if we fire $t_{89}, t_{10,1}, t_{9,10}, t_{12}, t_{10,1}, t_{34}, t_{23}, t_{12},$ and t_{56}, then v_3 will be deadlocked.

Because both circuits and cycle chains exist and are interactive, Theorem 9.5 should be applied to control the system. To avoid deadlock we need to keep $S(v_3) > 0$ and $g(w) = 0$ all the time. According to the control law, first we can fire $t_{89}, t_{10,1}, t_{9,10}, t_{12}, t_{10,1}, t_{34}, t_{23},$ and t_{12} so that V_1 moves to p_3, V_2 moves to p_1, and V_3 moves to p_4. As a result, there are three tokens in v_3 (one space available) and the direction $p_2 \to p_{11} \to p_6$ is active. Hence, V_4 is forbidden to enter p_6. Then by firing $t_{4,11}, t_{11,9}, t_{9,10},$ and $t_{10,1}$, V_3 reaches its destination. By this time, v_3 will not be deadlocked. We only need to avoid deadlock in the cycle chain. After that, we can fire $t_{34}, t_{4,11}, t_{11,6}, t_{2,11}, t_{67}, t_{11,6},$ and t_{78}. Consequently, V_1 and V_2 reach their destinations. Now the direction $p_2 \to p_{11} \to p_6$ is freed. Finally, we fire $t_{67}, t_{56}, t_{6,11},$ and $t_{11,2}$, and

V_4 reaches its destination. It should be noticed that in the above transition process, both V_1 and V_2 are in the cycle chain subnet w at the same time, but since $g(w) = 0$ holds, no deadlock is generated. Although there are five circuits, only one of them, i.e., v_3, needs to be identified to carry out this control law.

9.4 ROUTING OF AGV SYSTEMS BASED ON CROPN

Deadlock in AGV systems can be resolved at different levels. At the design level, the configuration of an AGV system can be carefully designed such that its management is simplified. At the planning and routing level, deadlocks in AGV systems can be predicted and avoided by extensive route preplanning (Kim and Tanchoco, 1991, 1993; Dowsland and Greaves, 1994; Huang et al., 1993; Krishnamurthy et al., 1993). However, these planning strategies are essentially open-loop control policies (Reveliotis, 2000) and based on deterministic timing of the vehicle travel. Because of the significantly stochastic nature of the manufacturing systems, these techniques are not robust. Deadlocks in AGV systems can also be detected and recovered by appropriate rerouting (Egbelu and Tanchoco, 1986), but they suffer from the disruption of vehicle rerouting.

The effective way to deal with the deadlock problem in AGV systems is deadlock avoidance. Since all the deadlock avoidance policies are closed-loop control and are implemented in real time, they are robust. In Reveliotis (2000), the banker's algorithm is used to solve the deadlock problem and at the same time determine the route for each vehicle to reach its destination from its source node. However, it does not consider the system performance. In Lee and Lin (1995), Yeh and Yeh (1998), and Wu and Zhou (2001b, 2004), to avoid deadlock in AGV systems, it is assumed that the route for each AGV is known and the task for deadlock avoidance is to guarantee that the mission of each AGV can be completed, or the AGVs can reach their destinations from their source nodes. Here, we decouple the upper-level planning from the lower-level logical control. The dispatching and routing problem can be solved at the upper level, called planning. At the planning level, one can concentrate on the system performance and provide a feasible initial state for the lower level, and at the lower level, it is supervisory control that guarantees the deadlock-free operation of the system. This control architecture for AGV systems is shown in Figure 9.14.

However, there is a gap between these two levels. Almost all the deadlock avoidance policies are developed based on certain system conditions. A required condition

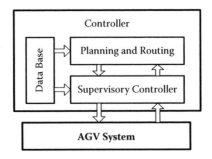

FIGURE 9.14 The control architecture for an AGV system.

should be satisfied at any state reached, including the initial state. We know that deadlocks result from the lack of resources (or buffer spaces). Consider part manufacturing processes in AMS, we can always assume that at the initial state all the resources are available to perform an operation. No matter what condition is required by a deadlock avoidance policy, it is satisfied at the initial state, and the system can be kept free of deadlocks by using a correct deadlock avoidance policy. However, this is no longer true for AGV systems, since AGVs always reside in the system and occupy some nodes (zones). Their distribution in the system changes from time to time. Such distribution becomes an initial state for the next missions of AGVs. Thus, sometimes, the condition required by a deadlock avoidance policy is not satisfied at initial states. The system may thus be deadlocked at the beginning, and the existing deadlock avoidance policies have to be extended in order to handle such situations.

AGV blocking can also happen in many multiple AGV systems. Assume that after V_1 reaches its destination and fulfills its mission, it stays there and waits for the next mission, and a V_1-occupying node is on the path of V_2. Then V_2 will be blocked by V_1. A deadlock control policy can avoid deadlocks in a system. It may not be able to avoid blocking, and this problem should be solved by the upper level (Reveliotis, 2000). Thus, AGV planning must be carefully done such that the condition required by a control policy is satisfied and there is no blocking in fulfilling the missions. These conditions motivate this work to bridge the gap between the two levels by appropriately routing and rerouting AGVs. To do so, assumptions made in the last section are adopted. The goal here is to route AGVs such that the traveling time is minimized, and at the same time the routing should be feasible in the sense of reachability.

9.4.1 PROBLEM DESCRIPTION

Before proceeding with our discussion, consider some initial situations for the AGV system. Replacing zones with nodes, the node relationship of the AGV system shown in Figure 9.1 can be shown as in Figure 9.15. Assume that V_{1-4} now stay in nodes 1, 2, 3, and 8 with their routes: V_1, $3 \to 2 \to 1 \to 3$; V_2, $2 \to 1 \to 3 \to 2 \to 4 \to 7 \to 8$; V_3, $8 \to 6 \to 3 \to 2 \to 5 \to 7 \to 8$; and V_4, $1 \to 3 \to 6 \to 8 \to 9 \to 10$, respectively. The CROPN is shown in Figure 9.16a. It can be seen that at M_0 circuit $v_1 = \{p_2, t_{21}, p_1, t_{13}, p_3, t_{32}, p_2\}$ is full of tokens with no leaving token, and the condition given in Theorem 9.1

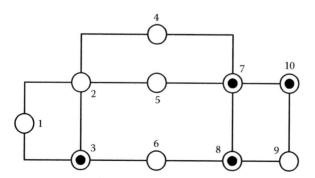

FIGURE 9.15 The configuration of an AGV system and initial AGV locations.

is violated. This part of the net is initially deadlocked. At the same time, AGVs V_2 and V_3 need to reach the same node 8; this is physically infeasible. However, in practice, it is possible that two kinds of materials are required to deliver to the same workstation. If V_{1-4} sit at nodes 3, 7, 8, and 10, respectively, but the routes are changed into V_1, $3 \rightarrow 6$ $\rightarrow 8 \rightarrow 9$; V_2, $7 \rightarrow 5 \rightarrow 2 \rightarrow 1 \rightarrow 3 \rightarrow 2 \rightarrow 4 \rightarrow 7$; V_3, $8 \rightarrow 6 \rightarrow 3 \rightarrow 2 \rightarrow 4 \rightarrow 7 \rightarrow 8$; and V_4, $10 \rightarrow 7 \rightarrow 8 \rightarrow 9 \rightarrow 10$, we can obtain the CROPN shown in Figure 9.16b. In it, there is a cycle chain $H = \{Y_2, Y_3\}$ formed by the routes of V_1 and V_3. In H, there are two tokens with opposite directions, and the condition given in Equation 9.2 is violated, i.e., the system will be deadlocked. With such initial states, a deadlock avoidance policy is helpless. One way to solve this problem is to reroute the AGVs.

First, we present the description of the problem for AGV routing and rerouting. The mission of an AGV can be stated as follows: An AGV situates in node s (denoted by n_s), the source node, and is required to pick up a material in node m (denoted by n_m), the middle node, and deliver the material to node d (denoted by n_d), the destination

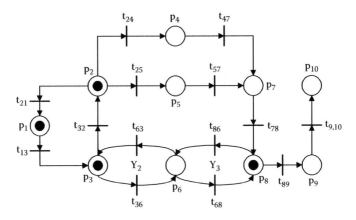

(a) V_1: 3-2-1-3; V_2: 2-1-3-2-4-7-8; V_3: 8-6-3-2-5-7-8; V_4: 1-3-6-8-9-10

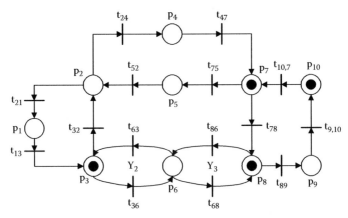

(b) V_1: 3-6-8-9; V_2: 7-5-2-1-3-2-4-7; V_3: 8-6-3-2-4-7-8; V_4: 10-7-8-9-10

FIGURE 9.16 The CROPNs that are infeasible in the sense of reachability.

node. To avoid blocking, an AGV may be ordered to go from n_d to n_e to release node n_d. Let $R_i = \{n_{is}, n_{i1}, n_{i2}, \ldots, n_{im}, n_{im+1}, n_{im+2}, \ldots, n_{id}, \ldots, n_{ie}\}$ denote the route on which V_i completes its mission. Note that R_i is an ordered set of nodes. It is possible that $n_{is} = n_{im}$ or $n_{im} = n_{id}$ in R_i. If for some time AGV V_i is idle and stays in n_s, then $n_{is} = n_{im} = n_{id}$. In this situation, we still treat it as a special route. Note that it is consistent with the definition of a route for AGV mission. If it is necessary for an idle V_i to release n_{is} such that other AGVs can go through, its route R_i may be expanded from n_{is} to n_{ie} for V_i.

If there is a lane between nodes i and j, then we say that these two nodes are adjacent. Because the configuration of an AGV system is fixed, we assume that the time needed for an AGV to travel from node i to j is known as a_{ij} and is the same for all AGVs. The time needed for AGV V_k to complete its mission is the sum of a_{ij} along path R_k. Assume that $R_k = \{n_1, n_2, \ldots, n_K\}$; then the time needed is denoted by $A_k = \Sigma_{i=1}^{K-1} a_{i,i+1}$. In this way, a strongly connected undirected network is formed with a_{ij} as the length between nodes i and j. In a CROPN, p_i corresponds to node i. Thus, we do not distinguish them.

Assume that there are n AGVs in the system and $U = \{1, 2, \ldots, n\}$ is the set of AGV indices. With these R_i, $i \in U$, we can obtain the CROPN by the modeling procedure presented in the last section. Let M_0 and Q_F denote the initial marking and the set of markings permitted by the control policy, respectively.

We assume that the task for each AGV is known, or for V_i, n_{is}, n_{im}, and n_{id} are known. The AGV routing problem is to find R_i for every $i \in U$ and

$$\text{Minimize } J = \sum_i A_i \tag{9.3}$$

such that $M_0 \in Q_F$

$$n_{ie} \neq n_{je}, \text{ if } i \neq j$$

$\exists$ fireable transitions in the CROPN, such that there is no blocking

We can use the following procedure to solve the problem.

Algorithm 9.2: Procedure to Solve Equation 9.3

Step 1: For V_i, $i \in U$, find the shortest path from n_{is} to n_{im} and the shortest path from n_{im} to n_{id}, and form the path $R_i = \{n_{is}, n_{i1}, n_{i2}, \ldots, n_{im}, n_{im+1}, n_{im+2}, \ldots, n_{id}\}$; for an idle AGV V_i, just set $R_i = \{n_{is}\}$.

Step 2: Construct all sub-CROPN for R_i's and unite them to obtain the CROPN.

Step 3: If $M_0 \in Q_F$, go to step 4; otherwise, reroute the AGVs such that $M_0 \in Q_F$.

Step 4: Expand the paths found in step 1 and modified in step 3 from n_{id} to n_{ie} if necessary, so that the last two conditions in Equation 9.3 are satisfied.

Step 1 finds the shortest path (shortest time) for each AGV and is of polynomial complexity (Aho et al., 1987). Step 2 takes $O(n|P|)$ time. Once step 3 is completed, the algorithm guarantees that no matter what is done in step 4, the final solution is correct, or the initial marking M_0 satisfies the condition required by the deadlock avoidance control policy. In step 4, we change only the stop nodes, not the initial ones. The following discussion focuses on steps 3 and 4.

9.4.2 AGV REROUTING

Considering the condition required by the control policy for deadlock avoidance in AGV systems in Sections 9.2 and 9.3, we know that if for every circuit, the condition given by Theorem 9.1 is satisfied, and at the same time the condition given by Theorem 9.4 is satisfied for every cycle chain, then the condition given by Theorem 9.5 must be satisfied. Thus, we only need to check every circuit and cycle chain to see if the condition is satisfied.

To violate Theorem 9.1's condition, AGVs need to reside at every node in a circuit, called a *full circuit*. At least three nodes are needed to form a circuit. Considering much fewer AGVs than the number of nodes, it is infrequent to have a full circuit. First, let us consider one full circuit only in the CROPN. In the CROPN model for AGV systems, to forbid an AGV to travel on the lane from node i to j, we need simply to remove transition t_{ij}.

Proposition 9.1: A full circuit v can be eliminated by the following algorithm:

Algorithm 9.3: Eliminate Full Circuit

1. Select a transition $t_{ij} \in T(v)$, ${}^{\bullet}t_{ij} = p_i \in P(v)$ and $t_{ij}{}^{\bullet} = p_j \in P(v)$.
2. Remove t_{ij} and find the shortest path from node i to j to form a new circuit that is not full in the remaining network.

The feasibility in step 2 is guaranteed since otherwise the number of AGVs must equal the number of nodes under the assumption that the network of an AGV system configuration is a strongly connected one in which at least two different paths exist between any two nodes. If multiple full circuits exist, we can remove them one by one using the above algorithm. It should be noticed that when we remove transition t_{ij} in v, we just disable the path from p_i to p_j, but not from p_j to p_i. In this way, we force the lane between p_i and p_j to be unidirectional.

Now we consider the condition given by Theorem 9.4. In a cycle chain, if the condition given by Theorem 9.4 is violated, we call this cycle chain a *conflict cycle chain*. Let $T(H)$ and $P(H)$ denote the set of transitions and the set of places in the cycle chain H.

Proposition 9.2: A conflict cycle chain formed by two AGVs disappears if any transition $t_{ij} \in T(H)$ is removed.

Proof: Assume that V_i and V_j form a conflict cycle chain in the CROPN and they locate in p_i and p_j, respectively. Let $P(V_i)$ and $P(V_j)$ be the sets of places on V_i and V_j, and between p_i and p_j, including p_i and p_j. Then $P(V_i) = P(V_j)$ holds. If a transition, say t_{ik}, is removed, then V_i needs to go another way to reach p_j. Even if V_i enters p_k later, $P(V_i) = P(V_j)$ will never hold. This implies that V_i and V_j will never form a conflict cycle chain again. ∎

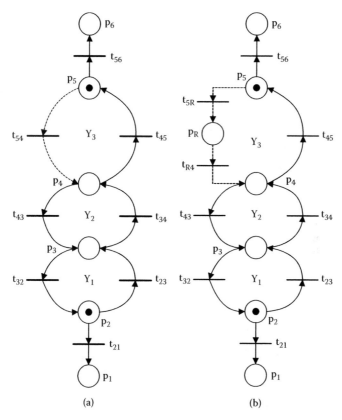

(a) (b)

FIGURE 9.17 A conflict cycle chain that is broken: (a) a conflict cycle chain $Y_1 - Y_2 - Y_3 = \{p_2, t_{23}, p_3, t_{34}, p_4, t_{45}, p_5, t_{54}, p_4, t_{43}, p_3, t_{32}, p_2\}$, and (b) the conflict cycle chain is broken as $\{p_2, t_{23}, p_3, t_{34}, p_4, t_{45}, p_5, t_{5R}, p_R, t_{R4}, p_4, t_{43}, p_3, t_{32}, p_2\}$.

In Figure 9.17a, a conflict cycle chain is shown. If transition t_{54}, denoted by a dash line, is removed, then this conflict cycle disappears, as shown in Figure 9.17b. It should be pointed out that p_R in Figure 9.17b is only a virtual place that stands for a segment of a path. Also, the route for the AGV in place p_5 may not enter place p_4, and in that case, the conflict cycle chain disappears with no question.

A cycle chain can be formed by multiple AGVs, and cycle chains can interact to form an interactive cycle chain. Like a single cycle chain, an interactive cycle chain can become conflicted if the condition given in Theorem 9.4 is violated. Such an interactive cycle chain is called a *conflict interactive cycle chain*. An interactive cycle chain can be mapped into an undirected network, a subnet in the network for the configuration of the AGV system. Consider the flow between two adjacent nodes. Let w and $G(w)$ denote an interactive cycle chain and the corresponding undirected network. Such a mapping is shown in Figure 9.18. If there are two transitions t_{ij} and t_{ji} between places p_i and p_j, then the set of transitions $\{t_{ij}, t_{ji}\}$ is mapped into an undirected arc (i, j). Let $N(t_{ij})$ denote the number of AGVs that will travel on t_{ij} in

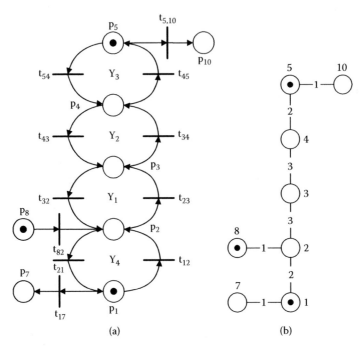

FIGURE 9.18 The mapping from (a) a cycle chain to (b) an undirected network.

the CROPN and $N(i, j) = N(t_{ij}) + N(t_{ji})$ be the flow on arc (i, j). In Figure 9.18b, the flow is put on the arcs. Given a cycle chain, arc (i, j) is maximal if $N(i, j) \geq N(i', j')$, $\forall i'$ and j'.

Note that a cycle chain formed by two AGVs is a special case of interactive cycle chains. It should also be pointed out that by Proposition 9.2, the broken interactive cycle chain may just be a part of another interactive cycle chain. With the result obtained above, we can present the following algorithm for breaking a conflict interactive cycle chain:

Algorithm 9.4: Breaking the Conflict Interactive Cycle Chain

1. Select a conflict interactive cycle chain w and map it into the corresponding undirected network $G(w)$, and find the edge (i, j) with maximal flow and AGVs that travel on (i, j) with different directions.
2. Select a transition between t_{ij} and t_{ji}, say t_{ij}, such that if t_{ij} is removed, the node network in the AGV system is still strongly connected. If both t_{ij} and t_{ji} are removable, we select t_{ij} such that $N(t_{ij}) \leq N(t_{ji})$, and remove t_{ij}.
3. Find the set of AGVs that will travel on the transition t_{ij}, say V_{cf}.
4. For any $V_i \in V_{cf}$, find the shortest path with t_{ij} removed.
5. If there is no more conflict interactive cycle chain, then stop; otherwise, go to step 1.

Proposition 9.3: Algorithm 9.4 terminates in finite steps.

Proof: From Proposition 9.2 we know that when a conflict interactive cycle chain is broken, at least one of the AGVs in the chain will not be in another conflict interactive cycle chain. Thus, the number of AGVs in such chains is reduced step by step, and finally the algorithm terminates. In the extreme case, when one direction for each lane is removed and the system is degraded into a unidirectional strongly connected system, there will be no cycle chain at all. In this extreme case, we only need to carry out the breaking step n times if there are n edges in the network for the AGV system. ■

In Algorithm 9.4, we need to carry out the computation for finding the edge with maximal flow and the shortest path in a network. It is well known that there are efficient algorithms to do so (Aho et al., 1987). Thus, the proposed algorithm is efficient.

In removing transitions, a node may become unreachable or may not be able to reach any other node. For example, in Figure 9.15, if both transitions t_{21} and t_{31} are removed, then node 1 is unreachable. Contrarily, if both transitions t_{12} and t_{13} are removed, node 1 cannot reach any other node. Such a situation is avoided in step 2. Notice that to break a cycle chain, we can remove any one of transitions t_{ij} and t_{ji}, and by appropriately selecting one of them, we can guarantee that the remaining network is still strongly connected. By assumption, the network for the configuration of an AGV system is strongly connected. This implies that every subnet (or a single node) has at least two lanes on which an AGV can reach or leave the subnet. For example, the shadow part shown in Figure 9.19 is a subnet of the network shown in Figure 9.15; there are three such lanes: (2, 4), (2, 5), and (3, 6). If transition t_{36} is removed for breaking some cycle chains, then when we come to lane (2, 5) and we can select t_{52} to be removed. In this way, the strong connectedness of the network can be guaranteed. In fact, with connectedness of a network in the configuration, in the extreme situation we can select only one direction for every lane, so that it becomes a unidirectional system and is still strongly connected, but there will be no cycle. In this way, we can break conflict interactive cycle chains by removing transitions, and at the same time the net is kept strongly connected.

In summary we can give the AGV rerouting algorithm as follows:

Algorithm 9.5: AGV Rerouting Algorithm

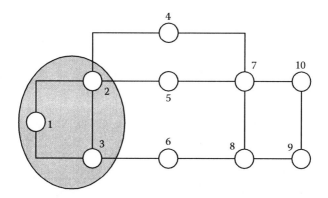

FIGURE 9.19 Strong connectedness of the network after breaking a cycle chain.

1. Eliminate the full circuit if there is one:
   a. For every edge on the full circuit, carry out Algorithm 9.3 and find the shortest paths for all the AGVs in the circuit
   b. Select the best one among the solutions found.
2. Perform Algorithm 9.4 to break the conflict interactive cycle chains.

9.4.3   ROUTE EXPANSION

To guarantee that the AGVs can reach their destinations, the blocking problem should be resolved. One way to do that is to expand the CROPN so that it is a strongly connected PN. Thus, when an AGV reaches its destination and blocks the other AGVs from reaching their destinations, the controller can move it to the other nodes. Because of the strong connectedness of the CROPN, an AGV can move anywhere. This guarantees the reachability of all the AGVs to their destinations. In this chapter, however, we expand the paths of only those AGVs that otherwise block others.

We use $P(R_j)$ to denote the set of places (or nodes) for the path of AGV V_j. A necessary condition for AGV V_i to block AGV V_j when V_i reaches n_{id} is $n_{id} \in P(R_j)$. Thus, a simple way is to expand route R_i so that $n_{id} \notin P(R_j)$ for any $j \neq i$. However, this may be infeasible. Consider the example in Figure 9.20. It is formed from the AGV reoutes V_1, $3 \rightarrow 2 \rightarrow 1 \rightarrow 3$; V_2, $7 \rightarrow 5 \rightarrow 2 \rightarrow 1 \rightarrow 3 \rightarrow 2 \rightarrow 4 \rightarrow 7$; V_3, $8 \rightarrow 6 \rightarrow 3 \rightarrow 2 \rightarrow 4 \rightarrow 7 \rightarrow 8$; and V_4, $10 \rightarrow 7 \rightarrow 5 \rightarrow 2 \rightarrow 3 \rightarrow 6 \rightarrow 8 \rightarrow 9$. We see that $n_{1d} \in P(R_2)$, $n_{1d} \in P(R_3)$, $n_{1d} \in P(R_4)$, $n_{2d} \in P(R_3)$, $n_{2d} \in P(R_4)$, and $n_{3d} \in P(R_4)$. This implies that it is not easy to make $n_{ie} \notin P(R_j)$.

If $n_{id} \in P(R_j)$ and V_j can go through node n_{id} before V_i does, then AGV V_i will never block V_j. By this observation, if $n_{id} \in P(R_j)$ and $n_{jd} \in P(R_k)$, and V_k can go through n_{jd} before V_j, and V_j can go through n_{id} before V_i, then no route expansion is needed.

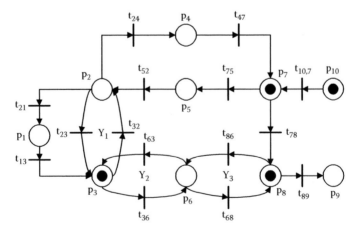

FIGURE 9.20 The CROPN for a set of AGV missions.

Algorithm 9.6: Route Expansion Algorithm

1. For all V_i and V_j, check if $n_{id} \in P(R_j)$, and order $V_1, V_2, \ldots, V_n$ such that if $n_{id} \in P(R_{i+1})$, then V_{i+1} can go through n_{id} before V_i.
2. In the obtained CROPN, test if the ordering given by the last step can be realized. If so, stop; otherwise, go to the next step.
3. Based on the CROPN, find an AGV ordering such that the previous ordering is least violated, and for V_i and V_j, $n_{id} \in P(R_j)$, and V_i goes through n_{id} before V_j, expand R_i to n_{ie} such that n_{ie} is not on any path of other AGVs, or it is on the paths of AGVs that go through n_{ie} before V_i.
4. If there exist two AGVs V_i and V_j such that $n_{id} = n_{jd}$, and if V_i reaches n_{id} before V_j, then expand the path of V_i as step 3 does.

All steps are of polynomial complexity. If $\{n_{id}\} \cap P(R_j) \neq \varnothing$, let $R_{ij} = R_i \cap R_j$ be the common path segment of AGVs V_i and V_j with n_{id} in it. We use p_i as the first place of R_{ij}; then the necessary condition for AGV V_j to go through n_{id} before V_i is that V_j can enter p_i before V_i. With this observation, we make the test in Algorithm 9.6 easier.

9.4.4 ILLUSTRATIVE EXAMPLES

In this subsection we use some examples to show the application of the proposed method. In the following we assume that the time needed for AGV travel between any two adjacent nodes is the same. Thus, for the shortest time we need to find the path with the least number of nodes.

Example 9.4

Consider the CROPN shown in Figure 9.16a, with V_1, $3 \rightarrow 2 \rightarrow 1 \rightarrow 3$; V_2, $2 \rightarrow 1 \rightarrow 3 \rightarrow 2 \rightarrow 4 \rightarrow 7 \rightarrow 8$; V_3, $8 \rightarrow 6 \rightarrow 3 \rightarrow 2 \rightarrow 5 \rightarrow 7 \rightarrow 8$; and V_4, $1 \rightarrow 3 \rightarrow 6 \rightarrow 8 \rightarrow 9 \rightarrow 10$, where circuit $v_1 = \{p_2, t_{21}, p_1, t_{13}, p_3, t_{32}, p_2\}$ is a full circuit. Assume that for V_1, n_{1m} and n_{1d} are p_1 and p_3; for V_2, n_{2m} and n_{2d} are p_4 and p_8; and for V_4, n_{4m} and n_{4d} are p_6 and p_{10}. By carrying out Algorithm 9.3, this full circuit is eliminated as follows:

Step 1: Removing transition t_{32} and carrying out Algorithm 9.2, we find $R_1 = \{3, 1, 2, 3\}$, $R_2 = \{2, 4, 7, 8\}$, and $R_4 = \{1, 3, 6, 8, 9, 10\}$.
Step 2: Removing t_{31} and carrying out Algorithm 9.2, we find $R_1 = \{3, 2, 3\}$, $R_2 = \{2, 4, 7, 8\}$, and $R_4 = \{1, 2, 3, 6, 8, 9, 10\}$.
Step 3: Removing t_{21} and carrying out Algorithm 9.2, we find $R_1 = \{3, 2, 3\}$, $R_2 = \{2, 4, 7, 8\}$, and $R_4 = \{1, 3, 6, 8, 9, 10\}$.
Step 4: Among the obtained solutions, the one found in step 3 is the best and we select that one.

With this solution, the resulting CROPN is shown in Figure 9.21. It can be seen that the full circuit is eliminated.

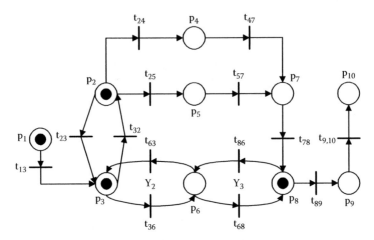

V_1: 3-1-3; V_2: 2-4-7-8; V_3: 8-6-3-2-5-7-8; V_4: 1-3-6-8-9-10

FIGURE 9.21 The CROPN after a full circuit is eliminated.

Example 9.5

Consider the CROPN shown in Figure 9.16b, with V_1, $3 \to 6 \to 8 \to 9$; V_2, $7 \to 5 \to 2 \to 1 \to 3 \to 2 \to 4 \to 7$; V_3, $8 \to 6 \to 3 \to 2 \to 4 \to 7 \to 8$; and V_4, $10 \to 7 \to 8 \to 9 \to 10$. $H = \{Y_2, Y_3\}$ is a conflict cycle chain. Assume that for V_3, n_{3m} and n_{3d} are p_5 and p_8, respectively. Applying Algorithm 9.4, by removing t_{86}, we find $R_3 = \{8, 7, 5, 7, 8\}$. The resulting CROPN is shown in Figure 9.22, where the conflict cycle chain disappears and there is no conflict cycle chain in it.

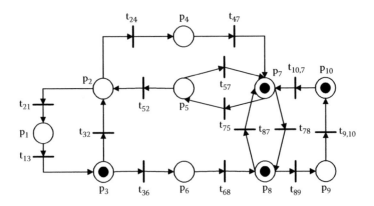

V_1: 3-6-8-9; V_2: 7-5-2-1-3-2-4-7; V_3: 8-7-5-7-8; V_4: 10-7-8-9-10

FIGURE 9.22 The CROPN after a conflict cycle chain is broken.

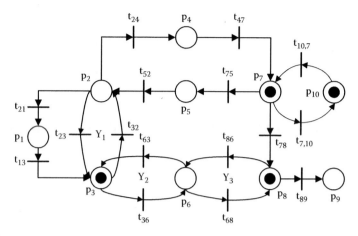

FIGURE 9.23 The CROPN after route expansion.

Example 9.6

Consider the CROPN shown in Figure 9.20. For this example we know $n_{1d} \in P(R_2)$, $n_{1d} \in P(R_3)$, $n_{1d} \in P(R_4)$, $n_{2d} \in P(R_3)$, $n_{2d} \in P(R_4)$, and $n_{3d} \in P(R_4)$. If V_4 can go through p_8 before V_3, V_3 can go through p_7 before V_2, and V_2 can go through p_3 before V_1, then no route expansion is needed. We see that $R_{12} = \{2, 1, 3\}$, $R_{13} = \{3, 2\}$, and $R_{23} = \{2, 4, 7\}$, where $R_{ij} = R_i \cap R_j$. To realize the above ordering, V_2 should enter p_2 before V_1, and V_3 should enter p_2 before V_2 and enter p_3 before V_1. This is impossible. For if V_2 enters p_2 first, then V_3 cannot move, and if V_3 enters p_2 first, V_2 cannot enter p_2 before V_1. It is easy to verify that if we exchange the positions of V_2 and V_3, then the ordering can be realized. Thus, as shown in Figure 9.23, we expand R_2 by adding p_{10} into it, because p_{10} is on R_4 and V_4 can go through p_{10} before V_2. The other routes do not need expansion, and there will be no blocking.

9.4.5 PERFORMANCE COMPARISON

As discussed in the introduction, there are several ways to resolve conflict in AGV systems. Our approach combines the routing and deadlock avoidance problems by providing a good AGV route to the lower-level deadlock controller. Thus, the methods used in the design level (Bozer Srinivasan, 1989, 1991, 1992; Sinriech and Tanchoco, 1995; Sinriech et al., 1996) are irrelevant to this study and no comparison is needed. The approaches for deadlock avoidance generally do not deal with the routing problem, but are based on the known routes to avoid deadlocks. There is work at the planning level that solves the routing problem and at the same time guarantees conflict-free operation (Egbelu and Tanchoco, 1986; Kim and Tanchoco, 1991, 1993; Dowsland and Greaves, 1994; Huang et al., 1993). However, these planning strategies are essentially open-loop control policies (Reveliotis, 2000) and based on deterministic timing of the vehicle travel. Because of the significantly stochastic nature of the manufacturing systems, these techniques are not robust. Thus, it is unnecessary to compare the performance of this study with theirs. Among the works in this field, the approach in Reveliotis (2000) presents a technique to avoid conflicts in AGV systems with feedback by providing the AVG routes. Thus, the work in Reveliotis (2000) in some sense is similar to ours, and it is meaningful to compare the performance of ours with that in Reveliotis (2000). We use two examples adapted from Reveliotis (2000) to do that.

The configuration of the AGV system considered for the examples is shown in Figure 9.24. According to Reveliotis (2000), the AGVs can only reside on the links or in the docking station that can hold all the AGVs. Thus, a link corresponds to a node in our model, and the configuration can be transferred to that shown in Figure 9.25.

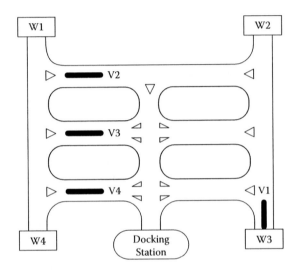

FIGURE 9.24 The configuration of the AGV system for the examples. (From Reveliotis, S. A. (2000). Conflict resolution in AGV systems, *IIE Trans.*, Vol. 32, No. 7, pp. 647–659. With permission.)

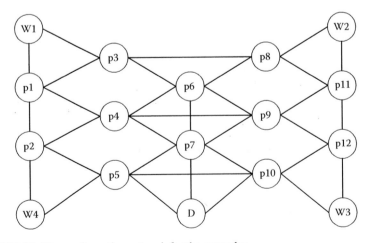

FIGURE 9.25 The configuration network for the examples.

Example 9.7

Consider the situation shown in Figure 9.24, where V_1 is in W_3, V_2 in p_3, V_3 in p_4, and V_4 in p_5. To complete the missions, V_1 should first reach W_1 and then D; V_2 first W_3, then D; V_3 first W_2, then W_3, and finally D; and V_4 first W_1, then W_2, and finally D. In Reveliotis (2000), a solution is obtained by the following routes: V_4, p_5 $\to p_2 \to p_1 \to W_1 \to p_1 \to p_2 \to p_5 \to p_{10} \to p_{12} \to p_{11} \to W_2 \to p_{11} \to p_{12} \to p_{10} \to D$; V_1, $W_3 \to p_{10} \to p_5 \to p_2 \to p_1 \to W_1 \to p_1 \to p_2 \to p_5 \to D$; V_3, $p_4 \to p_6 \to p_8$ $\to W_2 \to p_{11} \to p_{12} \to W_3 \to p_{10} \to D$; and V_2, $p_3 \to p_8 \to p_{11} \to p_{12} \to W_3 \to p_{10}$ $\to D$. Particularly, the algorithm requires that only after an AGV fulfills its mission can the other start its operation in the given order; in other words, the system can operate only sequentially.

Based on the configuration of the AGV system, it is reasonable to assume that the distance of arcs (p_5, p_{10}), (p_3, p_8), and (p_4, p_9) is 4 units, the distance of (W_1, p_3), (W_1, p_1), (p_1, p_3), (p_1, p_4), (p_2, p_4), (p_2, p_5), (W_4, p_5), (p_7, p_5), (D, p_5), (D, p_{10}), (p_7, p_{10}), (p_7, p_4), (p_7, p_9), (p_6, p_4), (p_6, p_3), (p_6, p_9), (p_6, p_8), (W_2, p_8), (W_2, p_{11}), (p_8, p_{11}), (p_9, p_{11}), (p_9, p_{12}), (p_{10}, p_{12}), (p_{10}, W_3), and (p_{12}, W_3) is 3 units, and the distance of (p_1, p_2) and (p_{11}, p_{12}) is 2 units, respectively. Then, by using the proposed method, we can find the routes: V_4, $p_5 \to p_2 \to p_1 \to W_1 \to p_3 \to p_8 \to W_2 \to p_{11} \to p_{12} \to p_{10} \to D$; V_1, $W_3 \to p_{10} \to p_5 \to p_2 \to p_1 \to W_1 \to p_1 \to p_2 \to p_5 \to D$; V_3, $p_4 \to p_6 \to p_8$ $\to W_2 \to p_{11} \to p_{12} \to W_3 \to p_{10} \to D$; and V_2, $p_3 \to p_8 \to p_{11} \to p_{12} \to W_3 \to p_{10}$ $\to D$. For this routing, based on Figure 9.25, we obtained the CROPN as shown in Figure 9.26. It is easy to verify that with the initial marking shown in Figure 9.26, conflict can be avoided by applying the policy presented in Section 9.2. In fact, for the AGVs to fulfill their missions, first for each AGV we can find a path such that V_1, $W_3 \to p_{10}$; V_4, $p_5 \to p_2$; and V_3, $p_4 \to p_8$, simultaneously. Second, for each AGV we can find a path such that V_4, $p_2 \to W_1$; V_1, $p_{10} \to p_1$; V_3, $p_8 \to W_2$; and V_2, $p_3 \to W_3$. Third, V_2, $W_3 \to D$; V_3, $W_2 \to W_3$; V_4, $W_1 \to p_8$; V_1, $p_1 \to W_1$, and then V_3, $W_3 \to D$; V_4, $p_8 \to W_2$; V_1, $W_1 \to D$; and finally V_4, $W_2 \to D$. It can be seen that

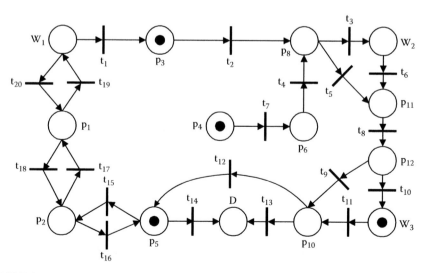

FIGURE 9.26 The Petri net model for Example 9.7.

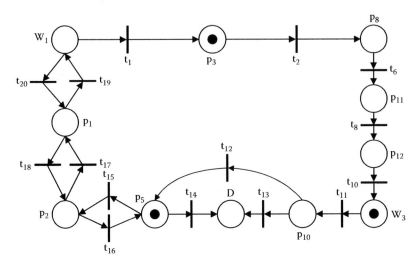

FIGURE 9.27 The CROPN for Example 9.8.

the total traveling distance obtained by our method is 95 units, which is shorter than the 106 units obtained by the method in Reveliotis (2000). Moreover, by our method, the missions of the AGVs can be fulfilled concurrently. Notice that V_4 starts its mission as early as other AGVs and is the latest one that reaches its destination without waiting anywhere during the mission. Thus, assume the traveling speed for the AGVs is f; then the total time for the AGVs to fulfill their missions is $29/f$ (the time taken by V_4), while it takes $106/f$ time units to fulfill their missions by the method in Reveliotis (2000). In other words, the total time for fulfilling all the missions by the proposed method is much shorter than that using the algorithm presented in Reveliotis (2000).

Example 9.8

Consider the situation that V_2 resides in W_1 and should go to W_3 and then to D, and at the same time V_1 resides in W_3 and should go to W_1 and then to D. As pointed out in Reveliotis (2000), the algorithm cannot find a solution, and it should be resolved by the performance-control level. However, it is easy to find a solution with the proposed method, $V_1, W_3 \rightarrow p_{10} \rightarrow p_5 \rightarrow p_2 \rightarrow p_1 \rightarrow W_1 \rightarrow p_1$ $\rightarrow p_2 \rightarrow p_5 \rightarrow D$ and $V_2, W_1 \rightarrow p_3 \rightarrow p_8 \rightarrow p_{11} \rightarrow p_{12} \rightarrow W_3 \rightarrow p_{10} \rightarrow D$. The corresponding CROPN is shown in Figure 9.27.

It should be pointed out that, in fact, the AGV banker's algorithm focused on the lower-level logical control to avoid conflicts, and it cannot consider the routing performance. Furthermore, because of the sequential nature of the banker's algorithm, it delays the completion of the missions. On the contrary, ours is a two-level approach. At the upper level we concentrate on the routing problem to improve the routing performance. At the lower level, we pursue an effective control policy. Thus, by combining them, we can obtain a better result.

9.5 SUMMARY

In an AGV system, the mission of each AGV is changed dynamically from time to time. Hence, a route is changed dynamically for each AGV. This disables deadlock avoidance techniques developed for part production processes where routes of parts are deterministic. Moreover, all deadlock avoidance methods require that at the initial state the system is not deadlocked. This is not true in an AGV system. Another problem is that when an AGV reaches its destination and stays there, it may block other AGVs from fulfilling their missions. Thus, it is very difficult to operate an AGV system in real time. Of course, AGV systems can be managed at the planning level. However, such an open-loop strategy is not robust to deal with events that occur in real time. Here, an effective approach is presented to solve this problem with the help of resource-oriented Petri nets.

REFERENCES

Aho, A. V., J. E. Hopcroft, and J. D. Ullman. 1987. *Data structures and algorithms*. Reading, MA: Addison-Wesley.

Banaszak, Z. A., and B. H. Krogh. 1990. Deadlock avoidance in flexible manufacturing systems with concurrently competing process flows. *IEEE Transactions on Robotics and Automation* 6:724–34.

Bozer, Y. A., and M. M. Srinivasan. 1989. Tandem configurations for AGV systems offer flexibility and simplicity. *Industrial Engineering* 21:23–27.

Bozer, Y. A., and M. M. Srinivasan. 1991. Tandem configuration for automated guided vehicle systems and the analysis of single vehicle loops. *IIE Transactions* 23:72–82.

Bozer, Y. A., and M. M. Srinivasan. 1992. Tandem AGV systems: A partitioning algorithm and performance comparison with conventional AGV systems. *European Journal of Operational Research* 63:173–92.

Dowsland, K. A., and A. M. Greaves. 1994. Collision avoidance in bi-directional AGV systems. *Journal of Operations Research Society* 45:817–26.

Egbelu, P. J., and J. M. A. Tanchoco. 1986. Potential for bidirectional guided-path for automated vehicles based systems. *International Journal of Production Research* 24:1075–97.

Fanti, M. P., B. Maione, S. Mascolo, and B. Turchiano. 1997. Event-based feedback control for deadlock avoidance in flexible production systems. *IEEE Transactions on Robotics and Automation* 13:347–63.

Hammond, G. 1986. *AGVS at work*. New York: Springer-Verlag.

Huang, J., U. S. Palekar, and S. G. Kapoor. 1993. A labeling algorithm for the navigation of automated guided vehicles. *Journal of Engineering for Industry* 115:315–21.

Kim, C. W., and J. M. A. Tanchoco. 1991. Conflict-free shortest bi-directional AGV routing. *International Journal of Production Research* 29:2377–91.

Kim, C. W., and J. M. A. Tanchoco. 1993. Operational control of bi-directional automated guided vehicle system. *International Journal of Production Research* 31:2123–38.

Koff, G. A. 1987. Automatic guided vehicle: Application, control, and planning. *Material Flows* 4:3–16.

Krishnamurthy, N. N., R. Batta, and M. H. Karwan. 1993. Developing conflict-free routes for automated guided vehicles. *Operations Research* 41:1077–90.

Lawley, M. A. 1999. Deadlock avoidance for production systems with flexible routing. *IEEE Transactions on Robotics and Automation* 15:1–13.

Lee, C.-C., and J. T. Lin. 1995. Deadlock prediction and avoidance based on Petri nets for zone-control automated guided vehicle systems. *International Journal of Production Research* 33:3249–65.

Malmbog, J. 1990. A model for the design of zone-control automated guided vehicle systems. *International Journal of Production Research* 28:1741–58.

Reveliotis, S. A. 2000. Conflict resolution in AGV systems. *IIE Transactions* 32:647–59.

Sinriech, D. and J. M. A. Tanchoco. 1995. An introduction to the segmented flow approach for discrete material flow systems. *International Journal of Production Research* 33:3381–410.

Sinriech, J. M. A. Tanchoco, and Y. T. Herer. 1996. The segmented bi-directional single-loop topology for material flow systems. *IIE Transactions* 28:4–54.

Wu, N. Q. 1999. Necessary and sufficient conditions for deadlock-free operation in flexible manufacturing systems using a colored Petri net model. *IEEE Transactions on Systems, Man, and Cybernetics C* 29:192–204.

Wu, N. Q., and M. C. Zhou. 2001a. Avoiding deadlock and reducing starvation and blocking in automated manufacturing systems. *IEEE Transactions on Robotics and Automation* 17:657–68.

Wu, N. Q., and M. C. Zhou. 2001b. Resource-oriented Petri nets in deadlock avoidance of AGV systems. In *Proceedings of 2001 IEEE International Conference on Robotics and Automation*, Seoul, Korea, pp. 64–69.

Wu, N. Q., and W. Q. Zeng. 2002. Deadlock avoidance in AGV system using colored Petri net model. *International Journal of Production Research* 40:223–38.

Wu, N. Q., and M. C. Zhou. 2004. Modeling and deadlock control of automated guided vehicle systems. *IEEE Transactions on Mechatronics* 9:50–57.

Wu, N. Q., and M. C. Zhou. 2007. Shortest routing of bi-directional automated guided vehicles avoiding deadlock and blocking. *IEEE/ASME Transactions on Mechatronics* 12:63–72.

Xing, K. Y., B. S. Hu, and H. X. Chen. 1996. Deadlock avoidance policy for Petri net modeling of flexible manufacturing systems with shared resources. *IEEE Transactions on Automatic Control* 41:289–95.

Yeh, M.-S., and W.-C. Yeh. 1998. Deadlock prediction and avoidance for zone-control AGVS. *International Journal of Production Research* 36:2879–89.

Zeng, L., H.-P. Wang, and S. Jin. 1991. Conflict detection of automated guided vehicles: A Petri net approach. *International Journal of Production Research* 29:865–79.

10 Control of FMS with Multiple AGVs

10.1 INTRODUCTION

A well-recognized way for manufacturers to keep agile is to implement flexible automation on the shop floor using advanced management, computer, and control technologies. The trend toward flexible automation has significantly increased the use of automated guided vehicles (AGVs) in material handling in automated factories for cost savings. The trend toward small to moderate volumes of many different board types in semiconductor manufacturing promotes the use of AGVs over conveyor systems. As a result, manufacturers have chosen AGVs to implement truly flexible material handling systems (MHS). On the other hand, managing AGVs and product routes becomes a challenging problem.

One of the challenging issues is deadlock resolution (Zhou and Fanti, 2004). Deadlock occurs due to limited production, storage, and transportation resources in flexible manufacturing systems (FMS), leading to a system wide standstill. As pointed out by Wu and Zhou (2004), deadlocks in FMS can be classified into two types. One is caused by the competition by parts for the manufacturing resources, including machines, buffers, and MHS. It occurs during the part processing. The other type is due to the competition for nodes and lanes by multiple AGVs in MHS. Deadlock resolution for these two types of deadlock issues has been studied extensively. However, the issues are studied separately. It is well known that either problem is intractable if the maximal permissive control policy is to be pursued when they are solved separately with single-capacity resources.

It seems that it is appropriate and straightforward to study these two types of deadlocks in FMS since part processing and delivering can be seen as two separate processes, and the tasks of these two processes are different. However, these two processes are often mutually dependent. Thus, if an MHS contains only one AGV, it is justified to treat these two processes separately. However, if it contains multiple AGVs, the situation differs in the sense of deadlock resolution. When material handling demands multiple AGVs, it makes sense and requires new research to solve the deadlock problem by considering these two processes together. In fact, the AGVs can be used as buffers for the part production processes, thereby helping resolve the deadlock issues caused by manufacturing resources.

Two examples are used to show that it is meaningful to study the deadlock resolution problem in considering a part production process and AGV system together.

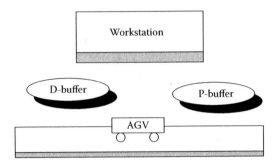

FIGURE 10.1 A deadlock situation in an AGV system.

Example 10.1

In Figure 10.1, an AGV carries a part for a workstation. When it reaches the workstation, it finds that the D-buffer is full, leading to a deadlock as treated in the existing literature. On the other hand, if another AGV is free at the moment, and can move away one of the parts in the D-buffer, then the deadlock can be resolved.

Example 10.2

Figure 10.2 shows that part A is processed by workstation m_1 and part B by workstation m_2. The next operation of part A requires workstation 2, and the next operation of part B requires workstation 1. This is a typical deadlock situation in the literature. However, in FMS with free AGVs, part A is unloaded and delivered to workstation 2 by one AGV, and part B by another, leading to no deadlock.

From these two simple examples, an AGV system, if well designed and operated, can enable a new and flexible way for deadlock resolution in FMS. Thus, an efficient deadlock control policy could be developed by taking this advantage to increase the manufacturing productivity. A challenging question is whether one can obtain a maximally permissive control policy with a computationally tractable algorithm for FMS with multiple AGVs.

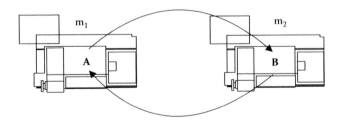

FIGURE 10.2 A deadlock situation in part processing.

An AGV system in FMS is assumed to be operated in a bidirectional path network and the system configuration is strongly connected. We extend the colored resource-oriented Petri net (CROPN) to model the systems. Based on the model and the results in Wu and Zhou (2005), a maximally permissive control policy is proposed in this chapter. This policy is shown to be computationally efficient regardless of whether the system is a single-capacity resource system. This can be viewed as a significant development in the area.

10.2 SYSTEM MODELING WITH CROPN

A model for the integrated system with part production process and AGVs considered together is needed. First, part production processes and an AGV system are modeled by CROPN as done in the previous chapters. Then these two models are integrated to describe the whole system.

An FMS that contains five workstations and an AGV system with two AGVs is shown in Figure 10.3. We first build a CROPN model for the AGV system. Consider the AGV system in the FMS shown in Figure 10.3. It contains two AGVs V_1 and V_2. Initially they are located in zones 4 and 8, respectively. For their next missions, the routes for these two AGVs are as follows: V_1, $4 \to 2 \to 3 \to 6 \to 8 \to 7 \to 9$ is to pick up a part at workstation 3 at zone 6 and then deliver to workstation 4 at zone 9, and V_2, $8 \to 6 \to 3 \to 1 \to 2 \to 4$ is to pick up a part at the load/unload station at zone 1 and deliver to workstation 1 at zone 4, respectively.

Based on the routing data and ROPN modeling method presented in Chapters 5 and 9, we obtain the subnets for V_1 and V_2 and the merged net as shown in Figure 10.4a–c, respectively. Here, we address the properties for the entire system, not only for the AGV system itself. Thus, we do not expand the model to form a strongly connected ROPN as done in Chapter 9.

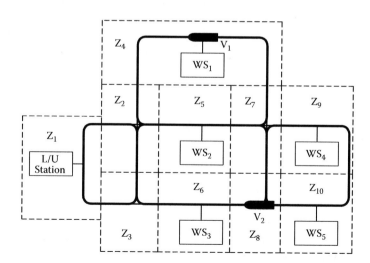

FIGURE 10.3 An FMS (RMS) containing five workstations and an AGV system with two AGVs.

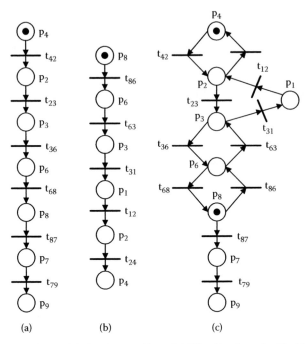

FIGURE 10.4 The PN models for the traveling of AGVs; (a) subnet for V_1, (b) subnet for V_2, and (c) merged net for both V_1 and V_2.

Now, with the model for the AGV system built, we give the CROPN model for the part production processes, as done in Chapters 7 and 8. For part processing, it is assumed as usual that before a part is moved to a machine for processing, the tools needed to process the part are already in the magazine of that machine. Therefore, tools do not contribute to deadlocks. Similarly, pallets and fixtures are irrelevant to deadlocks in this environment. Machines, buffers, and AGVs are the main resources in the production process and impact the occurrence of deadlocks. We take only machines and buffers into consideration.

There are buffers (input or output buffers, or both) associated with some machines, or there are multiple machines for some types of machines. Nevertheless, in the sense of deadlock resolution, the multiplicity of a type of machine or a machine with buffers just implies that it can hold multiple parts, and it is not necessary to distinguish the individual machines or the machine and its buffers. In this chapter, we model a machine by a single place, just like a node in the AGV system. This implies that the system considered here can be single capacity. However, to differentiate between the model for a machine in the part processing and a node in an AGV system, we use y and z to denote their place and transitions, respectively, instead of p and t. While $K(p) \equiv 1$ in an AGV system, $K(y) \geq 1$ holds in the part processing ROPN.

Assume that there are three types of parts to be processed in the five-workstation (machine) system, as shown in Figure 10.3. The processing routes for them are

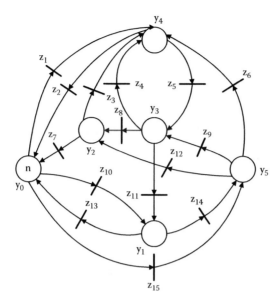

FIGURE 10.5 The ROPN for a part processing process.

A-part, $y_4 \rightarrow y_3 \rightarrow \{y_2, y_1\}$; B-part, $y_1 \rightarrow y_5 \rightarrow \{y_2, y_3\} \rightarrow y_4$; and C-part, $y_5 \rightarrow y_4 \rightarrow y_3 \rightarrow y_1$. Note that for A-part, $\{y_2, y_1\}$ means either y_2 or y_1 after y_3. For B-part, $\{y_2, y_3\}$ means either y_2 or y_3 after y_5. Then we can obtain the CROPN for the part production process according to the modeling method presented in the previous chapters. This model is shown in Figure 10.5.

Colors in the CROPN models for both the AGV system and part production process can be defined as done in the previous chapters and are omitted here. Both CROPN models for a part production process and AGV system describe a part of the FMS. The key is how to build an integrated model for the whole system by using these two models. Consider a transition in Figure 10.4 for the AGV system; it represents that an AGV travels from one zone to another. For example, transition t_{42} indicates V_1 traveling from zone 4 to zone 2, which physically describes an activity. On the contrary, a transition in the CROPN shown in Figure 10.5 describes the logic relation between two operations. For example, transition z_6 just indicates that after a part of type 3 is processed by machine 5 for its first operation, it must be delivered to machine 4 for its second operation. It does not describe which AGV delivers the part from machine 5 to 4, how the part is delivered, and by what route the part is delivered. It takes it for granted that the delivery job is done as expected. In the real manufacturing process, these activities should be specified, but done in real time. The CROPN in Figure 10.5 describes the competition for machines and buffers, but not for AGVs. In fact, the parts in processing compete not only for machines and buffers, but also for AGVs and traveling paths. The fact is that while an operation is known to be processed by a set of machines in advance, delivering a part from one

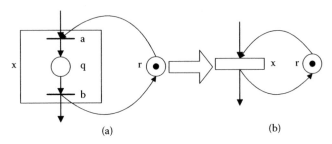

FIGURE 10.6 The macro-transition for material flow.

zone to another can be done by any AGV. Hence, modeling a delivering process can be very complex.

Consider the CROPN for part processing in Figure 10.5. Firing a transition, in fact, requires delivering a part from one zone to another, which can be described by a subnet that models the mission for an AGV shown in Figure 10.4. This reminds us of using a hierarchical way to integrate a part delivering process modeled in the CROPN for an AGV system into a model for a part processing process. Observe the subnets shown in Figure 10.4a and b. It is reasonable to model a subnet by a macro-place in the hierarchical model. Unfortunately, we cannot use a macro-place to replace a transition in the CROPN for the part production process in Figure 10.5, because otherwise it is not a PN. Thus, we use a macro-transition to describe a part delivering process by an AGV, as shown in Figure 10.6a. This macro-transition x contains two transitions a and b and place q. For convenience, we denote $x = \{a, q, b\}$, $x_i = \{a_i, q_i, b_i\}$, and $Q = \{q_i\}$. Firing transition a in x represents the use of an AGV in q, and firing b releases the AGV. A token in q represents an AGV delivering a part. Transition x can be treated just as an ordinary transition in the integrated CROPN. A token in place r represents an available AGV. The macro-transition can be pictured as a rectangle, as shown in Figure 10.6b.

With the macro-transitions, we can integrate the CROPN models for an AGV system and part processing process. Because a transition in the part processing CROPN, as shown in Figure 10.5, represents a task to deliver a part from one node to another by an AGV, to integrate the two models is to replace each transition shown in Figure 10.5 with a macro-transition straightforwardly. However, by doing so, it does not model the competition for AGVs by the parts, and this should be appropriately modeled if an effective control policy is wanted. Assume that there are N (>1) AGVs in the system; then we use N places named $r_1, r_2, \ldots,$ and r_N to model them. They are connected by $N - 1$ transitions $z_1, z_2, \ldots, z_{N-1}$, such that z_i is the output transition of r_i and the input transition of r_{i+1}, $i = 1, 2, \ldots N - 1$. Denote $R = \{r_1, r_2, \ldots, r_N\}$. In this way, N places are serially connected and form a direct place path. We let $K(r_i) = 1$ for any $i = 1, 2, \ldots,$ and N, and initially we put a token in each place, which implies that there are N AGVs available in the initial state. It should be pointed out that although we use N places to model the N AGVs, each place does not correspond to an individual AGV. The number N just means that there are N AGVs in total, and the serial structure guarantees the correct control logic presented later.

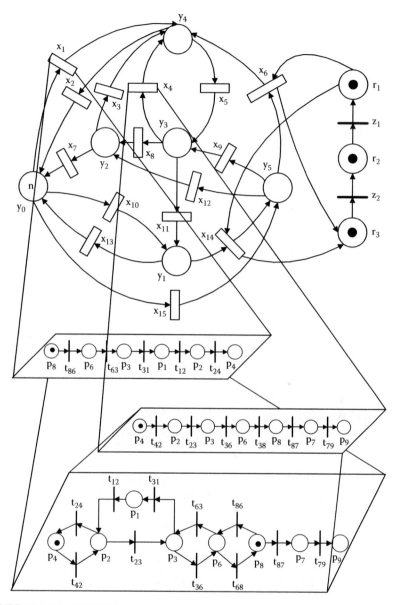

FIGURE 10.7 The CROPN for the integrated system.

The integrated CROPN model is shown in Figure 10.7, where every transition in the CROPN shown in Figure 10.5 is replaced by a macro-transition, and each macro-transition is the output transition of r_1 and the input transition of r_3. Among these arcs, only (r_1, x_6), (x_6, r_3), (r_1, x_{14}), and (x_{14}, r_3) can be seen in Figure 10.7, and the others are omitted for the sake of clarity. When a macro-transition is in firing,

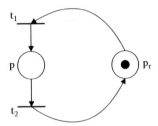

FIGURE 10.8 Another way to model the material delivery.

it corresponds to a subnet for an AGV to deliver a part from one node to another by some route in the AGV system. Firing x_4 may correspond to the subnet shown in Figure 10.4a, and firing x_1 to that shown in Figure 10.4b. The subnets then form the model for the tasks of the AGVs as shown in Figure 10.7.

One may argue that it may be better to model a subnet for a mission of an AGV by a macro-place, and add an input and output transition, respectively. Then a model for this process can be as shown in Figure 10.8. By replacing each transition in Figure 10.5 with the model shown in Figure 10.8, an integrated model may be obtained. Indeed, it can be done in this way. However, as discussed in Chapter 6, doing so may create some unnecessary ill-behaved siphon (IBS) and make the deadlock avoidance problem much more complicated and thus difficult. Such a problem can be avoided by using macro-transitions, as shown in Figure 10.6. At the same time, we can track AGVs when they are in their mission.

In Figure 10.7, there are three places r_1, r_2, and r_3 serially connected by z_1 and z_2, with a token in each place indicating that there are three AGVs in the system. In reality, the AGVs can act independently, or they can perform their missions concurrently. It seems that the structure of the subnet formed by r_1, r_2, r_3, z_1, and z_2 undermines the concurrency of the action for the AGVs, for z_1 and z_2 can fire only sequentially, but in fact, it does not. It should be noticed that it takes time to complete the firing of a macro-transition. During that time we can set transitions z_1 and z_2 as immediate ones. When a macro-transition fires, place r_1 is emptied since the fired macro-transition must be an output transition of r_1. Then z_1 is enabled and can fire immediately according to enabling and firing rules for finite capacity PN, thereby moving a token into r_1. Thus, another AGV represented by the token in r_1 can be used to fire another macro-transition immediately after firing of the previous macro-transition begins. In this way, the AGVs can act concurrently. In fact, our model controls only the firing orders of the macro-transitions (or the AGVs), but does not undermine the flexibility of the system. Nevertheless, it is critical to control the firing orders to make the system deadlock-free.

10.3 DEADLOCK AVOIDANCE POLICY

After the integrated CROPN model for the AMS with multiple AGVs is developed, we are ready to present the deadlock avoidance policy for the system. Deadlock in FMS is a phenomenon of the circular wait of jobs and is related to the liveness in the corresponding PN model. If the PN model of a system is live, the system is deadlock-free. Thus, to make an FMS deadlock-free is to develop a control policy under which its corresponding PN model is live. The sources that lead to deadlocks in FMS with multiple AGVs are: (1) some of the parts being carried by AGVs cannot reach the destinations because of their competition for limited zones, and (2) parts may be delivered to the destination machines for the next operations, but some of the machines are occupied and cannot be freed.

From Chapter 9, we know that for any missions of AGVs in an AGV system, deadlock can be avoided if the system is under control of the control policy presented there, and at the same time, initially the system is not deadlocked. In other words, in this condition, any part delivery can be fulfilled. If it is initially deadlocked, we can reroute the AGVs by using the method presented in Chapter 9 such that the missions can be fulfilled. Thus, it is reasonable to assume that a part can be delivered to any place by the AGV system that is under control. Then we can focus on the deadlock avoidance problem for a part production process with AGVs taken into consideration.

If parts can be delivered to destinations as required, the remaining work is to develop a control policy such that when a part is delivered to a machine, the machine is either free or to be freed. Of course, simple policies exist, such as only one part is allowed in the system. All policies developed by considering the part processing only can also be applied. However, such policies cannot be maximally permissive with tractable computation and do not take advantage of the multiplicity of AGVs. The goal here is to obtain the maximally permissive policy with tractable computation.

Observe the CROPN in Figure 10.7. If we treat each macro-transition just as an ordinary transition and the enabling and firing rules presented for a finite capacity PN are applied, then the deadlock problem is equivalent to the one for part processing without considering the AGVs. However, if we replace x_i in Figure 10.7 by the subnet shown in Figure 10.8, then the capacity for resources will be useless. Although the spaces in $Q = \{p\}$ may be interpreted as free resources, they are in fact not. Nevertheless, when macro-transition x_i is in firing, the token in q_i does represent that an AGV carrying a part is traveling. Thus, it is necessary to treat a macro-transition just as an ordinary transition, but sometimes the tokens in Q should be taken into account.

To make a PN live, control should be imposed to a PN. A CROPN is said to be a controlled CROPN if at least one transition in it is controlled. Here, in the CROPN for the FMS with multiple AGVs, all the macro-transitions are controlled.

Assume that $y_d \in b_i^{\bullet}$. Let $W(q_i) = \text{Min}(K(y_d) - M(y_d), M(q_i))$, where q_i and b_i are in macro-transition x_i. $W(q_i)$ gives the number of AGVs in place q_i that can be freed by using the free spaces in y_d. For example, in Figure 10.7, $y_4 \in b_6^{\bullet}$, assume that $K(y_4) = 2$, $M(y_4) = 1$, and $M(q_6) = 1$ at a marking M; then $W(q_6) = 1$, indicating that the AGV for firing x_6 can be freed by loading the part carried by the AGV into y_4. Further assume that y_h is a place and N' is the number of macro-transitions in a CROPN.

Let $\mu(M, y_h) = \min([K(y_h) - M(y_h) + \max(\sum_{i=1}^{N} M(r_i) + \sum_{j=1}^{N'} W(q_j) - 1, 0)], K(y_h))$ denote the potential available spaces for place y_h at marking M. For example, in the system modeled by the CROPN shown in Figure 10.7, assume that (1) there are two AGVs, V_1 and V_2, and $K(y_4) = 2$; (2) at marking M, V_1 is for firing x_2 and V_2 is in r_1. Hence, $M(r_1) = 1$, $M(r_2) = 0$, $M(q_2) = 1$, $M(q_j) = 0$, $\forall j \neq 2$, $N = 2$ and $N' = 15$; and (3) $M(y_4) = 2$. Thus, $[K(y_4) - M(y_4) + \max(\sum_{i=1}^{N} M(r_i) + \sum_{j=1}^{N'} W(q_j) - 1, 0)] = 1$ and $\mu(M, y_4) = 1$. Then, we present the control policy on the macro-transitions as follows.

M-policy: A macro-transition x_i in the CROPN for the FMS with multiple AGVs in marking M can fire if one of the conditions below is satisfied:

1. $M(y_1(C_i)) \geq 1$, $\forall y_1 \in {}^\bullet x_i$, with C_i being the color of x_i, and $\mu(M, y_2) \geq 1$, $\forall y_2 \in x_i{}^\bullet$ and $y_2 \notin R = \{r_1, r_2, \ldots, r_N\}$
2. $M(y_1(C_i)) \geq 1$, $\forall y_1 \in {}^\bullet x_i$, and macro-transition $x_j \in {}^\bullet y_1$, with $y_1 \notin R$ and $M(q_j) > 0$, where $q_j \in x_j$

From this control policy, some macro-transitions are controlled. The other transitions, including macro-transitions, can fire spontaneously according to the rules for colored PN. We explain the policy as follows:

1. If the output place y of a macro-transition x_i has one or more spaces, then x_i can fire, or if two or more AGVs are free or can be freed by unloading the parts into the corresponding destination places (where enough spaces are available), then any macro-transition can fire. For example, in Figure 10.7, if there is a free space in y_4 and a free AGV, transition x_1 can fire by using the free AGV. In this situation, only the information about y_4 being full or not is needed to decide if x_1 can fire, regardless of other places. If two AGVs are free, x_1 can fire even if y_4 is full.
2. When only one AGV is free, this AGV can only pick up the part represented by a token in place y that is the destination of one of the AGVs that are occupied by parts. After one of the tokens in y is removed, then one of the AGVs that are occupied can be freed. For example, in Figure 10.7, assume that there are two AGVs, V_1 and V_2, and at marking M, x_1 is firing by using V_1, V_2 is free, and y_4 is full. Then, V_2 can be used to fire only x_5.

Note that there are many PPCs in the CROPN shown in Figure 10.7. However, by the M-policy, it is unnecessary to identify these PPCs. Especially, we do not need to check whether these PPCs are full or not in real time in implementing the M-policy.

Theorem 10.1: The CROPN for AMS with multiple AGVs is live if it is controlled by the M-policy.

Proof: We assume, without loss of generality, that $K(y) = 1$, $y \neq y_0$, and there are two AGVs or $N = 2$. By condition 1, if $\exists y \in x^\bullet$ and $M(y) = 0$, then macro-transition x can fire when it is process-enabled. This condition is equivalent to rules for ordinary colored PN. It is easy to verify that by condition 1, every place $y \neq y_0$ can be full and at the same time one of the two AGVs is occupied. Then, if we can show when this marking is reached and the CROPN is still live, the theorem holds.

Assume that every $y \neq y_0$ is full, AGV V_1 is occupied, V_2 is free, and the part on V_1 is required to be delivered to place y_1. According to the M-policy, there is no macro-transition such that condition 1 is satisfied. Let $x_1 \in y_1{}^\bullet$ and $M(y_1(C_1)) > 0$, where C_1 is the color of x_1 (it is sure such x_1 exists). Then condition 2 for x_1 is satisfied and x_1 can fire. After firing x_1, y_1 is emptied and V_1 is freed and V_2 is occupied. Let $y_2 \in x_1{}^\bullet$, the part on V_2 must be delivered to y_2, and there exists $x_2 \in y_2{}^\bullet$ such that $M(y_2(C(x_2))) > 0$, or x_2 can fire according to condition 2. In this way, the macro-transitions on a route can fire one by one until a part is completed and delivered to y_0. After that the previous firing process can repeat, but may be in a different route, depending on the types of parts in the system. By releasing different parts into the system, every macro-transition can fire again. This shows the liveness of the controlled CROPN. ■

In the proof we assume that $K(y) = 1$ with $y \neq y_0$. This implies that the system is a single-capacity resource system. It is known that it is intractable to apply a maximally permissive control policy if the deadlock avoidance problem for part processing is considered independently (Lawley, 1999; Fanti et al., 1997). Thus, it shows that the M-policy is powerful to avoid deadlock. This policy takes advantage of the multiplicity of AGVs, and the number of parts in the system can be $\sum_i K(y_i) + N - 1$ with $i \neq 0$. From the proof we also know that when $N > 2$, the system is more flexible by applying this policy. Theorem 10.1 shows the sufficiency of the M-policy; the following theorem shows the optimality of the policy:

Theorem 10.2: The M-policy is maximally permissive.

Proof: We only need to show that if the conditions given by the M-policy are violated, then the CROPN can be deadlocked. Again assume that $K(y) = 1$, $y \neq y_0$, and there are two AGVs, or $N = 2$. Assume that V_1 carries a part to be delivered to y_1 and y_1 is full. At the same time, $y_2 \neq y_1$ is also full, $y_2 \in$ •x_1, $M(y_2(C_1)) > 0$, $y_3 \in x_1$• is full too, and V_2 is free. In this marking the CROPN is not dead, and from Theorem 10.1 it is also safe. According to the M-policy, this time x_1 cannot fire. It is easy to verify that if x_1 is fired by using V_2, then the system is dead, for after firing x_1, no AGV can be freed forever. ■

From the proof we can see why we model N AGVs by $r_1, r_2, \ldots,$ and r_N in a serial structure. Assume that, in Figure 10.7, instead of three places $r_1, r_2,$ and r_3, only one place r is used, and put three tokens in r. When all the places $(y \neq y_0)$ are full, then $x_1, x_{10},$ and x_{15} are enabled according to the policy, but after these three transitions start their firing at the same time, the system is dead. However, with our model, the transitions can start their firing only one by one; after two of the three start their firing, the third is not enabled. Consequently, the CROPN is kept live. Nevertheless, this does not present any restriction to the system. In reality, it needs time for an AGV to complete a task modeled by a macro-transition x_i. However, transitions z_1 through z_{N-1} are immediate. Thus, after one AGV is released to fire x_i, another can be used to fire x_j immediately if permissive by the control policy. Hence, in fact, the macro-transitions can fire concurrently.

Theorem 10.3: The computational complexity of the M-policy is $O(n^2)$, where n is the number of machines in the system.

Proof: To implement the M-policy, we just need to check the macro-transitions to see which transition is enabled by the M-policy. If there are n machines in the system, then there are $n + 1$ places, including the place for the load/unload station, but not the places for the AGVs. Hence, there are at most $(n + 1) \times n$ macro-transitions in the CROPN for the system. This implies that the computational complexity of the M-policy is $O(n^2)$. ■

The complexity of deadlock avoidance in AMS results from the fact that when the system enters a state that is not a deadlock, but no matter what the system acts, it will be deadlocked. Such a state is called unsafe. In the CROPN controlled by the M-policy, any nondeadlock state is safe. Thus, the M-policy is efficient, which results from the multiplicity of AGVs.

Notice that the M-policy does not state how to control the AGVs to deliver the parts to destinations as required. In general, controlling the AGVs is not a trivial job. Thus, the complexity of deadlock avoidance for the integrated system is bounded by the complexity in controlling the AGVs. However, it should be pointed out that the deadlock avoidance problem for part processing itself, without considering part delivery, is intractable (Lawley, 1999; Fanti et al., 1997). Thus, the approach presented in this chapter makes significant improvement in the deadlock resolution field. Furthermore, as pointed out in Chapter 9, all existing AGV systems reported in the literature can be efficiently controlled to be deadlock-free due to their strongly connected configuration layout and a small number of AGVs in them.

10.4   ILLUSTRATIVE EXAMPLE

Consider the CROPN shown in Figure 10.7. We assume that all the machines are single-capacity resources, or $K(y_i) = 1$, $i \in N_5$, and $N = 3$. Initially, $M_0(y_i) = 0$, $i \in N_5$, and $M_0(r_i) = 1$, $i = 1, 2$, and 3. Furthermore, assume that there are enough tokens in y_0. According to the M-policy, at M_0, x_1, x_{10}, and x_{15} are enabled, Thus we fire them sequentially, and a part of type 1 is put into y_4, a part of type 2 into y_1, and a part of type 3 into y_5. After the firing of these three macro-transitions, x_5, x_6, and x_{14}, two of x_1, x_{10}, and x_{15} are enabled. To fire x_5, x_6, and x_{14}, we need to wait for the completion of the parts in processing. Hence, we fire x_1 and x_{10} sequentially; two parts are held by two AGVs. At this time we cannot fire x_{15} according to the M-policy. Otherwise, the system will be deadlocked. After some time, x_5 can fire and a part of type 1 is put into y_3; then x_6 fires and a part of type 3 is put into y_4, and x_{14} fires, leading a part of type 2 into y_5. A part held by an AGV is put into y_1 and an AGV is freed. This time two AGVs are free. Thus, although y_5 is full, we can still fire x_{15}. By doing so, the AGV holds a part of type 3. Because the part of type 1 held by an AGV should be delivered to y_4, x_5 is enabled. We fire x_5, leading to y_4 being emptied and an AGV holding a part of type 3. Then the part of type 1 held by an AGV is put into y_4 and the AGV is freed. Then x_8 fires, the part of type 1 in y_3 is moved into y_2, and the part of type 3 held by an AGV is put into y_3. Again, two AGVs are free and x_1 is enabled and fires. As a result, an AGV holds a part of type 1. This time there are seven parts in the system: five of them are in processing and two of them are held by two AGVs. According to the M-policy, the following events can occur: x_{12} fires, y_5 is emptied, and the part of type 2 is held by an AGV $\rightarrow$ the part of type 3 held by an AGV is put into y_5 and the AGV is freed $\rightarrow$ the freed AGV then allows x_7 to fire and the completed part of type 1 is returned into y_0 $\rightarrow$ the part of type 2 held by an AGV is put into y_2 and two AGVs are free $\rightarrow$ x_{10} fires by using one of the free AGVs. In this way, we keep seven parts in the system for processing and the system is still live.

It should be noted that if the existing deadlock avoidance approaches for AMS are applied, the system can be deadlocked when the number of parts in the system reaches five. However, by using the M-policy presented in this chapter, it reaches seven and the system is still live. This shows the M-policy's power. In addition, from the example, the M-policy is easy to implement.

10.5 SUMMARY

An AMS is composed of workstations and an automated material handling system (MHS). The MHS may be composed of a number of robots, a single AGV, or multiple AGVs. The deadlock resolution issue is very important in AMS operation. In Chapters 7 and 8, effective deadlock avoidance policies are presented for part production processes. When the MHS in an AMS is composed of robots or a single AGV, the prior control policy presented can be used. Effective methods for deadlock resolution for an AGV system with multiple AGVs are presented in Chapter 9. With the results presented in Chapters 7 to 9, an integrated control policy is presented in this chapter for AMS with multiple AGVs. When the MHS contains multiple AGVs, the policy presented here is tractable and optimal. Thus, up to now, by using the ROPN modeling method, solutions for deadlock resolution in AMS are provided for commonly seen operational environments.

REFERENCES

Fanti, M. P., B. Maione, S. Mascolo, and B. Turchiano. 1997. Event-based feedback control for deadlock avoidance in flexible production systems. *IEEE Transactions on Robotics and Automation* 13:347–63.

Lawley, M. A. 1999. Deadlock avoidance for production systems with flexible routing. *IEEE Transactions on Robotics and Automation* 15:1–13.

Wu, N. Q., and M. C. Zhou. 2004. Modeling and deadlock control of automated guided vehicle systems. *IEEE Transactions on Mechatronics* 9:50–57.

Wu, N. Q., and M. C. Zhou. 2005. Modeling and deadlock avoidance of automated manufacturing systems with multiple automated guided vehicles. *IEEE Transactions on Systems, Man, and Cybernetics B* 35:1193–202.

Zhou, M. C., and M. P. Fanti, ed. 2004. *Deadlock resolution in computer-integrated systems.* New York: Marcel Dekker.

11 Control of FMS with Multiple Robots

11.1 INTRODUCTION

As discussed in the previous chapters, deadlocks in automated manufacturing systems (AMS) can be classified into two types. The machines in an AMS are versatile, and jobs of different types can be processed concurrently in the system. In the job production processes, jobs compete for manufacturing resources, including machines, buffers, and material handling devices, leading to one type of deadlock. The material handling systems (MHS) in an AMS may be an automated guide vehicle (AGV) system, or composed of a number of robots and conveyors. In an AGV system, AGVs compete for zones and lanes, leading to the other type of deadlocks. In Chapters 7 to 9, we deal with deadlock resolution problems for these two types of deadlocks separately by using the ROPN model presented in Chapter 5. The deadlock resolution problem for the integrated system with multiple AGVs as MHS is dealt with by using the ROPN model in Chapter 10. It shows that AGVs can be used as material handling devices as well as temporary buffers, in order to ease a deadlock resolution problem. Based on this observation, an effective deadlock control policy is presented in Chapter 10.

It is believed that if the MHS is composed of robots, the robots will contribute to deadlock in the AMS operation (Ezpeleta et al., 1995; Li and Zhou, 2004, 2008; Li et al., 2008). Thus, great attention is paid to studying this problem. In Chapter 5, it is shown that by modeling the system with ROPN, robots have no contribution to deadlock in AMS. The question is whether robots can be used as material handling devices as well as temporary buffers, just like AGVs presented in Chapter 10, in order to resolve deadlock issues. In this chapter, we will show that, in some sense, they can. Based on the results in Wu and Zhou (2007), this chapter presents the corresponding control policy by using an ROPN model.

11.2 MOTIVATION THROUGH EXAMPLE

From the results in Ezpeleta et al. (1995), Li and Zhou (2004, 2008), and Li et al. (2008), it seems that robots are sources of deadlock other than resources for easing deadlock resolution. That is not really true. To argue this, consider a simple example. The flexible manufacturing systems (FMS) in Figure 11.1 contains three workstations, m_1, m_2, and m_3, three robots, r_1, r_2, and r_3, and a load/unload area. Robot r_1 is used for loading and unloading parts and for part delivering between m_1 and m_3; r_2 and r_3 are used for part delivery between m_1 and m_2, and m_2 and m_3, respectively. Consider that a part type is to be processed in the system with a route $m_1 \rightarrow m_2 \rightarrow m_3 \rightarrow m_1$. Assume

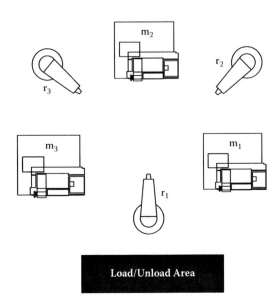

FIGURE 11.1 A simple example to show that a robot can be used as a buffer.

that at some time, m_1, m_2, and m_3 are processing a part for its first, second, and third operations, respectively. It is clear that there is a circular wait, or a deadlock if the robots in the system are not used as buffers. However, r_2 can pick up the part on m_1, and thus the part on m_3 can be delivered onto m_1 by r_1, the part on m_2 can be delivered onto m_3 by r_3, and the part held by r_2 can be dropped onto m_2. As a result, the deadlock can be avoided. This implies that as material handling devices in FMS, robots can act as buffer spaces too, so as to ease deadlock resolution in FMS as AGVs do.

It should be noticed that AGVs can travel on a built path. Thus, in general, an AGV system is designed in such a way that every AGV is able to visit any machine or workstation. Unlike an AGV, a robot in AMS may access only a very limited number of machines. Therefore, robots cannot be treated as central buffers. This means that the results presented in Chapter 10 are not applicable to the FMS with robots as MHS. This chapter discusses deadlock resolution problems in AMS with robots as MHS in treating robots as both material handling devices and buffers.

11.3 DEADLOCK CONTROL POLICY

To present the control policy, the CROPN model developed in Chapter 5 can be used directly here.

It is shown that if deadlock occurs in an ROPN, it must occur in a production process circuit (PPC) or in a strongly connected subnet formed by PPCs. Further, when a deadlock occurs, all the places for at least one PPC must be capacitated. When all of its places are capacitated, we say that a PPC is capacitated. We will show that when some PPCs are capacitated and a deadlock occurs in the ROPN, it may be avoided if some robots are treated as buffers. Let $T(v)$ be the set of transitions on PPC v, and $P(v)$ the set of places.

Definition 11.1: If $t_i \in T(v)$ is a transition on PPC v, there is a self-loop $\{p_{r1}, t_i\}$, and $R_i \neq \phi$ is the set of robots that can perform the tasks for transitions $T(v) - \{t_i\}$ with $r_1 \notin R_i$, t_i is said to be a buffering transition (BT) of v.

To explain the definition, consider the example below:

Example 11.1

The configuration of an AMS is shown in Figure 6.1. It is composed of four machines m_{1-4} and three robots r_{1-3}. Each machine can process two parts at a time. The working area is l_1, O_3, m_1, and m_3 for r_1; l_2, O_2, and m_{1-4} for r_2; and l_3, O_1, m_2, and m_4 for r_3. Assume that three types of parts, A, B, and C, are to be processed in the system with their routes: A-part, $m_1 \rightarrow m_3 \rightarrow m_4 \rightarrow m_2$; B-part, $m_4 \rightarrow m_1 \rightarrow m_2 \rightarrow m_4$; and C-part, $m_2 \rightarrow m_1 \rightarrow m_3 \rightarrow m_4$. The ROPNs without and with robot activity considered are shown in Figures 11.2 and 11.3, respectively.

There are five PPCs in the ROPN shown in Figure 11.2: $v_1 = \{p_1, t_{13}, p_3, t_{34}, p_4, t_{41}, p_1\}$, $v_2 = \{p_2, t_{24}, p_4, t_{42}, p_2\}$, $v_3 = \{p_1, t_{12}, p_2, t_{21}, p_1\}$, $v_4 = \{p_1, t_{12}, p_2, t_{24}, p_4, t_{41}, p_1\}$, and $v_5 = \{p_1, t_{13}, p_3, t_{34}, p_4, t_{42}, p_2, t_{21}, p_1\}$. For v_1, t_{13} can be served by r_1, and t_{34} and t_{41} can be performed by r_2 in the ROPN shown in Figure 11.3. Hence, t_{13} can serve as a BT according to Definition 11.1. For v_2, either t_{24} or t_{42} can be performed by either r_2 or r_3, or if t_{24} is performed by r_2, t_{42} can be performed by r_3 simultaneously, and vice versa. Thus, both t_{24} and t_{42} can serve as BT for v_2. For v_4, t_{24} can

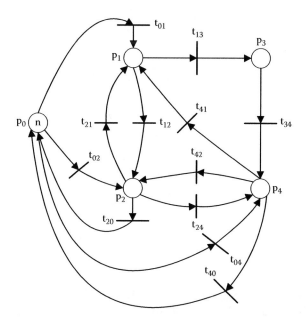

FIGURE 11.2 The ROPN of the production processes for Example 11.1.

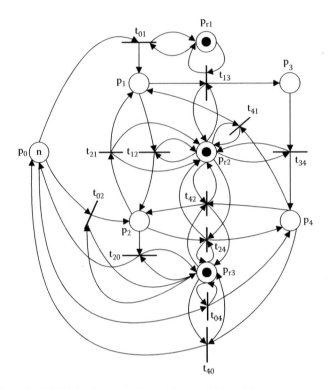

FIGURE 11.3 The ROPN for Example 11.1 with material handling processes modeled.

be performed by r_3, and t_{12} and t_{41} can be performed by r_2. Therefore, t_{24} is a BT for this PPC. For v_5, t_{42} can be performed by r_3, and t_{13}, t_{34}, and t_{21} can be performed by r_2, or t_{42} is a BT for this PPC. For v_3, both t_{12} and t_{21} must be performed by r_2. This violates the condition given in Definition 11.1. Hence, there is no BT in v_3.

Definition 11.2: Assume that the transfer task associated with a BT t should be performed by robot r, and t is process-enabled but not resource-enabled. Let robot r (1) pick up a part from t's input place, (2) wait for a free space in an output place of t, and (3) drop the part held into an output place once a free space is available. The above process is called t's R-firing. At step 1, t's R-firing is triggered. After step 3, t's R-firing is completed.

Notice that in a finite capacity PN, a transition can fire only if it is process and resource-enabled. It is regular transition firing. With R-firing, a transition t to be R-fired is just process-enabled at the beginning. By triggering t's R-firing, a free space in place $\bullet t$ is created, and then this free space can finally travel into $t^\bullet$ and t's R-firing can finish. By R-firing, the MHS resources are sufficiently used. With Definitions 11.1 and 11.2, we have the following lemma:

Lemma 11.1: Assume that a CROPN for an AMS contains only one PPC v and there is at least one BT for this PPC. If v is capacitated at a marking M, there exists a sequence of transition firings such that the PPC can be emptied.

Proof: Assume that there are n transitions and n places on v. Let $T(v) = \{t_1, t_2, \ldots, t_n\}$ and $P(v) = \{p_1, p_2, \ldots, p_n\}$ be the sets of transitions and places on v with $p_i = {}^\bullet t_i$, $i = 1, 2, \ldots, n$, $t_i{}^\bullet = p_{i+1}$, $i = 1, 2, \ldots, n-1$, and $t_n{}^\bullet = p_1$.

Case A: If a transition in $T_O(v)$, say t_k, is enabled, t_k can fire because t_k is not on the only PPC in CROPN. After firing t_k, one token is removed from v, and v is not capacitated. Assume that the set of robots serving for $T(v)$ is R_v. Because no transition in $T(v)$ is enabled, no robot in R_v needs to serve for the firing of any transition in $T(v)$ before the firing of t_k. Thus, any r in R_v is either free or serving for a firing of t that is not in $T(v)$. If r is serving for the firing of t, t must be enabled and r can be freed in a limited time, so that r can serve for the firing of a transition t_1 in $T(v)$ when t_1 is enabled. In other words, in both cases, the robots in R_v will be available for serving the firing of transitions in $T(v)$. This implies that v cannot be deadlocked after the firing of t_k. Thus, in a limited time, all the jobs in v can be completed and v will be emptied.

Case B: If no transition in $T_O(v)$ is enabled, an R-firing of a BT on v, say t_i, can be triggered. Afterwards, a free space in p_i is available, or t_{i-1} is both process and resource-enabled (just enabled hereafter). This time, as discussed above, the robots in R_i (Definition 11.1) are available for serving the firing of transitions in $T(v) - \{t_i\}$ in a limited time. Hence, there exists a sequence of firings $t_{i-1} \rightarrow t_{i-2} \rightarrow \ldots \rightarrow t_1 \rightarrow t_n \rightarrow t_{n-1} \rightarrow \ldots \rightarrow t_{i+2} \rightarrow t_{i+1}$. After the firing of t_{i+1}, a free space in p_{i+1} is available, and thus the R-firing of t_i can be completed. This process can be repeated until a transition in $T_O(v)$ is enabled, leading to the empty state of v. ∎

The fact that a PPC can be emptied implies that the PPC is deadlock-free. Thus, Lemma 11.1 shows that some deadlock in production processes can be avoided in a CROPN by taking robots as temporary buffers.

Let U denote a strongly connected subnet formed by multiple PPCs, T_U the set of transitions in U, and $T_O(U)$ the set of output transitions of U. A subnet U is said to be capacitated if every place in U is capacitated.

Definition 11.3: A PPC v is said to be process-enabled if all the transitions in $T(v)$ are process-enabled.

According to Definition 11.3, to make a PPC v process-enabled, every place on v must have at least one token. The following lemma gives the condition under which a strongly connected subnet formed by multiple PPCs is deadlock-free.

Lemma 11.2: Assume that a CROPN for an AMS contains only one strongly connected subnet U formed by multiple PPCs, and there is at least one BT for each PPC in U. If, at marking M, the subnet U is capacitated, there exists a sequence of transition firings such that the subnet U can be emptied.

Proof: If, in marking M, there are transitions in T_U and $T_O(U)$ that are enabled, and the firing of these transitions empties U, the lemma holds. If there is no transition in T_U and

$T_O(U)$ that are enabled at M, or after firing some transitions in T_U and $T_O(U)$ that are enabled at M, no transition in T_U and $T_O(U)$ is enabled, then in both cases, a deadlock occurs in U if robots cannot serve as buffers. As discussed above, if there is such a deadlock, it must occur in some PPCs. Thus, there is at least one PPC, say v, in U that is deadlocked under such an assumption. Because deadlock is a circular wait, every $t \in T(v)$ must be process-enabled, but not resource-enabled. This implies that v is process-enabled. By assumption, there is at least a BT on v, say t. Then, according to Lemma 11.1, by the R-firing of t, all the transitions in $T(v)$ can fire once in a limited time. The firing of transitions in $T(v)$ may lead to the enabling of some transitions in T_U and $T_O(U)$. Then, these transitions can fire. If these firings lead to the empty state of U, the lemma holds. If, after these firings, a deadlock in the sense that robots cannot serve as buffers in U occurs again, the previous firing process can be repeated. In this way, all the jobs in the subnet U can be completed and U can be emptied in a limited time. ■

Lemma 11.2 is the generalization of Lemma 11.1. A CROPN for an AMS with robots as MHS may contain more than one strongly connected subnet. Because these subnets do not interact with each other, we can deal with them one by one. Therefore, we have the following corollary.

Corollary 11.1: Assume that a CROPN for an AMS with robots as MHS contains more than one strongly connected subnet, and for each subnet the conditions given in Lemma 11.2 are satisfied. Then if, at a marking M, each subnet U_i is capacitated, there exists a sequence of transition firings such that each subnet U_i can be emptied.

Now we can develop a deadlock control policy for AMS with robots. Because of the limitation of the action area for the robots in the system, not every PPC in the corresponding CROPN has a BT. For example, $v_3 = \{p_1, t_{12}, p_2, t_{21}, p_1\}$ in Figure 11.3 has no BT. Let Π denote a strongly connected subnet formed by PPCs that have no BT, and T_Π and P_Π denote the sets of transitions and places in Π.

Definition 11.4: Job constraint (JC): At any marking M, there is at most one place in P_Π, say p_i, such that $M(p_i) = K(p_i)$, and for any other places $M(p_j) < K(p_j)$ with $j \neq i$.

Note that JC does not restrict any places that are not in any Π. In other words, those not in Π can be capacitated.

Lemma 11.3: If a CROPN for an AMS contains a strongly connected subnet Π and the JC condition is satisfied, there exists a sequence of transition firings that empties Π.

Proof: If some transitions in $T_O(\Pi)$ are enabled, the firing of any of these transitions does not violate JC. If no transition in $T_O(\Pi)$ is enabled and there is a place $p_i \in P_\Pi$ such that $M(p_i) = K(p_i)$, at least one of the output transitions of p_i, say t_i, must be enabled. The firing of t_i moves a token from p_i to $p_j \in P_\Pi$. Because $M(p_j) < K(p_j)$ before firing t_i, after its firing, JC is still satisfied. If no transition in $T_O(\Pi)$ is enabled and no place in P_Π is capacitated, the firing of any transition in T_Π will not violate JC. Thus, in this way, all the jobs in Π can be completed in a limited time and Π can be emptied. ■

Similarly, from Lemma 11.3, we know that if there is more than one strongly connected subnet Π in a CROPN, the system is deadlock-free if JC is met.

Definition 11.5: R-firing rule (RR): When a deadlock occurs in the sense of a production process without taking robot activity into account, R-fire a BT t of a PPC if t is process-enabled.

Combining Definitions 11.4 and 11.5, we present the following control policy.

Definition 11.6: The control policy JC$\wedge$RR is called robots-as-temporary-buffer policy (RTB-policy) for AMS with robots as MHS.

With the RTB-policy defined, we present the main result of this chapter.

Theorem 11.1: A CROPN of an AMS with robots as MHS is deadlock-free if the RTB-policy is applied.

Proof: We use U to denote the strongly connected subnet formed by PPCs with each PPC having a BT, Π to denote the subnet in which no PPC has BT, and Φ to denote the subnet in which some PPCs have a BT, but some do not. If the CROPN contains only U-subnets, it follows from Lemma 11.2 and Corollary 11.1 that it is deadlock-free. If it contains only Π-subnets, it follows from Lemma 11.3 that the CROPN is deadlock-free. It is clear that if the CROPN contains both U-subnets and Π-subnets, but they do not interact with each other, the CROPN is also deadlock-free.

Now we show that when the CROPN contains a Φ-subnet, it is still deadlock-free. By RTB-policy, if a transition in the CROPN is enabled in the sense of a production process without taking robot activity into account and JC, it can fire spontaneously. To show the CROPN is deadlock-free, we need only show that if deadlock occurs in the Φ-subnet in the sense of a production process without taking robot activity into account and JC, the deadlock can be resolved by RR. We first show that when there is a Π-subnet in Φ-subnet. When a deadlock occurs in Φ, no transition in Φ is enabled. Hence, no transition in Π is enabled. At this marking, if $p_i \in P_{\Pi}$ and $M(p_i) = K(p_i)$, then all the tokens in p_i must process-enable transitions in $T_O(\Pi)$. This implies that some tokens should go into Π, and at the same time, some tokens should go out of Π. However, these transitions are not resource-enabled. Thus, there are three cases: (1) there are PPCs out of Π that are process-enabled, (2) there are process-enabled PPCs with one place $p \in P_{\Pi}$ on each such PPC; and (3) the tokens in $p_i \in P_{\Pi}$ with $M(p_i) = K(p_i)$ process-enable transitions in $T_O(\Pi) \cap T_{\Phi}$, and, at the same time, transition $t_k \in T_{\Phi} - T_{\Pi}$ and $t_k^{\bullet} = p_j \in P_{\Pi}$ when T_{Φ} is the set of transitions in Φ.

Case 1: By RR, we can R-fire a BT for the process-enabled PPC as done in Lemma 11.2, and this does not violate RTB-policy.

Case 2: We assume that $p_d \in P_{\Pi}$ is on the process-enabled PPC. By R-firing a BT for this PPC, we first move a token from p_d, and then put a token into p_d. This implies that the RTB-policy is not violated either.

Case 3: This time, if we merge the Π-subnet into a place p, then there is a process-enabled PPC with p in it. Then we can R-fire a BT for this PPC such that a token is first removed from $p_i \in P_{\Pi}$, and then a token is put into $p_j \in P_{\Pi}$, and the RTB-policy is not violated.

Therefore, for any case we can make the transitions fire again. In this way, the tokens in Φ can be emptied.

If there are two or more Π-subnets in Φ, which do not interact with each other by definition, we can deal with them separately. Thus, tokens in such Φ can also be emptied. ∎

Similarly, it is easy to show that a CROPN containing two or more Φ-subnets that do not interact with each other is deadlock-free if the RTB-policy is applied. It follows from the proof of Theorem 11.1 that the RTB-policy is more permissive than the existing methods for AMS with robots as MHS, since in RTB-policy, robots are taken as temporary buffers, allowing the otherwise deadlocked production process without taking robot activity into account to be deadlock-free.

To implement the control policy we should (1) identify the set T_R; (2) identify all PPCs, Φ-subnets, and Π-subnets; (3) check JC; and (4) identify BTs for RR.

Cases 1 and 3: It is trivial.

Case 2: Since the number of PPCs in a ROPN grows exponentially, it is a problem in general cases. However, this is done off-line. More importantly, all the real FMS have a quite limited number of PPCs and can be handled using today's computing power to the best knowledge of the authors. After all the PPCs and BTs are identified, we obtain the Φ-subnets and Π-subnets.

Case 4: It is to identify the PPCs that are process-enabled, and it must be done on-line. Because there may be a large number of PPCs, we cannot check the PPCs one by one. We can simply check the BTs one by one. Note that the number of BTs is limited by the number of transitions. To check if a BT t is qualified for RR, we can do the following:

a. In a deadlocked Φ-subnet (in the sense of the CROPN after the removal of all robot places and JC), remove all the transitions that are not process-enabled.

b. Assume $^\bullet t = p_i$ and $t^\bullet = p_j$; find a direct path from p_j to p_i.

c. If such a path exists, t can R-fire; otherwise, it cannot. A polynomial complexity algorithm for finding such a path can be used (Kocay and Kreher, 2005).

Thus, the on-line control policy is computationally efficient.

It follows from the above discussion that all potential deadlocks caused by competing for robots in the PN models in Ezpeleta et al. (1995) and Li and Zhou (2004) are completely prevented by using a CROPN. It should also be noticed that a deadlock in a production process without taking robot activity into account is also a deadlock in the models in Ezpeleta et al. (1995) and Li and Zhou (2004). Such deadlocks can be prevented by restricting jobs released into the system in Ezpeleta et al. (1995) and Li and Zhou (2004). However, by our policy, many such deadlocks can be avoided by using the robots as temporary buffers, instead of restricting the number of jobs in the system. Thus, the proposed policy is therefore significantly more permissive than those in Ezpeleta et al. (1995) and Li and Zhou (2004).

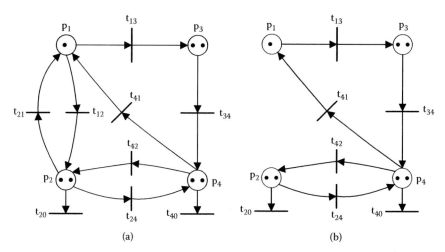

FIGURE 11.4 Illustration for the example: case 1.

11.4   ILLUSTRATIVE EXAMPLE

In this section, a simple example is used to show the application of the proposed policy.

Example 11.1 (continued)

Case 1: Assume that the capacity for all the machines is 2, and at a marking M, an A-part is in p_1, two B-parts are in p_2, an A-part and C-part are in p_3, and an A-part and B-part with its first operation performed are in p_4. The strongly connected subnet Φ at marking M is shown in Figure 11.4a.

At this marking, the CROPN is deadlocked in the sense of a production process without considering robot activity and JC. We can resolve it by RR. According to the RTB-policy, all the transitions in T_Π should be removed for checking the qualification of the BTs. As a result, t_{21} and t_{12} are removed. The other transitions in T_Φ are all process-enabled and cannot be removed. The remaining net is shown in Figure 11.4b. For t_{13}, a path $p_3 \rightarrow t_{34} \rightarrow p_4 \rightarrow t_{41} \rightarrow p_1$ from p_3 to p_1 is found. Thus, it is qualified to be a BT. For t_{42} and t_{24}, paths $p_2 \rightarrow t_{24} \rightarrow p_4$ and $p_4 \rightarrow t_{42} \rightarrow p_2$ are found, and thus are qualified to be BTs. By R-firing t_{13}, the B-part in p_4 moves into p_1, the C-part in p_3 moves into p_4, and the A-part in p_1 moves into p_3. Then the C-part in p_4 can fire t_{40} and go out of the subnet. Next, a B-part in p_2 goes into p_4 by firing t_{24}, and then goes out of the subnet. In this way, all the parts in the subnet can be completed.

Case 2: Assume that the capacity for all the machines is 1, and at a marking M, p_1 is empty, and there is an A-part in p_3, a B-part in p_2, and a B-part with its first operation performed in p_4. At this marking, besides t_{12} and t_{21}, transitions t_{13} and t_{42} are not process-enabled, and thus t_{12}, t_{21}, t_{13}, and t_{42} are removed. In this case, the only BT is t_{24}, and we cannot find a path from p_4 to p_2. Thus, according to the method discussed in Section 11.3, we merge p_1 and p_2 in Figure 11.5a into a single place, and the merged place is named p. The resulting net is shown in Figure 11.5b. Then we find a path

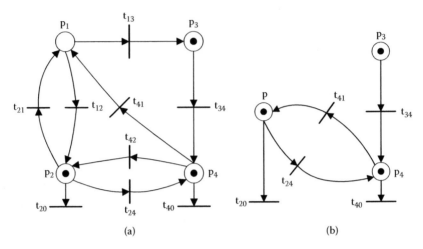

(a)   (b)

FIGURE 11.5 Illustration for the example: case 2.

$p_4 \rightarrow t_{41} \rightarrow p$ from p_4 to p. By R-firing t_{24}, the B-part is held by r_3, and thus the B-part can be moved into p_1 by firing t_{41} with r_2 without violating JC. When p_4 is emptied, r_3 drops the holding B-part into p_4, and then the B-part in p_4 fires t_{40} and goes out of the subnet, and the deadlock is resolved.

11.5   SUMMARY

An AMS contains a number of versatile machines (or workstations), buffers, and automated material handling systems (MHS). The MHS in an AMS can be an AGV system, or a system that is composed of a number of robots. For the AMS with an AGV system as its MHS, the problems of deadlock resolution for part processes and the AGV system can be solved either as two different problems or as an integrated system. As a result, a great number of techniques are presented. It is shown that AGVs can serve as both material handling devices and central buffers. Because deadlock is caused by the limitation of buffer spaces, treating AGVs as central buffers helps resolve deadlock problems.

For AMS with robots as MHS, the existing work treats the robots as material handling devices and shows that the robots are a kind of resource that contribute to deadlock. Here, we model the AMS with robots as MHS by using a CROPN. Contrary to the existing work, this shows that the robots yield no contribution to deadlock, and instead can help resolve deadlock by treating them as temporary buffers. Based on this model, a new deadlock control policy for such AMS is proposed and named the RBT-policy. It is more permissive than the existing ones. In addition, R-firing as a new concept is introduced into resource-oriented Petri nets. The essential difference between R-firing and regular transition firing is that while regular firing requires a free space representing a machine or a buffer, R-firing treats an available robot as a free space, which makes the policy more permissive. Finding a tractable maximally permissive control policy for AMS with robots as MHS remains an open problem.

REFERENCES

Ezpeleta, J., J. M. Colom, and J. Martinez. 1995. A Petri net based deadlock prevention policy for flexible manufacturing systems. *IEEE Transactions on Robotics and Automation* 11:171–84.

Kocay, W., and D. L. Kreher. 2005. *Graphs, algorithms and optimization.* Boca Raton, FL: Chapman & Hall/CRC.

Li, Z. W., and M. C. Zhou. 2004. Elementary siphons of Petri nets and their application to deadlock prevention in flexible manufacturing systems. *IEEE Transactions on Systems, Man, and Cybernetics A* 34:38–51.

Li, Z. W., and M. C. Zhou. 2008. Control of elementary and dependent siphons in Petri nets and their applications. *IEEE Transactions on Systems, Man, and Cybernetics A* 38:133–48.

Li, Z. W., M. C. Zhou, and N. Q. Wu. 2008. A Survey and comparison of Petri net-based deadlock prevention policy for flexible manufacturing systems. *IEEE Transactions on Systems, Man, and Cybernetics C* 38:173–88.

Wu, N. Q., and M. C. Zhou. 2007. Deadlock resolution in automated manufacturing systems with robots. *IEEE Transactions on Automation Science and Engineering* 4:474–80.

12 Control of Semiconductor Manufacturing Systems

Because of a growing need for a high level of automation, flexibility, and system utilization in semiconductor fabrication, there has been a trend toward the integration of process modules into a single equipment and the integration of several pieces of equipment into a more complex integrated system. Cluster tools and photolithography equipment (track system) are typical examples of integrated equipment used in semiconductor fabrication. Due to the high-throughput requirement and high-level concurrency in performing activities, one needs an effective methodology for their analysis and control. This chapter discusses some methods for their analysis, deadlock control, and scheduling by using a resource-oriented Petri net (ROPN).

12.1 MODELING, ANALYSIS, AND CONTROL OF CLUSTER TOOLS

Cluster tools provide a flexible, reconfigurable, and efficient environment for several manufacturing processes (Bader et al., 1990; Burggraaf, 1995). The benefit of using cluster tools includes improved yield and throughput, reduced contamination, better utilization of the floor space, and reduced human intervention (Singer, 1995). A cluster tool consists of several single-wafer processing modules (PMs) and a transportation module (TM) based on a wafer handling robot. Cluster tools have been increasingly used for etching, photolithography, chemical vapor deposition (CVD) processes, and even test processes. Some advanced cluster tools, for instance, some CVD cluster tools, can handle diverse wafer flow patterns concurrently. There is also a strict wafer residency time constraint such that a processed wafer at a PM should leave the PM's chamber within a specified time limit in order to prevent quality deterioration due to residual gases and heat. Thus, it is complicated to model and analyze the behavior of a cluster tool.

For the complex and concurrent systems, two types of properties are concerned: qualitative and quantitative. The former includes conflict, deadlock, blocking, and so on. Improper conflict resolution may disable the operation of the whole system. Thus, it is important to model the system and analyze its behavior to see whether it is conflict-free. When conflicts are detected, it is necessary to synthesize a conflict-free model to prevent them, or find a control policy to avoid them. The quantitative property concerned is temporal performance. Thus, a timed model should be built to evaluate the temporal performance for the system.

Traditionally, quantitative performance for cluster tools is analyzed by using timing diagrams that represent typical sequences of events (Perkinson et al., 1994, 1996; Venkatesh et al., 1997). However, as pointed out by Zuberek (2001), such an approach is highly dependent on the analyzed cluster tool and its properties, and it becomes quite complicated for tools that are complex. Furthermore, it can analyze the quantitative properties but not the qualitative ones. A timed Petri net (PN) is presented in Zuberek (2001) to analyze the performance of cluster tools. This chapter uses a PN to analyze the performance of cluster tools.

12.1.1 Cluster Tools

A cluster tool has one or two loadlocks (LLs), a number of PMs, and a single-blade or dual-blade robot, as shown in Figure 12.1. When a batch of wafers (e.g., twenty-five wafers) arrives, it is loaded into one of the LLs, which is then pumped down to a vacuum, and the wafers are thus ready to be processed. Then the robot performs

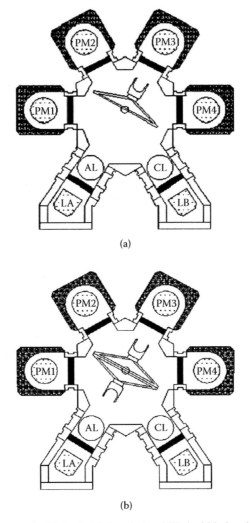

(a)

(b)

FIGURE 12.1 Cluster tools: (a) single-blade robot and (b) dual-blade robot.

loading, unloading, and movement of wafers for processing at PMs. A PM processes one wafer at a time within its chamber by a chemical process that is controlled by the control program for a recipe. When the processes of a wafer are completed, it comes back to the LL. After all the wafers in a batch are completed, the batch of the wafers is unloaded from the LL. If there are two LLs, while a batch of wafers in one of the LLs is being processed by the cluster tool, another batch of wafers can be loaded into the other LL. Thus, after one batch of wafers in an LL is completed, the wafers in the other LL can be processed continuously. Thus, at any time there are wafers to be processed.

A manufacturing process consists of a number of process operations, which together form a route. If a route has operations requiring different PMs, it is nonrevisiting. Otherwise, it is a revisiting route. To process a wafer, all operations in a route are assumed to be performed once. A PM may contain more than one chamber; thus, more than one wafer can be processed in such a PM at a time. Also, more than one PM can be configured in parallel to perform an identical operation to balance the workload. In this section, we deal with only one chamber PM and nonparallel situation. Note that a single-chamber and a nonparallel processing time that are equivalent to that achieved when performing an operation in the multichamber and parallel situations can be obtained (Perkinson et al., 1996).

A transportation module (TM) or a robot transfers wafers from one module to another. It performs three kinds of actions: picks wafers from a source module, rotates or moves wafers to a source or destination module, and places wafers into a destination module. The two blades of a dual-blade robot point in opposite directions, and both are able to carry a wafer. The two blades are tightly coupled by construction so that at any time only one of the two blades can pick or place a wafer from or into a module (Venkatesh et al., 1997).

12.1.2 ANALYSIS BY TIMED MG

A marked graph (MG) is a choice-free PN. If the sequence of activities for a process is deterministic, the process can be described by an MG. Then time can be associated with transitions or places, or both. For a given configuration of a cluster tool and operation mode, if the sequence of activities is deterministic, it can be modeled by an MG. Here a method for modeling and analysis of cluster tools is presented based on the work of Zuberek, 2001.

Consider a cluster tool with a single LL, a single-blade robot, and three PMs. Because the model for the steady-state behavior is the simplest one, we discuss it first. In steady state, every PM is processing one wafer (under the single-chamber PM's assumption), and the sequence of activities must be as follows: robot picks up the completed wafer at PM3 $\rightarrow$ moves to LL $\rightarrow$ drops the wafer into LL $\rightarrow$ moves to PM2 $\rightarrow$ picks up a completed wafer at PM2 $\rightarrow$ moves to PM3 $\rightarrow$ loads the wafer into PM3 $\rightarrow$ moves to PM1$\rightarrow$ picks up a completed wafer $\rightarrow$ moves to PM2 $\rightarrow$ loads the wafer into PM2 $\rightarrow$ moves to LL $\rightarrow$ picks up a wafer from LL $\rightarrow$ moves to PM1 $\rightarrow$ loads the wafer into PM1 $\rightarrow$ moves to PM3, waiting for the completion of the wafer in PM3. Then these activities are repeated, or these activities form an operation cycle. With this deterministic sequence of activities, a PN is obtained as shown in Figure 12.2 and Table 2.1. It can be seen that it is an MG.

In this PN, a transition represents an operation. Thus, it is straightforward to associate the time needed for an operation with the corresponding transition. Let $f(t)$

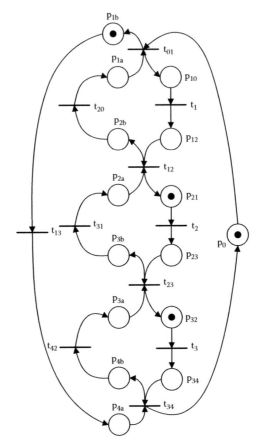

FIGURE 12.2 Marked graph for steady-state behavior of a cluster tool with a single-blade robot.

TABLE 12.1
Interpretation of Transitions in Figure 12.2

Transition	Operation
t_1	PM1 processes a wafer
t_2	PM2 processes a wafer
t_3	PM3 processes a wafer
t_{01}	Robot picks up a wafer from LL, moves to PM1, and loads the wafer into PM1
t_{13}	Robot moves to PM3
t_{34}	Robot picks up a wafer from PM3, moves to LL, and drops it into LL
t_{42}	Robot moves to PM2
t_{23}	Robot picks up a wafer from PM2, moves to the PM3, and loads it into PM3
t_{31}	Robot moves to PM1
t_{12}	Robot picks up a wafer from PM1, moves to the PM2, and loads it into PM2
t_{20}	Robot moves to LL

denote the time associated with transition t. In Figure 12.2, because p_0 is irrelative to the system cycle time, it and its related arcs can be removed from the net. Hence, only four circuits are identified: $v_1 = \{t_{01}, p_{10}, t_1, p_{12}, t_{12}, p_{2b}, t_{20}, p_{1a}, t_{01}\}$, $v_2 = \{t_{12}, p_{21}, t_2, p_{23}, t_{23}, p_{3b}, t_{31}, p_{2a}, t_{12}\}$, $v_3 = \{t_{23}, p_{32}, t_3, p_{34}, t_{34}, p_{4b}, t_{42}, p_{3a}, t_{23}\}$, and $v_4 = \{t_{01}, p_{1b}, t_{13}, p_{4a}, t_{34}, p_{4b}, t_{42}, p_{3a}, t_{23}, p_{3b}, t_{31}, p_{2a}, t_{12}, p_{2a}, t_{20}, p_{1a}, t_{01}\}$. According to the theory of timed PN, the cycle times for the circuits are $\tau_1 = f(t_1) + f(t_{01}) + f(t_{12}) + f(t_{20})$, $\tau_2 = f(t_2) + f(t_{12}) + f(t_{23}) + f(t_{31})$, $\tau_3 = f(t_3) + f(t_{23}) + f(t_{34}) + f(t_{42})$, and $\tau_4 = f(t_{01}) + f(t_{12}) + f(t_{23}) + f(t_{34}) + f(t_{13}) + f(t_{20}) + f(t_{31}) + f(t_{42})$. Then the cycle time for the system is $\tau_0 = \max(\tau_1, \tau_2, \tau_3, \tau_4)$. If $\tau_0 = \tau_4$, it is transport-bound; otherwise, it is process-bound.

Initially, when wafers are loaded into LL, PMs are idle. There is a transient process for the cluster tool to reach the steady state. However, the PN shown in Figure 12.2 cannot describe the initial transient behavior because the sequence of activities in the transient process is different from that in steady state. For the cluster tool to reach its steady state, the sequence of activities must be as follows: robot moves to LL → picks up a wafer from LL → moves to PM1 and loads the wafer into PM1, waiting for the completion of the wafer → picks up the wafer from PM1 → moves to PM2 and loads the wafer into PM2 → moves to LL and picks up a wafer from LL → moves to PM1 and loads the wafer into PM1 → moves to PM2 → picks up the wafer in PM2 → moves to PM3 and loads the wafer into PM3 → moves to PM1 and picks up the wafer in PM1 → moves to PM2 and loads the wafer into PM2 → moves to LL and picks up a wafer from LL → moves to PM1 and loads the wafer into PM1. Then the system is in the steady state. The PN shown in Figure 12.3 describes this process. However, this PN is not an MG; it cannot obtain the transient time by simply identifying circuits. In fact, the time needed for the transient process can be calculated by the operations given above.

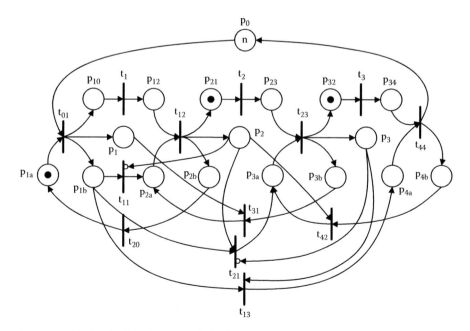

FIGURE 12.3 PN for initial transient behavior.

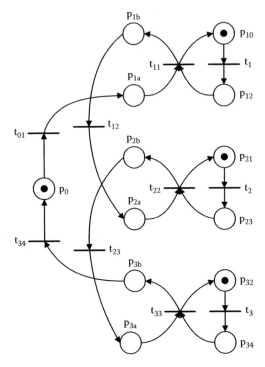

FIGURE 12.4 The PN for steady-state behavior of a cluster tool with a dual-blade robot.

Because there is only one LL, the wafer processing can be done only by batches. Thus, there exists a final transient process when there are three wafers remaining for processing in a batch. The PN in Figure 12.2 cannot be used to analyze the final transient behavior either, and a PN for the final transient process is needed. A PN similar to that for the initial transient process can be obtained.

Now consider a cluster tool with three PMs and a dual-blade robot in the steady state; the sequence of activities is different from that of the tool with a single-blade robot. For a cluster tool with a dual-blade robot, the sequence of activities is as follows: robot rotates, picks up a wafer from LL, and moves to PM1 $\rightarrow$ picks up the wafer in PM1 with one arm, rotates, and loads the wafer in the other arm into PM1 $\rightarrow$ moves to PM2 $\rightarrow$ picks up the wafer in PM2 with one arm, rotates, and loads the wafer in the other arm into PM2 $\rightarrow$ moves to PM3 $\rightarrow$ picks up the wafer in PM3 with one arm, rotates, and loads the wafer in the other arm into PM3 $\rightarrow$ moves to LL, rotates, and drops the wafer into LL. Then the activities are repeated with the same sequence. With this sequence of activities, an MG is obtained and is shown in Figure 12.4. The interpretation of transitions for Figure 12.4 is given in Table 12.2. In Figure 12.4, four circuits are identified: $v_1 = \{p_{10}, t_1, p_{12}, t_{11}, p_{10}\}$, $v_2 = \{p_{21}, t_2, p_{23}, t_{22}, p_{21}\}$, $v_3 = \{p_{32}, t_3, p_{34}, t_{33}, p_{32}\}$, and $v_4 = \{p_{1a}, t_{11}, p_{1b}, t_{12}, p_{2a}, t_{22}, p_{2b}, t_{23}, p_{3a}, t_{33}, p_{3b}, t_{34}, p_{p0}, t_{t01}\}$. Thus, we have $\tau_1 = f(t_1) + f(t_{11})$, $\tau_2 = f(t_2) + f(t_{22})$, $\tau_3 = f(t_3) + f(t_{33})$, and $\tau_4 = f(t_{01}) + f(t_{11}) + f(t_{12}) + f(t_{22}) + f(t_{23}) + f(t_{33}) + f(t_{34})$. The cycle time is $\tau_0 = \max(\tau_1, \tau_2, \tau_3, \tau_4)$. The sequence of

TABLE 12.2

Interpretation of Transitions in Figure 12.4

Transition	Operation
t_1	PM1 processes a wafer
t_2	PM2 processes a wafer
t_3	PM3 processes a wafer
t_{01}	Robot rotates, picks up a wafer from LL, and moves to PM1
t_{11}	Robot rotates, picks up the wafer in PM1, and loads it in the arm into PM1
t_{12}	Robot moves to PM2
t_{22}	Robot rotates, picks up the wafer in PM2, and loads it in the arm into PM2
t_{23}	Robot moves to PM2
t_{33}	Robot rotates, picks up the wafer in PM3, and loads it in the arm into PM3
t_{34}	Robot moves to the LL, rotates, and drops the wafer into LL

activities for initial and final transient processes can be analyzed, and PNs similar to those for the case of a single-blade robot cluster tool can be obtained.

By this method, the PN can only describe a process with a deterministic sequence of activities. To analyze the behavior of cluster tools, different models are needed for different processes, i.e., steady state, initial transient processes, and final transient process. Also, different models are needed for different configurations of cluster tools.

12.1.3 MODELING CLUSTER TOOLS BY CROPN

The behavior of cluster tools can also be modeled by CROPN. With a single CROPN model, quantitative and qualitative properties can be analyzed. Next, we present methods for modeling, analysis, and control of cluster tools by using CROPN.

In the CROPN modeling, PMs are treated as H-resources. Because a PM can process one wafer at a time, it is a single-capacity resource. Thus, a PM is modeled by a place p just as that in Chapter 5 for an H-resource with $K(p) = 1$.

A TM or robot is treated as a G-resource. Since a TM in a cluster tool may have a single blade or two blades, the model is different from that presented in Chapter 5 for a G-resource. We model the TM similar to that in Chapter 10 for the AGVs, and it is shown in Figure 12.5. In Figure 12.5, places r_1 and r_2 are used as resource places for the TM, and the capacity of either one is 1, or $K(r_1) = K(r_2) = 1$. If TM is a single-blade robot at the idle state, then there is a token in r_1. If it is a dual-blade robot at the idle state, then there is a token in both r_1 and r_2, respectively. A token in r_1 represents that a TM arm is ready to pick a wafer. Because r_1 can hold only one token, only one arm can pick a wafer at any time. Only when the token in r_1 is taken away can transition z fire, if there is a token in r_2. Its firing leads the token to r_1, and thus the other arm can pick a wafer. In this way, the action sequence performed by TM is correctly modeled, and at the same time, this model is applicable to a single- or dual-blade robot.

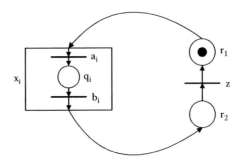

FIGURE 12.5 The PN model for TM.

To integrate the TM model into the CROPN, we define a macro-transition x_i, denoted by a box, to model the behavior of performing the actions of picking wafers, movement or rotation, and placing a wafer by TM. It contains two transitions, a_i and b_i, and a place, q_i. Transition a_i represents picking a wafer from a source module, and b_i placing a wafer into a destination module. A token in place q_i represents an arm of TM holding a wafer. Because an arm of the robot can hold only one wafer at a time, the capacity of q_i is also 1, or $K(q_i) = 1$.

Now the remaining module to be modeled is the LLs. To model them, the process of loading and unloading of wafer batches to and from a cluster tool should be modeled simultaneously. They are modeled by two places, p_{L1} and p_{L2}, for holding the raw wafers and the completed wafers, respectively. $K(p_{L1}) = K(p_{L2})$ is k and $2k$ for one and two LLs, respectively, where k is the batch size (twenty-five wafers). Place p_0 represents the load/unload station, and z_I and z_O represent wafer batch loading and unloading to and from the cluster tool. We assume, without loss of generality, that $K(p_0) = \infty$. The model is shown in Figure 12.6, where p represents the manufacturing process of wafers in a tool. If there are two LLs, $K(p_{L1}) =$

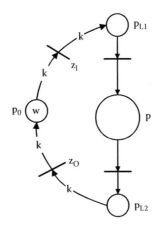

FIGURE 12.6 The PN model for LL.

$K(p_{L2}) = 2k$, after k wafers are completed and unloaded, i.e., z_O fires according to the behavior of the model, z_I can fire or k wafers can be loaded into the cluster tool. Thus, the manufacturing process can be done continuously. However, if there is only one LL, $K(p_{L1}) = K(p_{L2}) = k$, after k wafers are completed, p_{L1} is emptied. A batch of wafers can be loaded into the cluster tool (firing z_I) only after firing z_O; thus, there is an interruption between two batches of wafers. Therefore, it is meaningful to analyze the initial and final transient behaviors only for one LL case, but it is negligible for two LL cases.

With PN models for PM, TM, and LL, we can model the production processes of cluster tools. Suppose that all the wafers in a batch go through the same route {LL, PM_1, PM_2, ..., PM_k, LL}. As above, p_{L1} and p_{L2} represent the beginning and ending places for the LLs, respectively, and we use p_i to represent PM_i. Then put a macro-transition x_{01} between p_{L1} and p_1, a macro-transition x_i between p_i and p_{i+1}, and a macro-transition x_{k0} between p_k and p_{L2}.

For example, the PNs for $Q_1 = \{LL, PM_1, PM_2, PM_3, PM_4, LL\}$ and $Q_2 = \{LL, PM_1, PM_2, PM_3, PM_2, PM_3, PM_4, LL\}$ for a cluster tool with a two-blade robot and two LLs are shown in Figures 12.7 and 12.8, respectively. Q_1 is nonrevisiting,

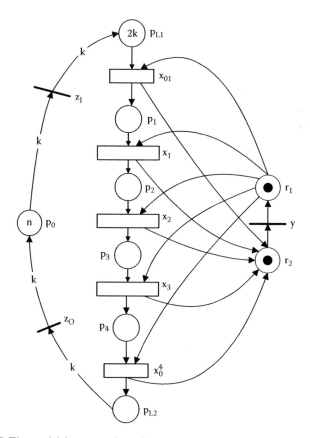

FIGURE 12.7 The model for a manufacturing process without revisiting.

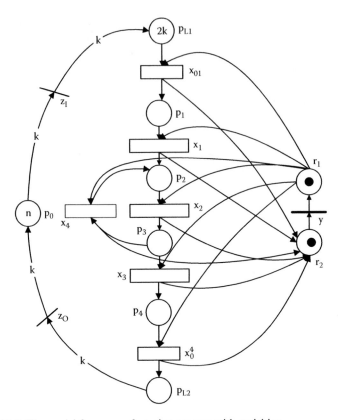

FIGURE 12.8 The model for a manufacturing process with revisiting.

but Q_2 is revisiting. The PN models shown in Figures 12.7 and 12.8 give the structure of the model; colors for the model can be defined based on the route of wafers to be processed according to the modeling method of ROPN. Because all the wafers are processed in the same route, colors for this model can be defined straightforwardly.

In these models, macro-transitions are used, and the enabling and firing rules for ordinary CROPN cannot be directly applied. Thus, we should define new enabling and firing rules. When there is no token in a place representing a PM, we say there is a space in it. Otherwise, if a place representing a PM has a token, we say it is full. When a place p representing a PM is full, while an arm is idle, this arm can be used to pick up the wafer in p, and the space in p is then released. The idle arm can thus be seen as a potential available space for p. Let $\chi(p) = [K(p) - M(p)] + \max[(M(r_1) + M(r_2) - 1), 0]$ denote the potential available spaces for p in M. Further, assume that x_2 is a macro-transition, $p \in {}^\bullet x_2$, and $x_1 \in {}^\bullet p$, and let $\delta(p) = M(p) + M(q_1)$, where q_1 is a place in Figure 12.5 corresponding to macro-transition x_1. We can define the following execution rules:

Definition 12.1: The enabling and firing rules are

1. An ordinary transition (i.e., nonmacro ones) in marking M is said to be enabled if it is both process and resource-enabled.
2. A macro-transition x in M is said to be enabled if one of the following conditions is satisfied:
   a. $M(p) \geq 1, \forall p \in {}^\bullet x$, and $\chi(p') \geq 1, \forall p' \in x^\bullet$
   b. $M(p) \geq 1$ and $\delta(p) \geq 2, \forall p \in {}^\bullet x$ that models a PM
3. When a macro-transition x_i is enabled, a_i fires first, then b_i.

Condition 1 is identical to the conditions for ordinary finite capacity PN. For a macro-transition, item a means that if its output place is full, but both arms of the dual-blade robot carry no wafer, then it is firable. Note that p_{L2} will never block the firing of a macro-transition, for it will never be full when there is a wafer in process. Item *b* says that when an arm of the dual-blade robot carries a wafer for a destination place p, then the output macro-transition x of p is enabled, no matter whether the output place of x is full or not. When the TM is a single-blade robot, condition 2 is reduced to condition 1. It is easy to check that by Definition 12.1, the behavior is correctly modeled. In this way, the behavior of the robot is exactly modeled. Assume that we take the box of the macro-transition away and model the behavior of the robot by two transitions and a place, then when an arm of the robot is free and there is a token in the input place of a_i, a_i can fire. This can result in some unnecessary conflict. However, if we model the behavior of the robot by an ordinary transition, then the implication is that the two arms cannot carry wafers simultaneously for a two-blade robot.

To analyze the quantitative properties of cluster tools, information time is necessary. It should be introduced into a CROPN model. To do so, the time duration may be associated with places (Lee and DiCesare, 1994), transitions (Zuberek, 2001; Marsan et al., 1984) and/or arcs (Zhou and Venkatesh, 1998). Here we associate the time duration with both transitions and places, leading to colored timed resource-oriented PN (CTROPN). We use $\tau(\bullet)$ to denote the time duration associated with a transition or a place, both of which are shown in Table 12.3. We assume, for the sake of simplicity, that the time needed for picking a wafer from an LL and a PM, and that for placing a wafer into an LL and a PM are the same. $\tau(q_i)$ contains the time for the robot to move, rotate, and wait. We use τ_m and τ_{rotate} to denote the time needed for moving between two adjacent modules and the time for rotating, respectively. We also assume that the time for the robot to move to the other module is $2\tau_m$. Note that in the processing, there may be revisiting. Thus, the time duration associated with a place representing a PM is dependent on the actual operation.

Observing the CTROPN developed for the cluster tools, we can make the following remarks:

1. CTROPN is very compact and concise. By CTROPN we model each PM by only one place, the robot by two places, and the LLs by two places; with colors introduced, the manufacturing processes can be well modeled. Thus, it is powerful in modeling.

TABLE 12.3

The Time Durations Associated with Transitions and Places

Symbol	Transition or Place	Actions	Time Duration
$\tau(z_I)$	Z_I	Put a batch into the loadlock and make it ready for processing	τ_{bload}
$\tau(z_O)$	Z_O	Unload a batch from the loadlock	$\tau_{bunload}$
$\tau(a_{01})$	a_{01}	Pick a wafer from the loadlock	τ_{pick}
$\tau(a_i), \tau(a_{k0})$	a_i, a_{k0}	Unload a wafer from a PM	τ_{pick}
$\tau(b_{01}), \tau(b_i)$	b_{01}, b_i	Load a wafer into a PM	τ_{place}
$\tau(b_{k0})$	b_{k0}	Place a wafer into the loadlock	τ_{place}
$\tau(y)$	y	Time delay between two arms to pick wafers	τ_{robot}
$\tau(p_{L1}), \tau(p_{L2})$	p_{L1}, p_{L2}		0
$\tau(p_i(op_i))$	p_i	The ith operation is processed in PM_i	τ_i
$\tau(q_i)$	q_i	An arm holding a wafer is in movement or waiting	
$\tau(r_1), \tau(r_2)$	r_1, r_2		0

2. CTROPN is generic since a cluster tool always consists of a number of PMs and a TM. The model structure is affected only by the number of process steps and revisit patterns.

3. It is easy to see that the CTROPN can exhibit the initial transient, steady-state, and final transient behaviors, and the differences among them are distinguished simply by the states (markings) of the model. This is due to the model structure and enabling rules.

4. CTROPN allows one to explore the flexibility of the event sequences. For example, for a dual-blade robot cluster tool, one may require the robot to pick a wafer from the loadlock while there is a wafer in PM_1 being processed, move to PM_1, and then do a swap operation at PM_1. It may be done another way. The robot may keep idle at PM_1 before the wafer in PM_1 is completed. After that, the robot unloads the wafer in PM_1 and loads it into PM_2. Then, the robot picks a wafer from the loadlock and loads it into PM_1. In this way, one arm of the robot can be used as a temporary buffer to satisfy residency constraints (Rostami et al., 2001).

5. The CTROPN can be used to analyze both qualitative and quantitative properties. We will do the analysis next.

12.1.4 ANALYSIS OF THE SINGLE-BLADE ROBOT CLUSTER TOOL

With the CTROPN for the cluster tools, we can now analyze their behavior, both qualitatively and quantitatively. In this section, we analyze the behavior of cluster tools with a single-blade robot. Among the properties we are most concerned with are whether the system is conflict-free for qualitative properties and with throughput for quantitative properties. Thus, based on the CTROPN, we only do the deadlock and throughput analysis.

12.1.4.1 Deadlock Analysis

First we examine the CTROPN in Figure 12.7. If the token in r_2 is taken away, then this model describes the manufacturing process without revisiting for single-blade robot cluster tools. By the modeling process of the CTROPN, the enabling and firing rules, we know that this net is conservative, neither token is destroyed, and none is created. At the same time, p_{L2} can hold all the tokens from p_{L1}, and only if k tokens in p_{L2} are taken away can k tokens be moved into p_{L1} again. Thus, p_{L2} will never block the manufacturing process, or the net is deadlock-free. Hence the system is deadlock-free.

Now we come to Figure 12.8. If the token in r_2 is taken away, then this model describes the manufacturing process with revisiting for single-blade robot cluster tools. There is a circuit $\{p_2, x_2, p_3, x_4, p_2\}$ in Figure 12.8. Assume that when a marking M is reached such that there are tokens in both p_1 and p_3, the token in p_3 represents the execution of the third operation, and next it will visit p_2. According to the enabling rules, at this marking, x_1 can fire and a token is moved into p_2. The net is now deadlocked. Thus, the system is not deadlock-free.

To overcome this problem, the CTROPN should be controlled. This can be done by introducing a control place c into the net such that c is the input place of transition x_1. Define $u = [K(p_2) + K(p_3)] - [M(p_2) + M(p_3)(d_4) + 1]$, where d_4 represents the color of transition x_4. Put u tokens in c at marking M, and require that only if $u \geq 1$, then x_1 can fire. The controlled net is shown in Figure 12.9. Notice that when the deadlock

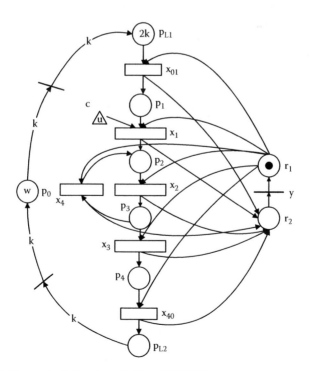

FIGURE 12.9 The controlled and conflict-free CTROPN.

situation above occurs, $u = 0$, or x_1 cannot fire and the net will not be deadlocked. It can be shown that this control can successfully avoid deadlock.

Proposition 12.1: The controlled CTROPN shown in Figure 12.9 is deadlock-free.

Proof: We assume that there is a token in each of the places p_1, p_3, and p_4, with the token in p_3 being for operation 3, and the robot is free. If we show that it is deadlock-free by starting from this marking, then the net is deadlock-free. To do this, we need only to find a firing sequence such that after the firing of any transition, any transition can fire again. In this marking, $u = 2 - 2 = 0$, or x_1 is disabled. Thus, only x_{40} and x_4 are enabled. We fire x_{40}, and the token in p_4 is moved into p_{L2}. Then with u being observed, we fire the following transition sequence: $x_4 \rightarrow x_2 \rightarrow x_1 \rightarrow x_{01} \rightarrow x_3 \rightarrow x_2$, and the system comes back to the initial marking. This means that the controlled net is deadlock-free. ∎

Although this control rule makes the system deadlock-free, it is a restrictive policy. Thus, the utilization of the PMs and the productivity of the system are limited due to the use of a single-blade robot.

12.1.4.2 Throughput Analysis for the Process without Revisiting

We can calculate the wafer output period to obtain the throughput. To calculate the period, we can analyze the sequences of occurrence of the events and the time durations associated with these events. Based on the CTROPN and from the enabling rules, we can analyze the sequences of events to obtain the period. For readability, in the following analysis, when we mention an event, we just use the action corresponding to the event, other than its transition or place.

At the initial state, there are tokens in p_{L1} and no token in any other place. Thus, only x_{01} is enabled. The transition firing sequence must be $x_{01} \rightarrow x_1 \rightarrow x_{01} \rightarrow x_2 \rightarrow x_1 \rightarrow x_{01} \rightarrow x_3 \rightarrow x_2 \rightarrow x_1 \rightarrow x_{01}$. Then the system will operate in the steady state. After it enters the steady state, according to the enabling and firing rules, the only enabled transition is x_{40}, so it can fire. After firing x_{40}, the only enabled one is x_3. After firing x_3, we can fire only x_2, and so on. Thus, the firing sequence must be $x_{40} \rightarrow x_3 \rightarrow x_2 \rightarrow x_1 \rightarrow x_{01}$. Then, this firing process is repeated. This means that the firing sequence is determined by the model and the operation sequence; it has no choice. Thus, we do not need to do reachability analysis for that. The evolution of the markings is shown in Table 12.4, when there is one LL. Note that O_{ij} represents the ith wafer being processed for the jth operation.

In the steady state, a cluster tool may operate in a process-bound or transport-bound region. In the first region, the period must be bound by a bottleneck PM, while the period is bound by the robot to transport the wafers in the second region.

When the system is in its steady state, there must be a token (wafer) in every place p_i representing PM_i, as shown in Table 12.4. Thus, the robot must move to p_4, and when the wafer in p_4 is completed, the robot picks it up and moves to p_{L2}. After the robot places the wafer into p_{L2}, it moves to p_3 and picks the wafer in p_3, then moves to p_4 and places it into p_4. Then, while the wafer is being processed in p_4, the robot repeats the similar events. Thus, the sequence of events for a processing cycle must be as follows: processing the wafer in p_i (τ_i) $\rightarrow$ the robot picks the wafer in p_i (τ_{pick}) $\rightarrow$ moves to p_{i+1}

TABLE 12.4

The Evolution of Markings for One LL System without Revisiting

Marking	Operation in p_1	Operation in p_2	Operation in p_3	Operation in p_4
M_1	O_{11}			
M_2	O_{21}	O_{12}		
M_3	O_{31}	O_{22}	O_{13}	
M_4	O_{41}	O_{32}	O_{23}	O_{14}
M_5	O_{51}	O_{42}	O_{33}	O_{24}
•			•	
•			•	
•			•	
M_{25}	$O_{25,1}$	$O_{24,2}$	$O_{23,3}$	$O_{22,4}$
M_{26}		$O_{25,2}$	$O_{24,3}$	$O_{23,4}$
M_{27}			$O_{25,3}$	$O_{24,4}$
M_{28}				$O_{25,4}$

$(\tau_m) \rightarrow$ places the wafer into p_{i+1} $(\tau_{place}) \rightarrow$ moves to p_{i-1} $(2\tau_m) \rightarrow$ picks a wafer in p_{i-1} $(\tau_{pick}) \rightarrow$ moves to p_i $(\tau_m) \rightarrow$ places the wafer into p_i (τ_{place}). Thus, we have

$$T_{process} = \max_i(\tau_i) + \tau_{pick} + \tau_m + \tau_{place} + 2\tau_m + \tau_{pick} + \tau_m + \tau_{place}$$

$$= \max_i(\tau_i) + 2\tau_{pick} + 4\tau_m + 2\tau_{place} \tag{12.1}$$

For each process period, the robot must transport a wafer from p_4 to p_{L2}, p_3 to p_4, p_2 to p_3, p_1 to p_2, and p_{L1} to p_1, and then it goes back to p_4. Thus, the sequence of events for the transportation must be as follows: the robot picks the wafer in p_4 $(\tau_{pick}) \rightarrow$ moves to p_{L2} $(\tau_m) \rightarrow$ places the wafer into p_{L2} $(\tau_{place}) \rightarrow$ moves to p_3 $(2\tau_m) \rightarrow$ picks a wafer in p_3 $(\tau_{pick}) \rightarrow$ moves to p_4 $(\tau_m) \rightarrow$ places the wafer into p_4 $(\tau_{place}) \rightarrow$ moves to p_2 $(2\tau_m) \rightarrow$ picks a wafer in p_2 $(\tau_{pick}) \rightarrow$ moves to p_3 $(\tau_m) \rightarrow$ places the wafer into p_3 $(\tau_{place}) \rightarrow$ moves to p_1 $(2\tau_m) \rightarrow$ picks a wafer in p_1 $(\tau_{pick}) \rightarrow$ moves to p_2 $(\tau_m) \rightarrow$ places the wafer into p_2 $(\tau_{place}) \rightarrow$ moves to p_{L1} $(2\tau_m) \rightarrow$ picks a wafer in p_{L1} $(\tau_{pick}) \rightarrow$ moves to p_1 $(\tau_m) \rightarrow$ places the wafer into p_1 $(\tau_{place}) \rightarrow$ moves to p_4 $(3\tau_m)$. Thus, the period can be calculated as

$$T_{transport} = 5\tau_{pick} + 5\tau_{place} + 16\tau_m \tag{12.2}$$

If $T_{transport} > T_{process}$, then the cluster tool analyzed is transport-bound. Otherwise, it is process-bound.

When the system reaches marking M_{25}, the final transient-behavior begins and the firing sequence is $x_{40} \rightarrow x_3 \rightarrow x_2 \rightarrow x_1 \rightarrow x_{40} \rightarrow x_3 \rightarrow x_2 \rightarrow x_{40} \rightarrow x_3 \rightarrow x_{40}$, and it leads to the initial state.

12.1.4.3 Throughput Analysis of a Process with Revisiting

When there is revisiting in a process route, the situation is more complicated, since its uncontrolled system is deadlock-prone. Thus, to analyze the performance correctly, we must do the analysis based on the controlled model shown in Figure 12.9.

TABLE 12.5

The Evolution of Markings for One LL System with Revisiting (as reflected in Figure 12.9)

Marking	Operation in p_1	Operation in p_2	Operation in p_3	Operation in p_4
M_1	O_{11}			
M_2	O_{21}	O_{12}		
M_3	O_{21}		O_{13}	
M_4	O_{21}	O_{14}		
M_5	O_{31}	O_{22}	O_{15}	
M_6	O_{31}		O_{23}	O_{16}
M_7	O_{31}	O_{24}		O_{16}
M_8	O_{41}	O_{32}	O_{25}	O_{16}
M_9	O_{41}		O_{33}	O_{26}
M_{10}	O_{41}	O_{34}		O_{26}
M_{11}	O_{51}	O_{42}	O_{35}	O_{26}
$\vdots$		$\vdots$	$\vdots$	
M_{72}	$O_{25,1}$		$O_{24,3}$	$O_{23,6}$
M_{73}	$O_{25,1}$	$O_{24,4}$		$O_{23,6}$
M_{74}		$O_{25,2}$	$O_{24,5}$	$O_{23,6}$
M_{75}			$O_{25,3}$	$O_{24,6}$
M_{76}		$O_{25,4}$		$O_{24,6}$
M_{77}			$O_{25,5}$	$O_{24,6}$
M_{78}				$O_{25,6}$

We can analyze the initial and final transient behaviors in a way similar to that in the case without revisiting. Here we present the analysis of steady-state behavior. From Table 12.5 we know that markings M_6, M_7, and M_8 form a steady-state cycle. Because p_2 and p_3 form a circuit, to analyze the processing cycle, we must consider p_2 and p_3 together. Based on Figure 12.9, to complete the processing cycle in the circuit formed by p_2 and p_3, the sequence of events must be as follows: operation 3 being processed in p_3 (τ_3) $\rightarrow$ when it is completed the robot picks it up (τ_{pick}) $\rightarrow$ moves to p_2 (τ_m) $\rightarrow$ places the wafer into p_2 (τ_{place}) $\rightarrow$ the wafer is processed for operation 4 in p_2 (τ_4) $\rightarrow$ the robot picks the wafer in p_2 (τ_{pick}) $\rightarrow$ moves to p_3 (τ_m) $\rightarrow$ places the wafer into p_3 (τ_{place}) (and the wafer is being processed in p_3 for operation 5 (τ_5)) $\rightarrow$ moves to p_1 ($2\tau_m$) $\rightarrow$ picks a wafer in p_1 (τ_{pick}) $\rightarrow$ moves to p_2 (τ_m) $\rightarrow$ places the wafer into p_2 (τ_{place}) (and the wafer is being processed in p_2 for operation 2 (τ_2)) $\rightarrow$ moves to p_3 (τ_m) $\rightarrow$ picks the wafer in p_3 (τ_{pick}) $\rightarrow$ moves to p_4 (τ_m) $\rightarrow$ places the wafer into p_4 (τ_{place}) $\rightarrow$ moves to p_2 ($2\tau_m$) $\rightarrow$ picks the wafer in p_2 (τ_{pick}) $\rightarrow$ moves to p_3 (τ_m) $\rightarrow$ places the wafer into p_3 (τ_{place}). At this time, a period is completed in the circuit and a wafer is completed. Notice that this firing sequence is the only possible sequence, and it is easy to obtain.

Notice that at M_8 operations 2 and 5 are simultaneously processed. Thus, in the sequence of events, the robot picks the wafer in p_2 (τ_{pick}) $\rightarrow$ moves to p_3 (τ_m) $\rightarrow$ places the wafer into p_3 (τ_{place}) (and the wafer is being processed in p_3 for operation 5 (τ_5)) $\rightarrow$ moves to p_1 ($2\tau_m$) $\rightarrow$ picks a wafer in p_1 (τ_{pick}) $\rightarrow$ moves to p_2 (τ_m) $\rightarrow$ places the wafer into p_2 (τ_{place})

(and the wafer is being processed in p_2 for operation 2 (τ_2)) → moves to p_3 (τ_m) → picks the wafer in p_3 (τ_{pick}) → moves to p_4 (τ_m) → places the wafer into p_4 (τ_{place}) → moves to p_2 ($2\tau_m$), the time needed is dominated by the process of one of the operations, as given by Equation 12.1. Thus, consider the fact that only one wafer is completed; we have

$$T_{23} = \tau_3 + \tau_4 + \max(\tau_2, \tau_5) + 4\tau_{pick} + 6\tau_m + 4\tau_{place} \tag{12.3}$$

For places p_1 and p_4, only one wafer is completed during this cycle, so we have

$$T_1 = \tau_1 + 2\tau_{pick} + 4\tau_m + 2\tau_{place} \tag{12.4}$$

$$T_4 = \tau_4 + 2\tau_{pick} + 4\tau_m + 2\tau_{place} \tag{12.5}$$

Therefore, the processing period is calculated by

$$T_{process} = \max(T_1, T_4, T_{23}) \tag{12.6}$$

From Equation 12.6 we know that if τ_2, τ_3, τ_4, and τ_5 are not much less than τ_1 and τ_4, then T_{23} will be much greater than T_1 and T_4, or the throughput is reduced greatly because of the revisiting process.

In a similar way, we can determine the sequence of events for the transportation cycle: robot picks the wafer in p_4 (τ_{pick}) → moves to p_{L2} (τ_m) → places the wafer into p_{L2} (τ_{place}) → moves to p_3 ($2\tau_m$) → picks the wafer in p_3 (τ_{pick}) → moves to p_4 (τ_m) → places the wafer into p_4 (τ_{place}) → moves to p_2 ($2\tau_m$) → picks the wafer (operation 2) in p_2 (τ_{pick}) → moves to p_3 (τ_m) → places the wafer into p_3 (τ_{place}) → picks the wafer in p_3 (τ_{pick}) → moves to p_2 (τ_m) → places the wafer into p_2 (τ_{place}) → picks the wafer in p_2 (τ_{pick}) → moves to p_3 (τ_m) → places the wafer (operation 5) into p_3 (τ_{place}) → moves to p_1 ($2\tau_m$) → picks a wafer in p_1 (τ_{pick}) → moves to p_2 (τ_m) → places the wafer into p_2 (τ_{place}) → moves to p_{L1} ($2\tau_m$) → picks a wafer in p_{L1} (τ_{pick}) → moves to p_1 (τ_m) → places the wafer into p_1 (τ_{place}) → moves to p_4 ($3\tau_m$). Thus, we obtain

$$T_{transport} = 7\tau_{pick} + 18\tau_m + 7\tau_{place} \tag{12.7}$$

Again, if $T_{transport} > T_{process}$, then the cluster tool analyzed is transport bound. Otherwise, it is process-bound. Note that the above analysis is based on the cluster tool's CTROPN model. The results vary with such a model.

12.1.5 ANALYSIS OF DUAL-BLADE ROBOT CLUSTER TOOLS

Now we come to analyze the behavior of dual-blade robot cluster tools based on the CTROPN to see how a dual-blade robot affects the behavior of cluster tools.

12.1.5.1 Deadlock Analysis

We have seen that if there is no revisiting for the manufacturing of wafers, a single-blade cluster tool is conflict-free in its operation. However, if revisiting exists in the manufacturing process, deadlock is possible. For a dual-blade cluster tool, if there is no revisiting for the manufacturing of wafers, it is easy to check that it is deadlock-free. But is it still deadlock-free if revisiting exists for a dual-blade robot cluster tool? The answer is positive and we have the following proposition.

Proposition 12.2: The CTROPN shown in Figure 12.8 is deadlock-free.

Proof: It is known that deadlock is a circular waiting, and it must occur in some circuits. In the CTROPN shown in Figure 12.8, there indeed is a unique circuit $v = \{p_2, x_2, p_3, x_4, p_2\}$. If we show that deadlock will never occur in v, then the proposition holds. Assume that there is a token in every place p_i, $i = 1, 2, 3, 4$, with the token in p_2 being processed for operation 2 and the token in p_3 for operation 3. Thus, the next visit for the token in p_2 is p_3, and the next visit for the token in p_3 is p_2. This means that the tokens are already in circular wait. Assume that this time the two arms of the robot are all free, or there is a token in both r_1 and r_2. According to enabling rule 2a, at this moment, transition $x_1(a_1)$ can fire and a token is moved from p_1 into $x_1(q_1)$. Notice that according to the enabling rules, b_1 cannot fire at this time. However, according to 2b, some time later x_2 (a_2) can fire and the token in p_2 is removed (corresponding to a wafer's pickup at PM_2). Then the sequence of events can be as follows: b_1 fires and a wafer is placed in p_2 for operation 2 → a_4 fires and the wafer in p_3 is picked up → b_2 fires and a wafer is placed in p_3 for operation 3 → a_2 fires and the wafer in p_2 is picked up → b_4 fires and a wafer is placed into p_2 for operation 4 → a_4 fires and the wafer in p_3 is picked up → b_2 fires and a wafer is placed into p_3 for operation 3 → a_2 fires and the wafer in p_2 is picked up → b_4 fires and a wafer is placed into p_2 for operation 4 → a_4 fires and the wafer in p_3 is picked up → b_2 fires and a wafer is placed into p_3 for operation 5 → a_2 fires and the wafer in p_2 is picked up → b_4 fires and a wafer is placed into p_2 for operation 4 → a_3 fires and the wafer in p_3 is picked up → b_2 fires and a wafer is placed into p_3 for operation 5 → a_{40} fires and the wafer in p_4 is picked up → b_3 fires and a wafer is placed into p_4 → b_{40} fires and the wafer is placed into p_{L2}. Now the two arms are all free again and the firing process can be repeated. Hence, the net is deadlock-free. ∎

It can be shown that no matter what the processing pattern (including multiple revisiting) is, the corresponding CTROPN will be deadlock-free for dual-blade robot cluster tools. Thus, there is a significant improvement in behavior when using a dual-blade robot.

12.1.5.2 Throughput Analysis for the Process without Revisiting

For the dual-blade robot cluster tools, if there is no revisiting for the manufacturing process, the throughput can be evaluated based on Figure 12.7. In the steady state, the unique transition firing sequence must be $x_{01} \rightarrow x_1 \rightarrow x_2 \rightarrow x_3 \rightarrow x_{04}$, and this process is repeated. It can be shown that the marking evolution is the same as in Table 12.4. However, the robot works in a different way. Consider a place p_i representing a PM and assume that x_{i-1} and x_i are input and output transitions of p_i, respectively. While a wafer is being processed in p_i, x_{i-1} (a_{i-1}) can fire and the robot with one wafer carried moves to p_i. When the wafer in p_i is completed, x_i (a_i) can fire and the wafer in p_i is picked up (τ_{pick}). Then the robot rotates to point to p_i (τ_{rotate}), b_{i-1} fires, and the token in $x_{i-1}(q_{i-1})$ is placed in p_i (τ_{place}) to be processed. Thus, the processing period can be calculated in a straightforward manner by

$$T_{process} = Max_i(\tau_i) + \tau_{pick} + \tau_{rotate} + \tau_{place} \tag{12.8}$$

Compared with the processing period given by Equation 12.1 for a single-blade robot cluster tool, there is some improvement, for the PM does not need to wait for the robot to carry a wafer to it.

In the steady state, we can think that every PM is processing a wafer. Thus, the sequence of events for the transportation is as follows: The robot picks a wafer in p_{L1} (τ_{pick}) → moves to p_1 (τ_m) → picks the wafer in p_1 (τ_{pick}) → rotates (τ_{rotate}) → places the wafer into p_1 (τ_{place}) → moves to p_2 (τ_m) → picks the wafer in p_2 (τ_{pick}) → rotates (τ_{rotate}) → places the wafer into p_2 (τ_{place}) → moves to p_3 (τ_m) → picks the wafer in p_3 (τ_{pick}) → rotates (τ_{rotate}) → places the wafer into p_3 (τ_{place}) → moves to p_4 (τ_m) → picks the wafer in p_4 (τ_{pick}) → rotates (τ_{rotate}) → places the wafer into p_4 (τ_{place}) → moves to p_{L2} (τ_m) → places the wafer into p_{L2} (τ_{place}). Thus, we have

$$T_{transport} = 5\tau_{pick} + 4\tau_{rotate} + 5\tau_{place} + 5\tau_m \qquad (12.9)$$

Compared with Equation 12.2, the number of movements of the robot is much less. Therefore, it represents a significant improvement.

12.1.5.3 Throughput Analysis of Process with Revisiting

Now we analyze the throughput for the manufacturing process with revisiting based on the model shown in Figure 12.8.

As done for a single-blade robot cluster, we should consider p_2 and p_3 together for the circuit. By analyzing the sequence of events based on Figure 12.8, we can obtain the sequence of markings, and the evolution of markings is shown in Table 12.6, where → represents a wafer picked up from or placed into an LL. It is shown that markings M_8 to M_{13} form a cycle. Notice that in each marking M_i, two operations are processed in p_2 and p_3 simultaneously. Thus, the time needed can be calculated by Equation 12.8, as shown in Table 12.6, where $\tau = \tau_{pick} + \tau_{place} + \tau_{rotate}$.

It follows from Table 12.6 that the time for the cycle is $2Max(\tau_2, \tau_3) + 2Max(\tau_4, \tau_5) + Max(\tau_2, \tau_5) + Max(\tau_4, \tau_3) + 6(\tau_{pick} + \tau_{place} + \tau_{rotate})$, and three wafers are completed during that time. Thus, we have

$$T_{23} = [2Max(\tau_2, \tau_3) + 2Max(\tau_4, \tau_5) + Max(\tau_2, \tau_5)]/3 + 2(\tau_{pick} + \tau_{place} + \tau_{rotate}) \qquad (12.10)$$

$$T_1 = \tau_1 + \tau_{pick} + \tau_{rotate} + \tau_{place} \qquad (12.11)$$

$$T_4 = \tau_4 + \tau_{pick} + \tau_{rotate} + \tau_{place} \qquad (12.12)$$

$$T_{process} = Max(T_1, T_4, T_{23}) \qquad (12.13)$$

The sequence of events for transportation for a cycle must be the following: The robot picks a wafer in p_{L1} (τ_{pick}) → moves to p_1 (τ_m) → picks the wafer in p_1 (τ_{pick}) → rotates (τ_{rotate}) → places the wafer into p_1 (τ_{place}) → moves to p_2 (τ_m) → picks the wafer in p_2 (τ_{pick}) → rotates (τ_{rotate}) → places the wafer into p_2 (τ_{place}) → moves to p_3 (τ_m) → picks the wafer in p_3 (τ_{pick}) → rotates (τ_{rotate}) → places the wafer into p_3 (τ_{place}) → moves to p_2 (τ_m) → picks the wafer in p_2 (τ_{pick}) → rotates (τ_{rotate}) → places the wafer into p_2 (τ_{place}) → moves to p_3 (τ_m) → picks the wafer in p_3 (τ_{pick}) → rotates (τ_{rotate}) → places the wafer into p_3 (τ_{place}) → moves to p_4 (τ_m) → picks the wafer in p_4 (τ_{pick}) → rotates (τ_{rotate}) → places the wafer into p_4 (τ_{place}) → moves to p_{L2} (τ_m) → places the wafer into p_{L2} (τ_{place}). Thus, we have

$$T_{transport} = 7\tau_{pick} + 7\tau_m + 7\tau_{place} + 6\tau_{rotate} \qquad (12.14)$$

TABLE 12.6

The Evolution of Markings for One LL and Dual-Blade Robot System with Revisiting

Marking	Operation in p_1	Operation in p_2	Operation in p_3	Operation in p_4	Time Bound
M_1	$\rightarrow O_{11}$				
M_2	$\rightarrow O_{21}$	O_{12}			
M_3	$\rightarrow O_{31}$	O_{22}	O_{13}		
M_4	$\rightarrow O_{41}$	O_{32}	O_{23}		
M_5	O_{41}	O_{14}	O_{33}		
M_6	O_{41}	O_{24}	O_{15}		
M_7	O_{41}	O_{34}	O_{25}	$O_{16}\rightarrow$	
M_8	$\rightarrow O_{51}$	O_{42}	O_{35}	$O_{26}\rightarrow$	$\mathrm{Max}(\tau_2, \tau_5) + \tau$
M_9	$\rightarrow O_{61}$	O_{52}	O_{43}	O_{36}	$\mathrm{Max}(\tau_2, \tau_3) + \tau$
M_{10}	$\rightarrow O_{71}$	O_{62}	O_{53}	O_{36}	$\mathrm{Max}(\tau_2, \tau_3) + \tau$
M_{11}	O_{71}	O_{44}	O_{63}	O_{36}	$\mathrm{Max}(\tau_4, \tau_3) + \tau$
M_{12}	O_{71}	O_{54}	O_{45}	$O_{36}\rightarrow$	$\mathrm{Max}(\tau_4, \tau_5) + \tau$
M_{13}	O_{71}	O_{64}	O_{55}	$O_{46}\rightarrow$	$\mathrm{Max}(\tau_4, \tau_5) + \tau$
M_{14}	$\rightarrow O_{81}$	O_{72}	O_{65}	$O_{56}\rightarrow$	
$\vdots$	$\vdots$	$\vdots$	$\vdots$		
M_{45}	$\rightarrow O_{24.1}$	$O_{23.2}$	$O_{22.3}$	$O_{21.6}$	
M_{46}	$\rightarrow O_{25.1}$	$O_{24.2}$	$O_{23.3}$	$O_{21.6}$	
M_{47}	$O_{25.1}$	$O_{22.4}$	$O_{24.3}$	$O_{21.6}$	
M_{48}	$O_{25.1}$	$O_{23.4}$	$O_{22.5}$	$O_{21.6}\rightarrow$	
M_{49}	$O_{25.1}$	$O_{24.4}$	$O_{23.5}$	$O_{22.6}\rightarrow$	
M_{50}		$O_{25.2}$	$O_{24.5}$	$O_{23.6}\rightarrow$	
M_{51}			$O_{25.3}$	$O_{24.6}$	
M_{52}		$O_{25.4}$		$O_{24.6}$	
M_{53}			$O_{25.5}$	$O_{24.6}\rightarrow$	
M_{54}				$O_{25.6}\rightarrow$	

Compared with the single-blade robot case, T_{23} in Equation 12.10 is less than that in Equation 12.3. With the single-blade robot cluster tools, if there is revisiting, to avoid deadlocks, some PMs must be idle at times. However, the dual-blade robot cluster tools are always deadlock-free therefore, their productivity must increase. In the dual-blade robot cluster tools, movements of the robot are also reduced as shown by Equations 12.7 and 12.14.

The results given by Equations 12.1, 12.2, 12.8, and 12.9 are the same as those obtained in Zuberek (2001), but here it is not necessary for us to identify the invariants as done in Zuberek (2001). By our model, we only need to identify the markings that form a steady-state cycle, and the time required for the cycle is calculated straightforwardly from the sequence of the events that occur in the cycle.

12.2 DEADLOCK AVOIDANCE IN TRACK SYSTEM

To avoid deadlocks in flexible manufacturing systems (FMS), restrictive policies are applied. The main idea of restrictive policies is to keep a certain number of resources in the systems idle, such that there is no circular waiting. If there are buffers for machines in the system considered, then we can just let some buffers be empty and keep the machines working. Thus, this will not affect the production rate too much. However, for a single-capacity system (SCS), such as semiconductor track systems, there is no buffer at all between the operations. Therefore, to avoid deadlocks in such systems by restrictive policies, some machines must be kept idle all the time. This will significantly affect the productivity, and the more conservative the control policy is, the lower the production rate is. As mentioned before, the least restrictive (or maximally permissive) policy for SCS is intractable. Hence, there is a strong motivation to develop policies that are less conservative than some existing ones, for example, the one proposed in Lawley (1999). The goal here is to find a deadlock avoidance control policy for a semiconductor track system that is not too conservative and computationally efficient. The results are based on Wu and Zhou (2007a).

12.2.1 SEMICONDUCTOR TRACK SYSTEM

A track system with coaters and developers performing photolithography processes is among the core equipment in semiconductor fabrication. It is clustered equipment composed of transfer robots and process modules such as spin coaters, hot or cool plates and spin developers. An example semiconductor track system is adapted from Yoon and Lee (2001), as shown in Figure 12.10. It contains thirty-one process modules and four robots. The process modules are composed of cool plates (CPs), hot plates (HPs or HHPs), post exposure bake hot plates (PEBs), edge exposures (EEs), spin coaters (SCs), spin developers (SDs), and low-pressure adhesions (LPAHs). At one end of the track system there are four cassette indexers. The opposite end is linked to an expose tool known as a stepper. The track system is organized as three cells, with one robot for each cell. These three robots transfer wafers from module to module in each cell, or from cell to cell. The fourth robot moves wafers between the track system and the expose tool.

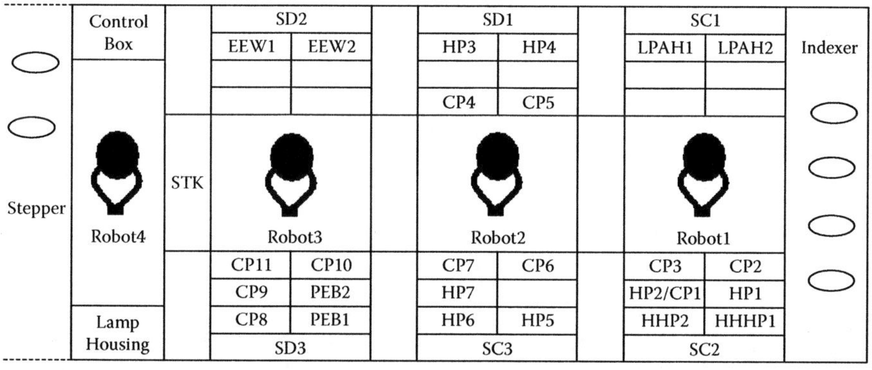

	Control Box		SD2			SD1			SC1		
			EEW1	EEW2		HP3	HP4		LPAH1	LPAH2	Indexer
						CP4	CP5				
Stepper	Robot4	STK	Robot3			Robot2			Robot1		
			CP11	CP10		CP7	CP6		CP3	CP2	
			CP9	PEB2		HP7			HP2/CP1	HP1	
	Lamp Housing		CP8	PEB1		HP6	HP5		HHP2	HHHP1	
			SD3			SC3			SC2		

FIGURE 12.10 The layout of a typical semiconductor track system.

Notice that although some machines have essentially identical capabilities, they can still be differentiated on the basis of their auxiliary equipment (Johri, 1993). Furthermore, even though two or more machines can perform the same set of operations, some may be more efficient than others with respect to certain operation subsets. Thus, each machine needs to be treated as a single one able to handle only one job at a time. This invalidates some deadlock avoidance policies, e.g., those developed under the assumption that a machine or cell is of multiple capacity (Lawley, 1999; Wu, 1999).

Processing flexibility is important in track systems. In such systems, there are two typical routes for deep-UV processing: top antireflective coating (TARC) and bottom antireflective coating (BARC). Table 12.7 shows an example of process routes for three wafer types. The first four and last seven steps are identical. While most operations have alternatives, some do not. Those operations with no alternatives make some policies too conservative, e.g., those in (Lawley, 1999).

As pointed out by Yoon and Lee (2001), the main characteristics of photoresist processes in a track system are that there is no reentrant flow for each wafer type. Thus, if only one type of wafer is processed in it at a time, the process flow is analogous to that in a flexible flow shop, generating no deadlock at all. However, because of the demand for increased levels of flexibility and the lengthy processing times of the underlying operations, it is required to conduct the concurrent processing of multiple wafer types (Reveliotis, 1999). Consequently, deadlock may occur.

TABLE 12.7
Process Flows of the Three Wafer Types

Operation	TARC 1	TARC 2	BARC
Op1	INDEXER	INDEXER	INDEXER
Op2	HHP1/HHP2	HHP1/HHP2	HHP1/HHP2
Op3	LPAH1/LPAH2	LPAH1/LPAH2	LPAH1/LPAH2
Op4	CP2/CP3	CP2/CP3	CP2/CP3
Op5	SC1/SC2	SC1/SC2	SC3
Op6	SC3	HP1/HP2	HP1/HP5
Op7	HP5/HP6	CP6/CP7	CP1/CP6
Op8	CP10/CP11	SC3	SC1/SC2
Op9	STEPPER	STEPPER	HP6/HP7
Op10	PEB1/PEB2	PEB1/PEB2	CP10/CP11
Op11	CP8/CP9	CP8/CP9	STEPPER
Op12	EEW1/EEW2	EEW1/EEW2	PEB1/PEB2
Op13	SD1/SD2/SD3	SD1/SD2/SD3	CP8/CP9
Op14	HP3/HP4	HP3/HP4	EEW1/EEW2
Op15	CP4/CP5	CP4/CP5	SD1/SD2/SD3
Op16	INDEXER	INDEXER	HP3/HP4
Op17			CP4/CP5
Op18			INDEXER

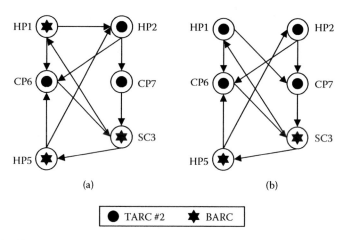

FIGURE 12.11 Deadlock situations in the track system.

Two deadlock states in the example track system are given in Figure 12.11. The routes for the two wafer types are as follows: TARC 2, HP1 or HP2 → CP6 or CP7 → SC3; and BARC, SC3 → HP1 or HP5 → CP6 or HP2. In Figure 12.11a and b, CP6 and HP2 are occupied by wafers of TARC 2 and SC3 is occupied by wafers of BARC. At the same time, no PM is idle. Thus, the process flow of wafer type TARC 2 is blocked by that of BARC, and vice versa. In fact, Figure 12.11a and b form a strongly connected graph. Hence, they are deadlock situations.

It should be pointed out that if the process flow of each wafer type is a flow shop in the strict sense, and there are opposite directions between different wafer types, then it is impossible to process these wafer types concurrently. It is the flexibility that makes the concurrent processing of multiple wafer types possible. Thus, if the system is well managed, high productivity and machine utilization can be achieved. This requires effective approaches that avoid deadlocks; otherwise, the concurrent processing of multiple wafer types will be disabled. The work of Yoon and Lee (2001) proposes an algorithm to identify the potential deadlock set for such a wafer flow in a track system.

12.2.2 MODELING BY ROPN

A wafer production process is essentially a part production process. The processing modules can be considered as manufacturing resources, just like machines in the general manufacturing system. As usual, they are treated as H-resources and each module is modeled by a place in ROPN. Because the wafers enter and are removed from the track system through the INDEXER, it is treated as a central storage modeled by p_0. It is reasonable to assume that the INDEXER has infinite capacity, or $K(p_0) = \infty$.

Based on the models for resources, the subnet models for TARC 1, TARC 2, and BARC of the example track system are obtained and shown in Figure 12.12. Each place corresponds to a resource, as shown in Table 12.8. Because of the one-to-one mapping between a resource and place, thereafter, we may refer to a place as a resource. To make the models more readable, we reduce a p to t arc, t, and

TABLE 12.8

Places Corresponding to the Resources in the PN Model

Resource	Place	Resource	Place
INDEXER	p_0	HHP1	p_1
HHP2	p_2	LPAH1	p_3
LPAH2	p_4	CP2	p_5
CP3	p_6	SC1	p_7
SC2	p_8	SC3	p_9
HP1	p_{10}	HP2	p_{11}
HP5	p_{12}	HP6	p_{13}
CP6	p_{14}	CP7	p_{15}
CP1	p_{16}	CP10	p_{17}
CP11	p_{18}	STEPPER	p_{19}
HP7	p_{20}	PEB1	p_{21}
PEB2	p_{22}	CP8	p_{23}
CP9	p_{24}	EEW1	p_{25}
EEW2	p_{26}	SD1	p_{27}
SD2	p_{28}	SD3	p_{29}
HP3	p_{30}	HP4	p_{31}
CP4	p_{32}	CP5	p_{33}

t to p' arc into a single arc from p to p', resulting in a P-net denoted by (P, A), where $A: P \times P \to N = \{0, 1\}$. Without loss of generality, we assume that initially there is no job in the modules, and there are enough jobs to be processed; i.e., in the initial state there are enough tokens in p_0 and all other paces are empty.

With the subnets we can obtain the union for the entire process according to the ROPN modeling method. The union of the three subnets shown in Figure 12.12 is given in Figure 12.13. For simplicity, we use p_x and p_y as macro-places to represent two subnets: one with p_{1-6} and another with p_{21-33}, where the same routings of three types are modeled. Colors can be defined based on the process route for each type of wafer leading to a CROPN.

12.2.3 DEADLOCK-FREE CONDITION FOR STRONGLY CONNECTED SUBNET

It follows from the results in Chapter 7 that deadlock must occur in production process circuits (PPCs). There are many PPCs in the CROPN of the overall track system as shown in Figure 12.13a. To mention a few, $v_1 = \{p_7, t_1, p_9, t_3, p_{12}, t_{27}, p_{14}, t_{30}, p_7\}$, $v_2 = \{p_8, t_2, p_9, t_3, p_{12}, t_{27}, p_{14}, t_{31}, p_8\}$, $v_3 = \{p_7, t_1, p_9, t_3, p_{12}, t_{26}, p_{16}, t_{28}, p_7\}$, $v_4 = \{p_8, t_2, p_9, t_3, p_{12}, t_{26}, p_{16}, t_{29}, p_8\}$, $v_5 = \{p_7, t_{11}, p_{10}, t_{16}, p_{15}, t_{20}, p_9, t_3, p_{12}, t_{26}, p_{16}, t_{28}, p_7\}$, $v_6 = \{p_7, t_{12}, p_{11}, t_{18}, p_{15}, t_{20}, p_9, t_3, p_{12}, t_{26}, p_{16}, t_{28}, p_7\}$, and $v_7 = \{p_9, t_3, p_{12}, t_{27}, p_{14}, t_{19}, p_9\}$. PPCs may be interactive, and interactive PPCs form strongly connected (SC) subnets. To avoid deadlock in a part production process, we should avoid deadlock in every strongly connected subnet. The deadlock avoidance control policies presented in Chapters 7 and 8 may be applied here. However, they are mainly suitable to multiple-capacity systems. Thus, a different control policy is needed here.

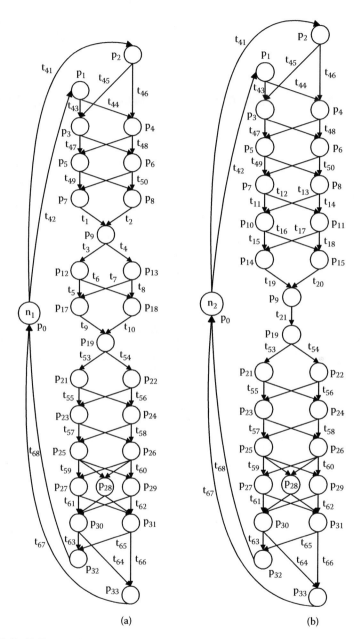

(a) (b)

FIGURE 12.12 P-nets of the process flows of three types of wafers.

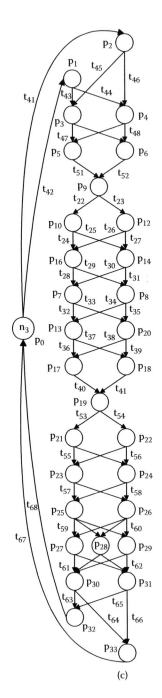

(c)

FIGURE 12.12 Continued.

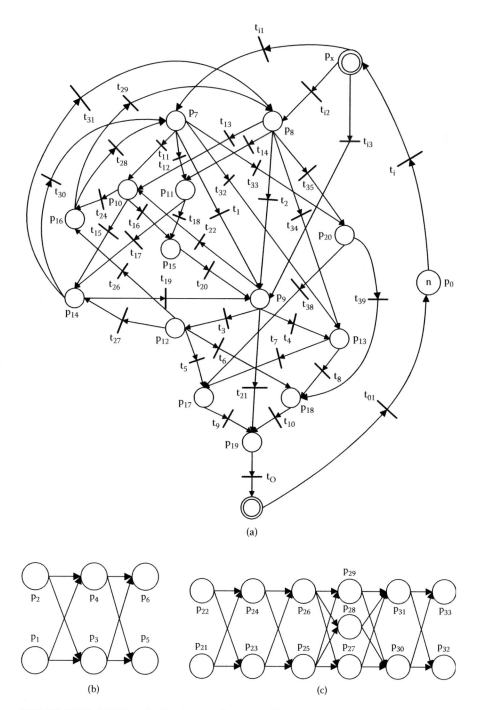

(a)

(b)

(c)

FIGURE 12.13 (a) PN model for the overall system, (b) p_x, and (c) p_y.

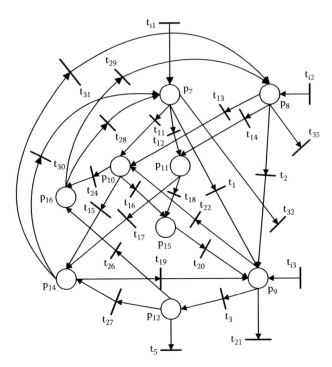

FIGURE 12.14 Strongly connected subnet obtained from the CROPN shown in Figure 12.13a.

In the CROPN of the example track system shown in Figure 12.13a, there is one strongly connected subnet, as shown in Figure 12.14. Besides the subnet, we also keep the SC's input and output transitions in Figure 12.14. However, if a place in the subnet has more than one input (output) transition, we merge them into one for conciseness.

Let $P(v^n)$ and $T(v^n)$ denote the sets of places and transitions in v^n, respectively. v^n's input transition set is denoted by $T_I(v^n) = \{t: {}^\bullet t \notin P(v^n) \text{ and } t^\bullet \in P(v^n)\}$, and output set by $T_O(v^n) = \{t: {}^\bullet t \in P(v^n) \text{ and } t^\bullet \notin P(v^n)\}$. We assume, in this section, that all transitions in $T_I(v^n)$ are process-enabled and all in $T_O(v^n)$ are resource-enabled. This implies that there are always wafers to be processed by the resources corresponding to $P(v^n)$, and when the operations of a wafer are completed in v^n, it can be removed. Moreover, let $\theta(p)$ denote a token in p, $\Psi(\theta(p))$ the set of colors of $\theta(p)$, and $\Psi(T_1)$ a set of colors corresponding to transitions in T_1.

Definition 12.2: If there is a subset $\Psi_1 \neq \varnothing$, $\Psi_1 \subseteq \Psi(\theta(p))$, $p \in P(v^n)$, and $\exists\, T_1 \subseteq T_O(v^n)$, $\ni \Psi_1 = \Psi(T_1)$, token $\theta(p)$ is a potential leaving token of v^n; otherwise, it is a cycling one.

A potential leaving token can be removed from v^n. For example, in Figure 12.14, if a token in p_9 representing TARC 1 can fire t_3 or t_{21}, we may fire t_{21} to force it to leave the subnet. Thus, it is a potential leaving token.

Assume that token $\theta(p)$ in p, $p \in P(v^n)$, represents the ith operation of a wafer being executed by resource p. The resources that can be used to perform the remaining operations after operation i are called downstream resources, denoted by set $R(\theta(p))$. To avoid deadlocks in v^n, we are greatly interested in $R(\theta(p))$ and $P(v^n)$.

We use $D(\theta(p)) = R(\theta(p)) \cap P(v^n)$ to denote the downstream places (resources) of token $\theta(p)$ in v^n. Assume that k operations (except the operation being performed in p) remain to be processed for $\theta(p)$, $p \in P(v^n)$, and are numbered by 1, 2, ..., and k. Let $V = \{1, 2, ..., k\}$ and $P_i = D_i(\theta(p)) \subseteq D(\theta(p))$, $i \in V$, be the set of places able to perform operation i. We use $H(\theta(p))$ to denote the set of the remaining processing paths of token $\theta(p)$. In marking M, token $\theta(p_i)$ is the token in p_i. If after firing some transitions it reaches p_j, we write $\theta(p_i \to p_j)$. Note that token $\theta(p_i \to p_j)$ is now in p_j.

Definition 12.3: Assume that $p_i \in P(v^n)$ and $p_j \in P(v^n)$, $p_j \in D(\theta(p_i))$ at marking M, and $P_i \subseteq D(\theta(p))$, $i \in V$, is the set of places representing the resources that are able to perform operation i. Token $\theta(p_j)$ is a consistent token of $\theta(p_i)$, if

1. Token $\theta(p_j)$ is a potential leaving token; or
2. Assume that there are n operations in $H(\theta(p_i \to p_j))$ and k operations in $H(\theta(p_j))$, respectively; then the following conditions are satisfied:
   a. $n \geq k$.
   b. $D_f(\theta(p_j)) = D_f(\theta(p_i \to p_j))$, for $f = 1, 2, ..., k$; in other words, $H(\theta(p_j)) \subseteq H(\theta(p_i \to p_j))$.
   c. $\exists\, p_d \in D_k(\theta(p_i))$ such that $\theta(p_j \to p_d)$ is a potential leaving token.

Otherwise, it is called an inconsistent token.

By saying that $\theta(p_j)$ is a consistent token of $\theta(p_i)$, we mean that tokens $\theta(p_j)$ and $\theta(p_i)$ go in similar directions and $\theta(p_j)$ goes first; token $\theta(p_i)$ can use the resources released by $\theta(p_j)$, but $\theta(p_j)$ will never require the resources released and currently occupied by $\theta(p_i)$. Thus, $\theta(p_j)$ will not block $\theta(p_i)$. To explain this, consider the simple sub-net in Figure 12.15 and the process routes in Table 12.9. In Figure 12.15a, $\theta(p_6)$ in $D(\theta(p_1)) = \{p_3, p_4, p_6\}$ is not a potential leaving token. Furthermore, since $H(\theta(p_6)) = \{p_1 \to p_2, p_1 \to p_3, p_5 \to p_2, p_5 \to p_3\} \not\subset H(\theta(p_1 \to p_6)) = \{p_3 \to p_6, p_4 \to p_6\}$, $\theta(p_6)$ is an inconsistent token of $\theta(p_1)$ at this marking. Similarly, $\theta(p_6)$ is an inconsistent token of $\theta(p_2)$. However, $\theta(p_3)$ is a consistent token of both $\theta(p_1)$ and $\theta(p_2)$ since it is a potential leaving token at this marking. In Figure 12.15b, $\theta(p_2)$ is a consistent token of $\theta(p_7)$, for

1. There are two operations in both $H(\theta(p_7 \to p_2)) = \{p_3 \to p_6, p_4 \to p_6\}$ and $H(\theta(p_2)) = \{p_3 \to p_6, p_4 \to p_6\}$.
2. $D_1(\theta(p_7 \to p_2)) = \{p_3, p_4\} = D_1(\theta(p_2))$ and $D_2(\theta(p_7 \to p_2)) = \{p_6\} = D_2(\theta(p_2))$, or $H(\theta(p_7 \to p_2)) = \{p_3 \to p_6, p_4 \to p_6\}$.
3. $\theta(p_2 \to p_6) \in D_2(\theta(p_2))$ is a potential leaving token.

Similarly, for token $\theta(p_3)$

1. There is a single operation in both $H(\theta(p_7 \to p_3))$ and $H(\theta(p_3))$.
2. $D_1(\theta(p_7 \to p_3)) = \{p_6\} = D_1(\theta(p_3))$.
3. $\theta(p_3 \to p_6) \in D_1(\theta(p_2))$ is a potential leaving token.

Hence, $\theta(p_3)$ is also a consistent token of $\theta(p_7)$. By Definition 12.3, a consistent token $\theta(p_j)$ of $\theta(p_i)$, $p_j \in D(\theta(p_i))$ can travel on any path in $H(\theta(p_i \to p_j))$ and finally leaves v^n in $H(\theta(p_i \to p_j))$. Thus, we have the following result:

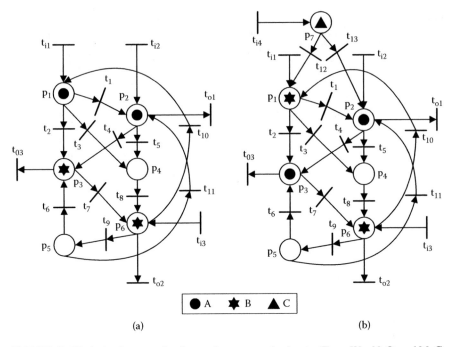

(a)   (b)

FIGURE 12.15 A simple example of strongly connected subnets. (From Wu, N. Q. and M. C. Zhou (2007). Real-time deadlock-free scheduling for semiconductor track systems based on colored-timed Petri nets, *OR Spectrum*, Vol. 29, No. 3, 421–443. With permission.)

Proposition 12.3: A consistent token of $\theta(p_i)$ will never block token $\theta(p_i)$.

Definition 12.4: A token $\theta(p)$ in v^n at marking M is said to be blocked if $\exists\, i \in V, \ni$ each place in $p_i \in D_i(\theta(p))$ is occupied by its inconsistent tokens.

For example, in Figure 12.15a, $\theta(p_6)$ is an inconsistent token of both $\theta(p_1)$ and $\theta(p_2)$. Thus, both $\theta(p_1)$ and $\theta(p_2)$ are blocked by $\theta(p_6)$. Similarly, in Figure 12.15b, $\theta(p_2)$ and $\theta(p_3)$ are blocked by $\theta(p_6)$. At the same time, $\theta(p_6)$ is blocked by $\theta(p_2)$ and $\theta(p_3)$. When deadlock occurs in v^n at a marking M, some tokens block each other and circular wait occurs. A token can be blocked by its inconsistent tokens, but not by its consistent ones.

Definition 12.5: A token $\theta(p)$ in v^n at marking M is said to be free if $\forall\, i \in V, \exists$ one place in $P_i = D_i(\theta(p)) \subseteq D(\theta(p))$ that is empty or has a consistent token.

TABLE 12.9
Process Routes for the Simple Example

Type	Operation 1	Operation 2	Operation 3	Operation 4
A	p_1/p_2	p_3/p_4	p_6	
B	p_6	p_1/p_5	p_2/p_3	
C	p_7	p_1/p_2	p_3/p_4	p_6

For example, $\theta(p_6)$ in Figure 12.15a is free because p_5 is empty and $\theta(p_3)$ is a consistent token.

Lemma 12.1: If token $\theta(p)$, $p \in P(v^n)$, is a free token in marking M, then there exists a firing order of transitions such that $\theta(p)$ can leave v^n.

Proof: Assume that k operations (except the operation being processed in p) remain to be processed for $\theta(p)$, $p \in P(v^n)$, and they are numbered by $1, 2, \ldots$, and k. Let $N_k = \{1, 2, \ldots, k\}$ and $P_i = D_i(\theta(p)) \subseteq D(\theta(p))$, $i \in N_k$, be the set of modules able to process operation i. Starting from the last operation k, $P_k = D_k(\theta(p)) \subseteq D(\theta(p))$ is the set of modules for processing operation k. By assumption, there is at least one place $p_d \in P_k$, such that p_d is empty or token $\theta(P_d)$ is a consistent token of $\theta(p)$. If p_d is empty, then we can check operation $k - 1$. If $\theta(p_d)$ is a consistent token, then by Definition 12.3, it must be a potential leaving token of subnet v^n. Thus, by firing a transition in $p_d{}^\bullet \cap T_O(v^n)$, $\theta(p_d)$ leaves v^n and the resource p_d is released. Now we check operation $k - 1$ and assume, without loss of generality, that $P_{k-1} = D_{k-1}(\theta(p)) \subseteq D(\theta(p))$ are all occupied and the consistent token $\theta(p_{d-1})$ with $p_{d-1} \in P_{k-1}$ is in place p_{d-1}. If it is a potential leaving token, let it release p_{d-1}. If not, by Definition 12.3, after the completion of operation $k - 1$, $\theta(p_{d-1})$ can be processed by any resource in P_k. Thus, we fire the transition from p_{d-1} to p_d, so p_{d-1} is released and $\theta(p_{d-1} \to p_d)$ becomes a potential leaving token and is allowed to leave the subnet. By doing so operation by operation, token $\theta(p)$ can finally leave subnet v^n. ∎

Definition 12.6: Assume that token $\theta(p)$ in v^n in marking M is blocked, and $P_i = D_i(\theta(p)) \subseteq D(\theta(p))$ is occupied by inconsistent tokens. If for any such P_i there exists a place $p_k \in P_i$ that is occupied by a free token, then $\theta(p)$ is said to be pseudo-blocked.

For example, in Figure 12.15a, in this marking both $\theta(p_1)$ and $\theta(p_2)$ are blocked by $\theta(p_6)$, but $\theta(p_6)$ is a free token. Thus, both $\theta(p_1)$ and $\theta(p_2)$ are pseudo-blocked. However, in Figure 12.15b, $\theta(p_2)$ and $\theta(p_3)$ are not pseudo-blocked because $\theta(p_6)$ is not a free token.

Theorem 12.1: A subnet v^n in the CROPN for a track system is live if in any marking M for any token $\theta(p)$, $p \in P(v^n)$, $\theta(p)$ is a free token or is pseudo-blocked.

Proof: By assumption, initially there is no token in the subnet and there are enough tokens to be released into the subnet. When tokens are released into the subnet and the condition is satisfied, it is sure that some tokens must be free tokens; otherwise, some token will be non-pseudo-blocked. Thus, from Lemma 12.1, there exists a firing order of transitions such that all the free tokens can leave the subnet. These firings only release resources previously occupied and do not make any token blocked. By Definition 12.6, the leaving of free tokens makes the pseudo-blocked tokens free. Therefore, finally all the tokens in v^n can be removed from the subnet and the subnet returns to the initial marking. Then by releasing different wafers, any transitions in the subnet can be fired. Thus, the subnet is live. ∎

A live PN is deadlock-free. Hence, the condition presented in Theorem 12.1 guarantees that a strongly connected subnet is deadlock-free. It should be pointed out that this condition is just a sufficient one. A necessary and sufficient condition for such systems remains unknown.

It should also be pointed out that the deadlock control policy proposed here takes advantage of routing flexibility. In fact, if there is no routing flexibility and two types of wafers flow in the system in opposite directions, to avoid deadlocks, only one type of wafer can be released into the system at a time. Because of the routing flexibility, by using the proposed policy we can carefully select different routes for different types of wafers based on the state of the system.

In Lawley (1999), the deadlock avoidance problem is studied for single-capacity systems (SCS) with routing flexibility, and a control policy is presented. The control policy is as follows. Assume that every processing stage of every part type has n processing alternatives. Then, the system is deadlock-free if one of the following conditions is satisfied: (1) at most n parts are allowed in the system at one time, or (2) at most $2n - 1$ parts are allowed in the system at one time, coupled with single-step look-ahead for SCs in the system. For the track system studied here, some operations can be processed by only one module, or $n = 1$. Thus, by the control policy in Lawley (1999), only one part is allowed in the system at one time. It is very conservative and unaccepted in practice. Notice that by Theorem 12.1, we do not make any restriction on the number of wafers in the system. Hence, the control policy presented here is much less conservative than the one in Lawley (1999). This can be shown by an illustrative example in the next section. A deadlock avoidance policy is also presented in Yoon and Lee (2004). However, it cannot avoid the second-level deadlocks, and if such a deadlock occurs, deadlock detection and recovery technique should be used. Fortunately, our proposed policy can avoid this type of deadlock as well.

12.2.4 IMPLEMENTATION OF THE DEADLOCK-FREE CONDITION

Based on the deadlock-free condition for a strongly connected subnet, we discuss the problem of implementing the deadlock-free condition for an overall semiconductor track system. We obtain the following control law:

Theorem 12.2: A strongly connected subnet v^n in the CROPN for a track system is deadlock-free if (1) initially there is no token in the set of place $P(v^n)$, and (2) $t \in T_I(v^n)$ and $t \in T(v^n)$ are control-enabled only if their firing yields a marking in subnet v^n that satisfies Theorem 12.1's condition.

By this law, we take the advantage of routing flexibility and let the wafers with the same direction go through the subnet on a path, and the other resources can be used by other wafers with a different direction. Thus, this control law does not restrict the number of wafer types. It avoids deadlocks by restricting paths for the wafers to go through. Therefore, it is flexible and effective.

Computational complexity is an important issue in implementing a deadlock control law. An effective control law must be computationally tractable. To implement the control law presented here, one needs to identify all strongly connected subnets in a CROPN. There exist efficient algorithms to find these subnets (Cormen et al., 1990). Furthermore, they can be found off-line once all types of wafers to be processed are known. Hence, this does not present any burden to implement the control law.

Next, we need to check whether firing a transition violates the condition in Theorem 12.1, and this must be done on-line. In other words, we need to check whether each token in v^n is free, pseudo-blocked, or non-pseudo-blocked. To do that for one token in $p \in P(v^n)$, we need to check every $p_i \in D(\theta(p))$ to see whether token $\theta(p_i)$ is a consistent token or an inconsistent token of $\theta(p)$. Let N be the upper bound of the number of operations of a wafer type to be processed in the subnet, and K is the upper bound of the number of possible options for each operation. To check each token, the complexity is $N \times K$. Assume that $|P(v^n)| = M$. Then the computational complexity is given by $O(M \times N \times K)$. This implies that this control law is computationally efficient, considering the relatively small K. For example, in the subnet shown in Figure 12.14, the routes for TARC 1, TARC 2, and BARC are $\{p_7/p_8, p_9\}$, $\{p_7/p_8, p_{10}/p_{11}, p_{14}/p_{15}, p_9\}$, and $\{p_9, p_{10}/p_{12}, p_{14}/p_{16}, p_7/p_8\}$, with the number of operations being two, four, and four, respectively. Thus, $N = 4$ and $K = 2$, for this example. From Figure 12.14 we know $M = 9$.

Notice that K is already known; we need to know M and N. However, we can identify the subnets off-line, or M can be obtained off-line. After identifying the subnets we can identify N for each subnet. This guarantees the effectiveness of the real-time control.

It follows from Chapter 7 that if each strongly connected subnet is live, the overall system is live. Thus, to avoid deadlock for the track system, we need to apply this control policy to each subnet only.

12.2.5 ILLUSTRATIVE EXAMPLE

We use the track system presented in Section 12.2.1 to show the application and power of the presented deadlock control law. The system is modeled by CROPN, as shown in Figure 12.13. As can be seen in Figure 12.14, there is only one strongly connected subnet. Thus, we only need to show the control process in that subnet.

Step 1: A wafer TARC 2 is released to place p_7, and at the same time a wafer BARC is released to p_9. At this time, $\theta(p_7)$ is pseudo-blocked, but $\theta(p_9)$ is a free token, or the condition is satisfied.

Step 2: $\theta(p_7)$ moves to p_{10} and $\theta(p_9)$ moves to p_{12}, and at the same time, a wafer TARC 2 is released to place p_8, and a wafer BARC is released to p_9. Then $\theta(p_9)$ and $\theta(p_{12})$ are free tokens, and $\theta(p_8)$ and $\theta(p_{10})$ are pseudo-blocked. Condition in Theorem 12.1 is satisfied.

Step 3: $\theta(p_{12})$ moves to p_{14}, $\theta(p_9)$ moves to p_{12}, $\theta(p_{10})$ moves to p_{15}, and $\theta(p_8)$ moves to p_{11}. At the same time, a wafer TARC 1 is released to place p_8, and a wafer BARC is released to p_9. Then $\theta(p_{14})$, $\theta(p_{12})$, and $\theta(p_9)$ are free tokens, and $\theta(p_8)$, $\theta(p_{15})$, and $\theta(p_{11})$ are pseudo-blocked.

Step 4: $\theta(p_{12})$ moves to p_{16}, $\theta(p_9)$ moves to p_{12}, and $\theta(p_8)$ moves to p_9. At the same time, a wafer TARC 1 and a wafer TARC 2 are released to p_7 and p_8, respectively. Then $\theta(p_7)$, $\theta(p_8)$, $\theta(p_{15})$, $\theta(p_9)$, and $\theta(p_{11})$ are free tokens, and $\theta(p_{12})$, $\theta(p_{14})$, and $\theta(p_{16})$ are pseudo-blocked.

Step 5: $\theta(p_9)$ leaves the subnet, $\theta(p_8)$ moves to p_{10}, and $\theta(p_7)$ moves to p_9. At the same time, two wafers TARC 1 are released to p_7 and p_8, respectively. Then $\theta(p_7)$, $\theta(p_8)$, $\theta(p_{15})$, $\theta(p_9)$, $\theta(p_{10})$, and $\theta(p_{11})$ are free tokens, and $\theta(p_{12})$, $\theta(p_{14})$, and $\theta(p_{16})$ are pseudo-blocked.

Step 6: At this time we can move all the tokens out of the subnet if necessary, and the subnet returns to the initial state.

It should be noticed that after step 5, all the places in the subnet are full of tokens, but the subnet is still live. By applying the control law, three types of wafers can be processed in the system simultaneously. It is easy to verify that if only one type of wafer is allowed to be processed at a time, some modules will be kept idle. This has shown that although the presented control law is based on a sufficient deadlock-free condition and computationally effective, it is not very conservative and is certainly acceptable for practical use. From this example, we also know that with the control policy, when there are tokens in the strongly connected subnet, the tokens can be removed from the subnet, or t_5, t_{21}, t_{32}, and t_{35} in $T_O(v^n)$ can fire. When the subnet is emptied, wafers of various types can be released to the subnet for processing, or t_{i1}, t_{i2}, and t_{i3} in $T_I(v^n)$ can fire. Therefore, the entire system is live.

12.3   DEADLOCK-FREE SCHEDULING OF A TRACK SYSTEM

To increase productivity and utilization in FMS, deadlock resolution and scheduling problems may be solved together. Although with a deadlock avoidance technique it is difficult to take the productivity into consideration, such a technique does provide an effective support to generate a deadlock-free schedule. Thus, to overcome the problem in scheduling a track system in semiconductor fabrication, we solve it in a hierarchical way. We use a deadlock avoidance technique as a lower level controller. It monitors the state of the system on-line and decides if an event that dispatches a job to a module is safe. This decision is sent to the upper layer, where the scheduling problem is solved. When a job is to be sent to a module for processing according to a scheduling algorithm, the deadlock controller checks if this event is safe. If so, it is sent to the module and otherwise disabled. The architecture for solving this problem

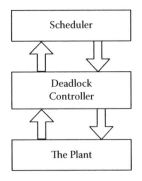

FIGURE 12.16 The architecture of scheduling a track system. (From Wu, N. Q. and M. C. Zhou (2007). Real-time deadlock-free scheduling for semiconductor track systems based on colored-timed Petri nets, *OR Spectrum*, Vol. 29, No. 3, 421–443. With permission.)

is shown in Figure 12.16. Based on the CROPN model and the deadlock avoidance control policy, some dispatching rules are used to schedule the system. The results here are based on Wu and Zhou (2007b).

In semiconductor fabrication, wafers are put in cassettes for transportation and to keep them free of dust. Thus, wafers are loaded into and unloaded from a track system in cassettes. Only when all the operations for the wafers in a cassette have been processed can the cassettes with the processed wafers be unloaded. However, this does not mean we can only process the wafers cassette by cassette. In fact, there is more than one cassette indexer (four in the example system shown in Figure 12.10). Hence, multiple cassettes can be loaded into a track system and wafers in these cassettes can be processed concurrently. This implies that the processing in a track system needs to be done continuously without interruption. Thus, we need to schedule such a repeated fabrication process with the objective to increase the system productivity. By a repeated fabrication process, we mean that the wafers can be processed cassette by cassette with the same number and types of wafers repeatedly. However, from the viewpoint of fabrication detail, it is not necessarily a repeated process, for the processing paths for the wafers in a cassette may not be able to follow exactly the ones for the wafers in the previous cassette.

12.3.1 Dispatching Rules

To develop scheduling rules for a track system, time information should be introduced into the CROPN. In Petri nets, time can be associated with places or transitions. In our CROPN, each place corresponds to a resource in a track system, and a token in a place represents a wafer being processed by a module. Thus, it is straightforward to associate the time with places in our model. Let wp_i denote wafer type i, and wp_{ij} the jth operation of wp_i-type wafers. Further, assume that module k is modeled by p_k in the model. $wp_{ij}(p_k)$ denotes wp_{ij} being processed on module k, and $f_{ij}(p_k)$ denotes the time needed for processing the jth operation of a wp_i-type wafer on module k.

Definition 12.7: If a token in p_k at marking M represents wp_{ij} being processed, we define the color of p_k as $C(p_k)(M) = w_{ij}$, and the time associated with p_k is $\tau(p_k)(M) = \tau(wp_{ij}(p_k)) = f_{ij}(p_k)$ at M.

To explain Definition 12.7, consider the marking shown in Figure 12.17. We use wp_1, wp_2, and wp_3 to denote wafer types TARC 1, TARC 2, and BARC, respectively, and assume that it takes eight and three time units for modules p_7 and p_{13} to process a TARC 1 wafer, two and six for p_9 and p_{14} to process a TARC 2 wafer, and five and four for p_7 and p_{16} to process a BARC wafer. Assume that the wafer types in the places at the marking shown in Figure 12.17 are as demonstrated in Table 12.10. Then, we obtain the colors and time associated with the places in this marking in Table 12.10.

With the model, we can discuss the scheduling problem. It is known from the above discussion that there are multiple wafer types to be processed in a track system concurrently with alternative modules for most of operations. Although there is no reentrant flow in processing the operations for each single wafer type, there are many operations for each type (fourteen, fourteen, and sixteen for TARC 1, TARC 2, and BARC, respectively). Thus, it is well known that the scheduling problem even for a

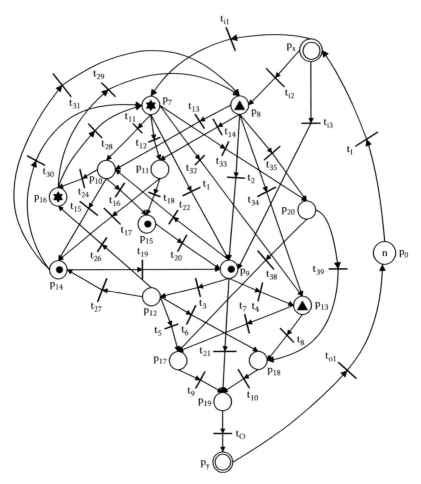

FIGURE 12.17 The CROPN model of a track system for illustrating time information.

TABLE 12.10

The Colors and Time Associated with the Places in the Marking Shown in Figure 12.17

Place	Wafer Type	$C(p_i)(M)$	$\tau(p_i)(M)$
p_7	BARC	w_3	5
p_8	TARC 1	w_1	8
p_9	TARC 2	w_2	2
p_{13}	TARC 1	w_1	3
p_{14}	TARC 2	w_2	6
p_{16}	BARC	w_3	4

known set of wafers to be processed is NP-complete. Of course, if we can schedule the production process such that after the processing of the wafers in one cassette, the production process for the wafers in the next cassette exactly follows that of the previous cassette, then this production process will be repeated for a set of wafers with a fixed ratio of different types and sequences. In fact, the production process is not repeated because of the alternative modules for processing the operations. For example, for the track system given in Section 12.2, when the production begins with all the modules idle, we can release two wafers to the system at time zero. However, when the last wafer in the cassette is processed by p_1 and another wafer is still being processed at p_2, we can release only one wafer from the next cassette to the system. Therefore, we are to schedule a system with continuous processing. Thus, we need a releasing rule for the problem.

Another important characteristic of a track system is that there are alterative modules for most of the operations, and some may be more efficient than others with respect to certain operation subsets. Thus, with wafers of different types to be processed in the system concurrently, operations from different wafer types may choose the most efficient module to be processed. Thus, the productivity can increase.

From the track system given in Section 12.2, we know that some modules can be used for multiple types of wafers, while others for only one type of wafer. Therefore, from this point of view, we can increase the productivity and utilization by processing multiple wafer types concurrently in the system.

With the above discussion, we present the scheduling rules as follows. Note that we use $N_x = \{1, 2, \ldots, x\}$ given a positive integer x. Also, we assume q, g, h, and d are positive integers in the following rules.

1. **Releasing rule:** Assume that there are q types of wafers to be processed in the track system, and in a cassette there are all the types, with the number of wafers being $n_1 - n_q$ to be processed, respectively. Let S be the common divisor of $n_1, n_2, \ldots,$ and n_q and $a_i = n_i/S$, $i \in N_q$. Assume that after some time, $b_1 - b_q$ wafers of types $1 - q$ remain in the system respectively. Let $\gamma_i = b_i/a_i$, $i \in N_q$ and $\gamma = \max\{\gamma_i, i \in N_q\}$. If $\gamma_i = \gamma$ and a module for the first operation of the wafers then a wafer of type i is released to the system.

2. **Dispatching rules:** Assume that $\theta(p_k)$ in p_k represents the jth operation of wp_i that is just completed by p_k at marking M, and there are h operations remaining to finish wp_i. Let $P_b = \{p_i, i \in N_q\} = D_b(\theta(p_k))$ and $f_b = \min\{f_{i(j+b)}(p_z), z \in N_q\}$, $b \in N_h$, and $f_i = \sum_{b \in N_h} f_b$ as the minimal remaining processing time.

   a. If token $\theta(p_k)$ in p_k represents the $(j-1)$th operation of wp_i that is just completed by p_k, and the jth operation can be processed by any one of the available modules $p_i, i \in N_d$ and $f = \min\{\tau(p_i)(M'), i \in N_d\} = \min\{f_{ij}(p_k), k \in N_d\}$. Then if $f = f_{ij}(p_y)$, this wafer will be sent to p_y for the jth operation.

   b. Assume that token $\theta(p_k)$ in p_k represents the $(j-1)$th operation of wp_i that is just completed by p_k, and token $\theta(p_c)$ in p_c represents the $(b-1)$th operation of wp_a that is just completed by p_k. Then the available module to process the next operation for both wafers is p_d and $f_{ij}(p_d) < f_{ab}(p_d)$, and the wafer just completed in p_k is sent to p_d for processing.

c. If there is a tie in situation 2, we send the wafer with maximal f_i to p_d for the processing of the next operation. If it is still a tie, we select the wafer with the lower index value.

The deadlock controller works as follows: If a transition fires and a token $\theta(p_k)$ in marking M is moved to p_j to become $\theta(p_j)$ according to rules 1, 2, and 3, such that M becomes M' and M' satisfies the control policy given by Theorem 12.2, then this firing is executed; otherwise, it is not.

With the releasing rule we keep the same mix of wafer types with respect to the number of wafers of different types, and this releasing process can be repeated one cassette after another. In this way, the processing time needed for the wafers in a cassette will not change too much from cassette to cassette. Thus, we keep a reasonable wafer mix in the system and output. With the dispatching rules, we attempt to process the wafers by their most efficient modules while keeping as many modules working as possible to maximize the concurrency.

By combining these rules with the deadlock avoidance policy presented in Section 12.2, we can keep the system free of deadlock. If a module fails and this failure does not cause a deadlock in that marking according to the control policy, the system can operate continuously.

12.3.2 ILLUSTRATIVE EXAMPLE

Assume that the three types of wafers are to be processed in the track system given in Section 12.2 according to the processing flows in Table 12.7 and processing times in Table 12.11. Each cassette contains twenty-five wafers, with the mix of ten TARC 1, ten TARC 2, and five BARC (Yoon and Lee, 2001).

By using the proposed deadlock avoidance policy, different types of wafers can be processed in the system concurrently. According to the releasing rule, we release the wafers as follows: one BARC, one TARC 1, one TARC 2, one TARC 1, and one TARC 2, and this releasing process is repeated until all the wafers are released. Then the dispatching rules and deadlock controller are used to complete all the jobs. It takes 126 time units to complete all the operations of twenty-five wafers in the cassette. Nineteen wafers have been observed in the system at time 65 under the proposed policy driven by the proposed rules.

Now consider other deadlock avoidance and scheduling approaches.

1. One way to keep the system free of deadlock is to process wafers type by type. Thus, the wafers are released into the system as five BARC, ten TARC 1, and ten TARC 2. Notice that in the processing routes the wafers of TARC 1 and TARC 2 always go in the same direction and can enter the strongly connected subnet concurrently. However, the last wafer of BARC must leave the strongly connected subnet before a wafer of TARC 1 enters it. In this way, it takes 147 time units to complete operations. Notice that if we continue to process wafers in another cassette, then all the wafers of TARC 1 and TARC 2 must leave the strongly connected subnet before the wafers of BARC can enter the strongly connected subnet. In addition, this result is obtained with the strongly connected subnet being identified. The set of

TABLE 12.11

Processing Time for the Operations

Operations	TARC 1 Modules	TARC 1 Processing Time (time units)	TARC 2 Modules	TARC 2 Processing Time (time units)	BARC Modules	BARC Processing Time (time units)
Op_1	P_0	0	P_0	0	P_0	0
Op_2	P_1/P_2	7/4	P_1/P_2	4/6	P_1/P_2	4/3
Op_3	P_3/P_4	5/3	P_3/P_4	5/9	P_3/P_4	5/1
Op_4	P_5/P_6	6/3	P_5/P_6	2/7	P_5/P_6	5/9
Op_5	P_7/P_8	7/8	P_7/P_8	5/5	P_9	1
Op_6	P_9	1	P_{10}/P_{11}	7/9	P_{10}/P_{12}	9/3
Op_7	P_{12}/P_{13}	9/1	P_{14}/P_{15}	7/6	P_{16}/P_{14}	6/4
Op_8	P_{17}/P_{18}	6/1	P_9	2	P_7/P_8	5/4
Op_9	P_{19}	2	P_{19}	2	P_{13}/P_{20}	3/2
Op_{10}	P_{21}/P_{22}	5/8	P_{21}/P_{22}	2/1	P_{17}/P_{18}	8/5
Op_{11}	P_{23}/P_{24}	7/5	P_{23}/P_{24}	5/4	P_{19}	2
Op_{12}	P_{25}/P_{26}	6/3	P_{25}/P_{26}	5/4	P_{21}/P_{22}	1/7
Op_{13}	$P_{27}/P_{28}/P_{29}$	6/4/6	$P_{27}/P_{28}/P_{29}$	1/4/3	P_{23}/P_{24}	3/5
Op_{14}	P_{30}/P_{31}	6/3	P_{30}/P_{31}	1/9	P_{25}/P_{26}	2/5
Op_{15}	P_{32}/P_{33}	6/9	P_{32}/P_{33}	2/9	$P_{27}/P_{28}/P_{29}$	7/3/3
Op_{16}					P_{30}/P_{31}	9/1
Op_{17}					P_{32}/P_{33}	3/5

Source: (From Wu, N. Q. and M. C. Zhou (2007). Real-time deadlock-free scheduling for semiconductor track systems based on colored-timed Petri nets, *OR Spectrum*, Vol. 29, No. 3, 421–443. With permission.)

modules that involve the potential deadlocks identified in Yoon and Lee (2001) is in fact the set of places in the strongly connected subnet. However, no deadlock avoidance policy is presented in Yoon and Lee (2001).

2. If the policy of Lawley (1999) is used, the condition given in Theorem 7 in Lawley (1999) should be applied, for the system here is an SCS. Thus, it allows $2n - 1$ (n is the number of processing alternatives) jobs for processing at a time in the system. For our system, there are no alternatives for some operations, and thus $n = 1$. Hence, only one job is allowed in the whole system at any time. It will take a much longer time to complete the wafers in a cassette. By using those policies (Fanti et al., 1997), the number of jobs allowed in the system must be less than the number of places in the strongly connected subnet of our model. In the example, the number of places is nine. Thus, at most eight wafers can concurrently be processed in the system at any time. In this way, the time required for completing the wafers will be longer than that for using the proposed approach.

3. If the deadlock control policy in Yoon and Lee (2004) is used, second-level deadlock, such as that shown in Figure 12.18, may occur. In Figure 12.18, places p_7 and p_{10} are occupied by tokens of TARC 2, and p_9 and p_{12} by tokens of BARC; this is not a deadlock state, but the system will be deadlocked no

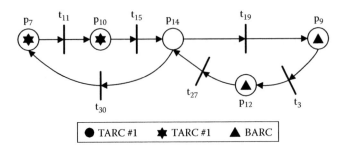

FIGURE 12.18 A potential second-level deadlock situation.

matter whether t_{15} or t_{27} fires. Then, a recovery strategy (by moving a wafer from the system) should be used. However, by our policy, such a situation will never occur.

12.4   SUMMARY

Cluster tools and a track system are typical examples of integrated equipment used in semiconductor fabrication. They are reconfigurable manufacturing systems and are difficult to manage. ROPN has been shown to be an integrated way to model both of them for analysis and control. Based on CROPN models, efficient approaches for analysis and control are proposed.

For some wafer fabrication processes, there is a strict wafer residency time constraint. It requires that when a wafer is completed in a PM, it should be taken away in a limited time; otherwise, the wafer will be scrapped. In this case, the scheduling problem of cluster tools is complicated, and the schedulability and optimal scheduling problem remain open. Some results can be found in Wu et al., 2008. The proposed method for track system scheduling is just a heuristic; its optimal scheduling problem is open too. Thus, further research is needed to develop new approaches for these problems.

REFERENCES

Bader, M. E., R. P. Hall, and G. Strasser. 1990. Integrated processing equipment. *Solid State Technology* 33:149–54.

Burggraaf, P. 1995. Coping with the high of wafer fabs. *Semiconductors International* 38:45–50.

Cormen, T. H., C. E. Leiserson, and R. E. Rivest. 1990. *Introduction to algorithms*. New York: MIT Press/McGraw-Hill.

Fanti, M. P., B. Maione, S. Mascolo, and B. Turchiano. 1997. Event-based feedback control for deadlock avoidance in flexible production systems. *IEEE Transactions on Robotics and Automation* 13:347–63.

Johri, P. K. 1993. Practical issues in scheduling and dispatching in semiconductor wafer fabrication. *Journal of Manufacturing Systems* 12:474–485.

Lawley, M. A. 1999. Deadlock avoidance for production systems with flexible routing. *IEEE Transactions on Robotics and Automation* 1:1–13.

Lee, D. Y., and F. DiCesare. 1994. Scheduling FMS using Petri nets and heuristic search. *IEEE Transactions on Robotics and Automation* 10:123–32.

Marsan, M. A., G. Conte, and G. Balbo. 1984. A class of generalized stochastic Petri nets for the performance evaluation of multiprocessor systems. *ACM Transactions on Computer Systems* 2:93–122.

Perkinson, T. L., R. S. Gyurcsik, and P. K. MacLarty. 1996. Single-wafer cluster tool performance: An analysis of the effects of redundant chambers and revisitations sequences on throughput. *IEEE Transactions on Semiconductor Manufacturing* 9:384–400.

Perkinson, T. L., P. K. MacLarty, R. S. Gyurcsik, and R. K. Cavin III. 1994. Single-wafer cluster tool performance: An analysis of throughput. *IEEE Transactions on Semiconductor Manufacturing* 7:369–73.

Reveliotis, S. A. 1999. Accommodating FMS operational contingencies through routing flexibility. *IEEE Transactions on Robotics and Automation* 15:3–19.

Rostami, S., B. Hamidzadeh, and D. Camporese. 2001. An optimal periodic scheduler for dual-arm robots in cluster tools with residency constraints. *IEEE Transactions on Robotics and Automation* 17:609–18.

Singer, P. 1995. The driving forces in cluster tools development. *Semiconductors International* 38:113–18.

Venkatesh, S., R. Davenport, P. Foxhoven, and J. Nulman. 1997. A steady-state throughput analysis of cluster tools: Dual-blade versus single-blade robots. *IEEE Transactions on Semiconductor Manufacturing* 10:418–23.

Wu, N. Q. 1999. Necessary and sufficient conditions for deadlock-free operation in flexible manufacturing systems using a colored Petri net model. *IEEE Transactions on Systems, Man, and Cybernetics* C 29:192–204.

Wu, N. Q., C. Chu, F. Chu, and M. C. Zhou. 2008. A Petri net method for schedulability and scheduling problems in single-arm cluster tools with wafer residency time constraints. *IEEE Transactions on Semiconductor Manufacturing* 21: 224–237.

Wu, N. Q., and M. C. Zhou. 2007a. Deadlock modeling and control of semiconductor track systems using resource-oriented Petri nets. *International Journal of Production Research* 45:3439–56.

Wu, N. Q., and M. C. Zhou. 2007b. Real-time deadlock-free scheduling for semiconductor track systems based on colored timed Petri nets. *OR Spectrum* 29:421–43.

Yoon, H. J., and D. Y. Lee. 2001. Identification of potential deadlock set in semiconductor track systems. In *Proceedings of 2001 IEEE International Conference on Robotics and Automation*, Seoul, Korea, 1820–25.

Yoon, H. J., and D. Y. Lee. 2004. Deadlock-free scheduling of photolithography equipment in semiconductor fabrication. *IEEE Transactions on Semiconductor Manufacturing* 17:42–54.

Zhou, M. C., and K. Venkatesh. 1998. Modeling, simulation and control of flexible manufacturing systems: A Petri net approach. Singapore: World Scientific.

Zuberek, W. M. 2001. Timed Petri nets in modeling and analysis of cluster tools. *IEEE Transactions on Robotics and Automation* 17:562–75.

13 Modeling and Control of Assembly/ Disassembly Systems

13.1 INTRODUCTION

The deadlock problem in another class of flexible manufacturing systems (FMS) is characterized by a nonsequential resource requirement. Its solution requires new techniques (Roszkowska and Wojcik, 1993; Roszkowska, 2004; Fanti et al., 1998; Hsieh, 2004). The typical example of such an FMS is a flexible assembly system (FAS). In FAS, there are base components of the products and parts to be mounted on the former. Often, the former and pallets are presented and delivered separately with the latter. The assembly of a complex product can in many cases be divided into a series of subassemblies. Both base components and parts to be mounted on the former take space in the assembly process, and they cannot be finished independently. The transportation task can also be divided into the transport of base components through the assembly, parts into the system, and semifinished parts for further assembly. Common transport devices may be used for some of these tasks. The material flow in this system can be abstracted as a fork/join pattern (Paik and Tcha, 1995; Roszkowska, 2004). An assembly/disassembly material flow is formed.

Fanti et al. (1997) studied the deadlock avoidance problem in FAS by using digraphs and synthesizing a supervisory controller. Hsieh (2004) studied the deadlock avoidance problem for the assembly systems with joint material flow using Petri nets and presented a deadlock avoidance algorithm. With this algorithm, deadlock is avoided by restricting the maximal number of parts that can be routed into the system. An assembly process is realizable if and only if there exists a feasible execution sequence (Roszkowska and Wojcik, 1993). By using Petri nets, Roszkowska and Wojcik (1993, 2004) studied FAS with fork/join material flow, proved that the realizability of the process was NP-complete, and proved that it was NP-hard to find the maximally permissive deadlock avoidance policy. With these conclusions, a control policy is proposed to determine the reserved buffer spaces for all the buffers such that there are enough spaces for the assembly to continue until its completion. This chapter discusses the deadlock avoidance problem in FAS based on Wu et al. (2008). The system is modeled by a resource-oriented Petri net (ROPN), and an efficient deadlock avoidance policy is proposed.

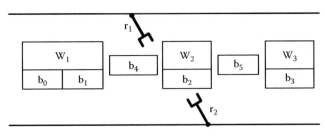

FIGURE 13.1 Schematic structure of a flexible assembly system. (Adapted from Roszkowaska, 2004.)

13.2 A FLEXIBLE ASSEMBLY SYSTEM

The FAS is shown in Figure 13.1. It has of two material handling devices, robot 1 (r_1) and robot 2 (r_2), and three workstations, w_1, w_2, and w_3. There are an input buffer b_0 and an output buffer b_1 for workstation w_1, buffer b_2 for w_2, and b_3 for w_3. Between them, b_4 can be accessed by w_1 and w_2, and b_5 can be accessed by w_2 and w_3. The robots handle the transport of materials from/to the system input/output and between the buffers. Robot r_1 delivers trays with parts and unpalletized subassemblies, while robot r_2 handles pallets with base components. In an assembly process, trays with parts and subassemblies are held in b_4 or b_5, and pallets with base components are put onto b_0, b_1, b_2, and b_3 for assembly. When a workstation performs an operation, it may take the parts or subassemblies in a tray at b_4 or b_5 and mount them onto the base components in workstation buffers, b_{0-3}. For example, when a base component is in b_0, w_1 takes parts from b_4 and mounts them onto it, then puts it into b_1 after completion. Notice that the workstations themselves cannot hold any extra component, and they just perform assembly operations one at a time.

Assume that two products, say A-product and B-product, should be assembled in the system concurrently. Their assembly processes are shown in Figure 13.2, where *tray* denotes trays with parts and subassemblies, and *base* denotes base components. Take A-part as an example for explanation. After robot r_1 takes a tray with parts to be mounted onto the base into b_4, robot r_2 transports a base component into b_0, and then w_1 mounts the parts in b_4 onto the base in b_0, and the finished one is moved into b_1, while the tray with other parts remains in b_4. Then robot r_2 can deliver the base component into b_2 and w_2 performs an operation on it, and after that it remains in b_2. It can then be delivered into b_3 by r_2. If, at the same time, the tray in b_4 is delivered into b_5 by r_1, and other parts from the central storage are released into the system and put into b_5 by r_1, then w_3 is ready to perform its assembly. After that, the finished product remains in b_3 and the tray remains in b_5, respectively. They are ready to be delivered out of the system, and the buffer spaces are released once delivered. Notice that when w_3 does its assembly, it takes two parts from b_5: one is delivered to b_5 from b_4, and the other from the central storage.

From the assembly processes shown in Figure 13.2, it is obvious that the material flow in the system is of the fork/join pattern, or assembly/disassembly form,

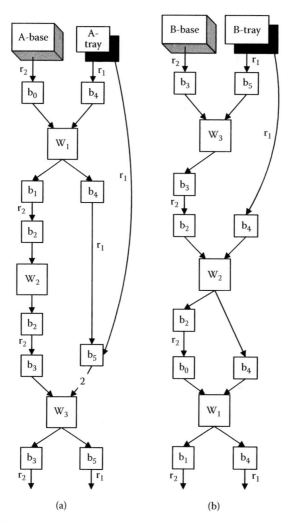

FIGURE 13.2 The assembly processes of two products.

and the resource requirement for processing an operation is complex. Two deadlock examples are presented as follows:

Base flow: Assume that the capacity of buffers b_{0-3} is 1, and the capacity of b_{4-5} is more than 1. There are A-bases in b_0 and b_1, respectively, and two A-trays in b_4, and at the same time there is a B-base in b_3. Now the A-base in b_1 or the B-base in b_3 can move into b_2 according to Figure 13.2. Either ends up with a deadlock.

Assembly: Assume that both buffers b_0 and b_4 have a capacity of one, and b_4 holds an A-tray and b_0 holds a B-base. No operation can occur at w_1, leading to system deadlock eventually, if not immediately. Finally, part flow can result in deadlock as well, to be illustrated later.

13.3   R-POLICY

Before developing a CROPN model and deadlock avoidance control policy for FAS, we discuss a control policy proposed by Roszkowaska (2004), called the R-policy for short. Her Petri net model is constructed based on a process-oriented modeling method.

The system is modeled by a PN, as shown in Figure 13.3, where place b_i represents buffer b_i. It models the assembly process for each product by a subnet, and all the subnets together form the PN model of the entire system. For example, subnets in Figure 13.3a and b describe the assembly processes of A-product and B-product, respectively. The part flows are as follows:

A-base: $t_2 \rightarrow b_0 \rightarrow t_3 \rightarrow b_1 \rightarrow t_5 \rightarrow b_2 \rightarrow t_6 \rightarrow b_2 \rightarrow t_7 \rightarrow b_3 \rightarrow t_9 \rightarrow b_3 \rightarrow t_{11}$
A-part: $t_1 \rightarrow b_4 \rightarrow t_3 \rightarrow b_4 \rightarrow t_4 \rightarrow b_5 \rightarrow t_9 \rightarrow b_5 \rightarrow t_{10}$ and $t_8 \rightarrow b_5 \rightarrow t_9 \rightarrow b_5 \rightarrow t_{10}$
B-base: $t_{13} \rightarrow b_3 \rightarrow t_{14} \rightarrow b_3 \rightarrow t_{15} \rightarrow b_2 \rightarrow t_{17} \rightarrow b_2 \rightarrow t_{18} \rightarrow b_0 \rightarrow t_{19} \rightarrow b_1 \rightarrow t_{21}$
B-part: $t_{12} \rightarrow b_5 \rightarrow t_{14}$ and $t_{16} \rightarrow b_4 \rightarrow t_{17} \rightarrow b_4 \rightarrow t_{19} \rightarrow b_4 \rightarrow t_{20}$

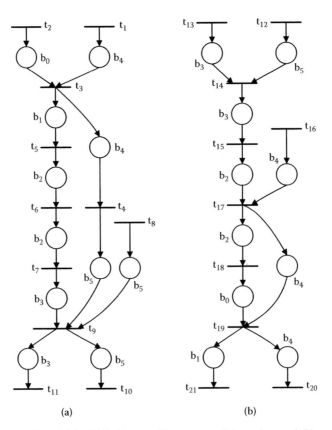

(a)   (b)

FIGURE 13.3 PN model for FAS: (a) assembly process of A-product and (b) assembly process of B-product. (Adapted from Roszkowaska, 2004.)

Then the two subnets in Figure 13.3a and b form the PN model for the example. This PN model defines the realizability problem of the assembly processes.

Definition 13.1: The PN is realizable if and only if there exists a firing sequence σ of transitions such that $M_0[\sigma > M_0$, and each $t \in T$ occurs in σ exactly once.

Definition 13.1 implies that if one product of every product type is put into the system for assembly, the assembly can be completed and the system comes back to its initial state. Based on Definition 13.1, the realizability problem can be stated as follows: Given the PN model for FAS and the capacity for each buffer b_i, is the PN realizable? It is proved that this problem is NP-complete in the strong sense. That is, to determine the minimal capacity for each buffer is NP-complete. Based on this result, an algorithm is presented to determine the minimal buffer capacity necessary for the control policy presented in Roszkowaska (2004). In the following definition, $I(p, t)$ and $O(p, t)$ represent the number of input arcs from p to t and output arcs from t to p, respectively.

Definition 13.2: For a PN of FAS, let $T^N = \{t \in T | \; |{}^\bullet t| > 1 \vee |t^\bullet| > 1\}$ be the set of transitions, called nodes, that have more than one input place or more than one output place; $P^N = \bigcup_{t \in T^N} {}^\bullet t \cup t^\bullet$ be the set of input and output places of all $t \in T^N$, called node parts; and $\overline{P}^N = P - P^N$ be the set of buffers used for storing nonnode parts. Denote $\alpha(b) = \text{Max}_{t \in T^N} (\text{Max}(0, O(b,t) - I(b,t)))$. The minimal buffer capacity is calculated as $v(b) = v'(b) + v''(b)$, where

$$v'(b) = \begin{cases} 1, b \in \overline{P}^N - P^N \\ \alpha(b), b \in P^N - \overline{P}^N \\ \alpha(b) + 1, b \in P^N \cap \overline{P}^N \end{cases} \tag{13.1}$$

$$v''(b) = \sum_{t \in T^N} I(b,t) \tag{13.2}$$

It is shown that with the capacity $v(b)$ for each buffer b, the PN is realizable. Transition $t \in T^N$ is an assembly or disassembly transition. Then $\max(0, O(b, t) - I(b, t))$ is the difference of output tokens from b and input tokens to b when $t \in T^N$ fires. It represents the buffer space requirements for buffer b for an assembly operation. By Equation 13.1, if buffer b is not the input or output buffer of an assembly/disassembly transition for all product types, $v'(b) = 1$; if buffer b is an input or output buffer of an assembly/disassembly transition for all product types, $v'(b) = \alpha(b)$, where $\alpha(b) = \max(0, O(b, t) - I(b, t))$ is the largest space requirement among all the assembly operations; if b is an input or output buffer of an assembly/disassembly transition for some product types, but it is not for other types, $v'(b) = \alpha + 1$. $I(b, t)$ represents the buffer space requirement for buffer b for the assembly of a product type. Thus, by Equation 13.2, $v''(b)$ is the sum of space requirements for buffer b by all the product types.

Example 13.1

For the PN shown in Figure 13.3, $T^N = \{t_3, t_9, t_{14}, t_{17}, t_{19}\}$, $P^N = P$, $\overline{P^N} = \{b_2\}$, $\overline{P^N} - P^N$ $= \varnothing$, $P^N - \overline{P^N} = \{b_0, b_1, b_3, b_4, b_5\}$, and $P^N \cap \overline{P^N} = \{b_2\}$. Thus:

$v'(b_0) = \alpha(b_0) = \max(\max(0, 0\text{-}1), \max(0, 0\text{-}1)) = 0$, $v''(b_0) = 2$, $v(b_0) = 0 + 2 = 2$

$v'(b_1) = \alpha(b_1) = \max(\max(0, 1\text{-}0), \max(0, 1\text{-}0)) = 1$, $v''(b_1) = 0$, $v(b_1) = 1 + 0 = 1$

$v'(b_2) = \alpha(b_2) = 1 + \max(0, 1\text{-}1) = 1 + 0 = 1$, $v''(b_2) = 1$, $v(b_1) = 1 + 1 = 2$

$v'(b_3) = \alpha(b_3) = \max(\max(0, 1\text{-}1), \max(0, 1\text{-}1)) = 0$, $v''(b_3) = 2$, $v(b_3) = 0 + 2 = 2$

$v'(b_4) = \alpha(b_4) = \max(\max(0, 1\text{-}1), \max(0, 1\text{-}1), \max(0, 1\text{-}1)) = 0$, $v''(b_4) = 3$, $v(b_4) = 0 + 3 = 3$

$v'(b_5) = \alpha(b_5) = \max(\max(0, 1\text{-}2), \max(0, 0\text{-}1)) = 0$, $v''(b_5) = 3$, $v(b_5) = 0 + 3 = 3$

In FAS, deadlock occurrence is due to competition for the finite buffer spaces. Thus, in operating an FAS, a restrictive use of buffer spaces can avoid deadlock. Based on the idea of buffer space reservation, a deadlock avoidance control policy is proposed in Roszkowaska (2004).

Definition 13.3: Let $P_{in}^N = {}^{\bullet}T^N$ be the set of node input parts. For each marking M and each buffer b, buffer space reservation is defined as $\eta(M, p) = v'(b) + |\{p|p \in P_{in}^N \wedge M(p) = 0\}|$.

$\eta(M, p)$ contains two parts. $v'(b)$ is a constant and determined by the structure of PN. It is known that if a buffer b is in $P^N \cap \overline{P^N}$ at least one space should be reserved at any marking M. The other part, $|\{p|p \in P_{in}^N \wedge M(p) = 0\}|$, is determined by the marking dynamically. It says that if a raw part is needed in buffer b for the assembly operation of a product type, and it is not yet at marking M, then a space in buffer b must be reserved for that operation.

Definition 13.4: For each transition $t \in T$, define:

1. TZ(t), operation zone, the set of transitions that includes t and its successor t' such that t' is not a node, and no transition t'' that succeeds t and precedes t' is a node. If t is not a node, then its operation zone consists of all transitions that lie on the path from t to that nearest node, or if t has no node, the successor of t that is an output transition in that PN. If t is a node, then TZ(t) captures t and all zones TZ(t') of its immediate nonnode successor t'.
2. PZ(t), part zone, a set of places that is the union of the output parts of all transitions $t' \in$ TZ(t).
3. BZ(t), buffer zone, a set of places (buffers) that includes all places used to store parts $p \in$ TZ(t), except the output place of t, if t is a nonnode, and

except the output places of t that are used to store exclusively node parts, that is, places $p \in P^N - \bar{P}^N$ if t is a node.

4. $QZ(t)$, node zone, a set of places that comprises all places $p \in PZ(t)$ such that there exists a path in PN from p to a node.

Example 13.2

In the PN in Figure 13.3, some zones can be identified as:

$TZ(t_1) = \{t_1\}$, $PZ(t_1) = \{b_4\}$, $BZ(t_1) = \varnothing$, $QZ(t_1) = PZ(t_1)$

$TZ(t_2) = \{t_2\}$, $PZ(t_2) = \{b_0\}$, $BZ(t_1) = \varnothing$, $QZ(t_2) = PZ(t_2)$

$TZ(t_3) = \{t_3, t_4, t_5, t_6, t_7\}$, $PZ(t_3) = \{b_1, b_2, b_2, b_3, b_4, b_5\}$, $BZ(t_3) = \{b_2, b_3, b_5\}$, $QZ(t_3) = PZ(t_3)$

$TZ(t_4) = \{t_4\}$, $PZ(t_4) = \{b_5\}$, $BZ(t_4) = \varnothing$, $QZ(t_4) = PZ(t_4)$

$TZ(t_5) = \{t_5, t_6, t_7\}$, $PZ(t_5) = \{b_2, b_2, b_3\}$, $BZ(t_5) = \{b_2, b_3\}$, $QZ(t_5) = PZ(t_5)$

Based on Definitions 13.3 and 13.4, and the buffer capacity determined by Definition 13.2, a deadlock control policy is proposed in Roszkowaska (2004), called R-policy for short.

R-policy: In a PN for FAS developed above, firing transition t at marking M changes M to M'. Then, t can fire if and only if at least one of the two following conditions holds, $\forall b \in {}^\bullet t \cup t^\bullet$:

1. $\forall b \in {}^\bullet t \cup t^\bullet$, $\mu(M', b) \le K(b) - \eta(M', b)$; and
2. a. $\forall p \in QZ(t)$, $M(p)=0$
   and
   b. $\forall b \in BZ(t)$, $\mu(M', b) < K(b)$

Note that $\mu(M', b)$ represents the number of parts in buffer b at marking M', for a buffer may correspond to several places in the PN. For example, there are three places for buffer b_5 in Figure 13.4a. $K(b)$ is the capacity of buffer b, and $\eta(M', b)$ is defined in Definition 13.3.

In fact, condition 1 contains two parts, as explained after Definition 13.3. For condition 2, (a) implies that t in a subnet of the PN can fire only $\forall p \in QZ(t)$, $M(p)=0$ in the subnet, and (b) says that after firing t, any b cannot be full, or there is at least a free space.

Example 13.3

A marking is shown in Figure 13.4 that is reachable by applying the R-policy. At this marking, the PN is still live. However, no more product can be released into the system for assembly before some products are completed and leave the system. Otherwise, the R-policy is violated.

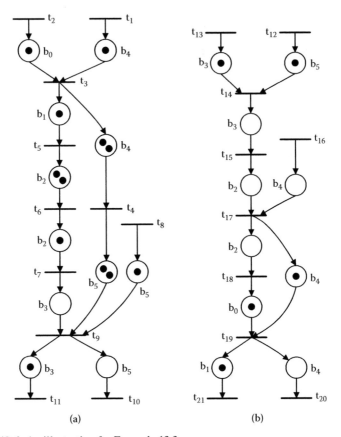

(a)   (b)

FIGURE 13.4 An illustration for Example 13.3.

13.4   MODELING FAS BY CROPN

Because the assembly/disassembly process in FAS is different from the sequential part production processes, the CROPN model for FAS is different from that for part production processes. We discuss the CROPN modeling for FAS in detail.

13.4.1   MODELS FOR RESOURCES

In AMS, when deadlock occurs, it must be in such a situation that the manufacturing processes block each other, such that there is no path for any part to go through the system. It is the limited quantity of resources that causes system deadlock. If modeled by CROPN, only some resources will make a contribution to system deadlock, but others will not. Thus, in modeling the system, different resources should be treated in different ways. Similarly, for FAS we deal with buffers differently than we do workstations and robots. The former is treated as H-resources and the latter as G-resources.

We model an H-resource in FAS by an H-place, as shown in Figure 13.5a. A token in an H-place represents a part occupying a space in the place (buffer), no matter whether

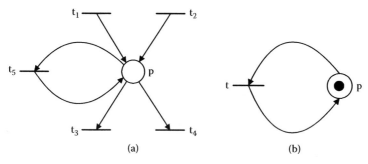

FIGURE 13.5 The PN models for (a) H-resource and (b) G-resource.

it is just sitting there or being processed by some other resources. The input and output transitions (e.g., t_{1-4}) of an H-place p represent assembly at workstations (e.g., t_5) or delivering by robots. Multiple input transitions (e.g., t_{1-2}) to p model the fact that the resource is competed by multiple processes. Multiple output transitions (e.g., t_{3-4}) describe the choices for the next operation. Unlike sequential manufacturing processes, in assembly processes, while a part is held in a buffer, it may be processed by some other device, e.g., workstation. This is equivalent to the part being processed and coming back immediately. The self-loop $\{p, t_5, p\}$ in Figure 13.5a describes such an assembly process.

We model a G-resource by a G-place, as shown in Figure 13.5b. Unlike an H-place, a token in a G-place represents that the G-resource is available for performing an operation. Firing transition t in Figure 13.5b implies that the resource is performing an operation. After its firing, the token comes back and the resource becomes available again. In this way, the dynamic behavior of the resources is captured, and we will see that this makes the model simple.

13.4.2 MODELS FOR INDIVIDUAL PRODUCTS

We can now develop the model for assembly processes. We first model an assembly process for each product to obtain a subnet. First, we present the PN models for the basic operations in FAS.

The first one is material delivery by a robot. When a robot delivers a part or base, the latter must be moved from one location to another, i.e., a buffer to another in FAS or a place to another in its net model. With the G-resource model, this process can be modeled by PN, shown in Figure 13.6. Firing t implies that a part is moved from p_1 to p_2 by robot r.

There are four situations for the operations performed by a workstation, as modeled in Figure 13.7.

1. A workstation processes a base component in a buffer without using any part, as shown in Figure 13.7a. After it, the base remains in the buffer.
2. A base component is in a buffer. A workstation uses a part from another buffer, i.e., p_1 in Figure 13.7b, to perform the operation. After it, the base remains in its buffer, i.e., p_2 in Figure 13.7b, but the buffer holding the part, i.e., p_1, is emptied.

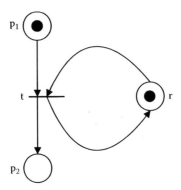

FIGURE 13.6 The PN model for material delivery by robot.

3. A base component is in buffer 1, i.e., p_1 in Figure 13.7c. A workstation uses a part in buffer 2, i.e., p_2, with a tray to perform the operation. After the completion, the base is put into buffer 3, i.e., p_3, to free buffer 1, but buffer 2 is still occupied by the tray.
4. A workstation uses a part in buffer 2, i.e., p_2 in Figure 13.7d, to perform an operation on the base component in buffer 1, i.e., p_1. After it, the base and the tray remain in buffers 1 and 2, respectively.

Note that place w models a G-resource, and others model H-resources. These PN models are called primitives.

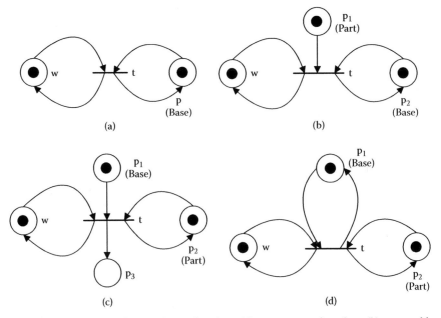

FIGURE 13.7 The PN primitives by workstations: (a) w processes a base in p; (b) w assembles all the parts in p_1 to a base in p_2; (c) w assembles some parts in p_2 to a base in p_1 and moves it to p_3; and (d) w assembles some parts in p_2 to a base in p_1, and after that they remain in p_1 and p_2.

Consider the primitive shown in Figure 13.7d. It indicates that when the assembly is completed, the base and part still occupy the buffer spaces. However, if this assembly operation is the last for a product, then the completed product can be moved away immediately, and the base and parts will no longer occupy buffer spaces. Thus, in this case, the model in Figure 13.7d can be replaced by the one shown in Figure 13.8.

Now, with the primitives for resources and basic operations, we present ROPN for individual products. For easy understanding, we use b_i, $i \in N_5$, to name the places corresponding to buffers, w_{1-3} for three workstations, and r_{1-2} for two robots. We use a place p_0 to represent the central storage for FAS, which hosts all the raw base components, parts, and final products. In this way, the PNs for assembling products A and B are obtained as shown in Figures 13.9 and 13.10, named A- and B-subnets for short.

In the A-subnet, self-loops $\{t_1, r_2\}$, $\{t_5, r_2\}$, $\{t_8, r_2\}$, $\{t_2, r_1\}$, $\{t_6, r_1\}$, and $\{t_9, r_1\}$ correspond to the primitive shown in Figure 13.6, representing that a base or tray

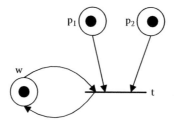

FIGURE 13.8 The equivalent PN model of Figure 13.7d for the last assembly operation.

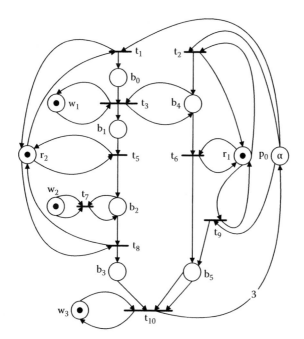

FIGURE 13.9 A-subnet for assembling product A.

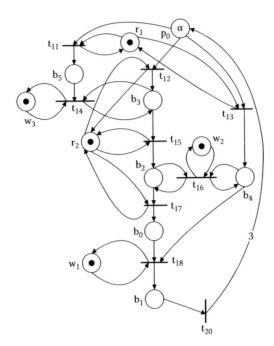

FIGURE 13.10 B-subnet for assembling product B.

is delivered by r_1 or r_2; $\{w_2, t_7, b_2\}$ to Figure 13.7a, meaning that w_2 performs an operation on a base in b_2; $\{w_1, t_3, b_0, b_1, b_4\}$ to Figure 13.7c, meaning that w_1 assembles a part from b_4 onto the base in b_0 and moves it into b_1; and $\{w_3, t_{10}, b_3, b_5\}$ to Figure 13.8b, meaning that w_3 assembles a part from b_5 onto the base in b_3 as the last assembly step. In Figure 13.9, there are two arcs between b_5 and t_{10}, since firing t_{10} takes two different A-parts from b_5; the weight of arc (t_{10}, p_0) is 3, representing that the finished A-product has a base and two parts.

In the B-subnet, $\{t_{11}, r_1\}$, $\{t_{13}, r_1\}$, $\{t_{12}, r_2\}$, $\{t_{15}, r_2\}$, and $\{t_{17}, r_2\}$ correspond to the primitive shown in Figure 13.6, $\{w_3, t_{14}, b_5, b_3\}$ to Figure 13.7b, $\{w_1, t_{18}, b_0, b_1, b_4\}$ to Figure 13.7c, and $\{w_2, t_{16}, b_2, b_4\}$ to Figure 13.7d. Because t_{18} represents the last assembly step, an arc (t_{18}, p_0) goes to p_0 directly, not to b_4, meaning that a tray can be moved out of the assembly process immediately. The weight of arc (t_{20}, p_0) is 3, representing that two parts are mounted onto a base for product B.

It is shown in Wu (1997) that by modeling the G-resources in this way, deadlock resulting from processes performed by G-resources can be eliminated. Thus, in the sense of deadlock avoidance, these G-places and their associated arcs can be removed from the model. By removing them, Figures 13.9 and 13.10 are reduced to Figure 13.11.

13.4.3 ROPN FOR THE WHOLE SYSTEM

With the subnets for the individual products, we can obtain the ROPN for the whole system by merging the subnets according to the rule presented in Chapter 5. The CROPN for the whole FAS is obtained and shown in Figure 13.12. Thereafter, an

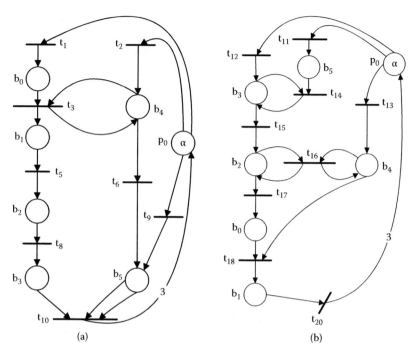

FIGURE 13.11 The reduced PNs for the products.

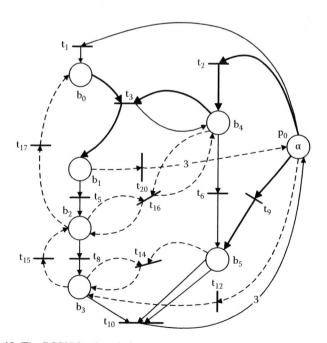

FIGURE 13.12 The ROPN for the whole system.

ROPN for FAS is referred to as the merged one from the reduced subnets. We use solid, dashed, and bold-solid lines to denote the part flow for the assembly processes of products A and B, and the shared path flow for clarity. The part flows are as follows: A-base component, $p_0 \to t_1 \to b_0 \to t_3 \to b_1 \to t_5 \to b_2 \to t_8 \to b_3 \to t_{10} \to p_0$; A-part, $p_0 \to t_2 \to b_4 \to t_3 \to b_4 \to t_6 \to b_5 \to t_{10} \to p_0$ and $p_0 \to t_9 \to b_5 \to t_{10} \to p_0$; B-base component, $p_0 \to t_{12} \to b_3 \to t_{14} \to b_3 \to t_{15} \to b_2 \to t_{16} \to b_2 \to t_{17} \to b_0 \to t_3 \to b_1 \to t_{20} \to p_0$; B-part, $p_0 \to t_9 \to b_5 \to t_{14} \to b_3$ and $p_0 \to t_2 \to b_4 \to t_{16} \to b_4 \to t_3 \to p_0$.

The ROPN obtained is compact, but it cannot describe the part flow exactly. For example, when a token is in place b_2, one may not know which transition, t_8, t_{16}, or t_{17}, it enables. This problem can be solved by introducing colors into ROPN, resulting in colored ROPN (CROPN). In Chapter 5, when we define colors for ROPN, each transition is given only one color. Here, for FAS, this is not enough; each transition may have more than one color, and the colors are defined next.

Notice that when we obtain the ROPN by merging the subnets, we allow only one transition between two places and in the same direction. Consequently, some transitions come from only one subnet, while others are from more than one subnet. In ROPN, if a transition $t \in T$ is from a single subnet, then we call t a nonshared transition. If, however, it is from $k \geq 2$ subnets, we call it a k-shared transition, or multishared one. For example, transitions t_1, t_5, t_6, t_8, t_{10}, t_{14}, t_{15}, t_{17}, and t_{20} are nonshared transitions, while t_2, t_3, and t_9 are two-shared.

Definition 13.5: $C(t_i) = \{c_i\}$ is the only color for a nonshared transition $t_i \in T$, and $C(t_i) = \{c_{iA}, c_{iB}, \ldots, c_{iK}\}$ for a multishared transition if it is from subnets $A, B, \ldots$, and K.

A multishared transition represents the assembly of multiproducts. The assembly of different products is thus distinguished by the transition's colors.

Definition 13.6: If transition $t_i \in p_i^\bullet$, the tokens in place p_i enabling t_i have a color f_j that is associated with color c_i.

Note that, with color f_j, we use j to identify a component and its operation that should be processed by transition t_i. With the definitions of the enabling and firing rules for colored PN and Definition 13.5, Equations 3.1 and 3.2 in Chapter 3 can be used in our CROPN. In our model, shown in Figure 13.12, transitions t_1, t_5, t_6, t_8, t_{10}, t_{14}, t_{15}, t_{17}, and t_{20} are single colored, but t_2, t_3, and t_9 have two colors: c_{2A} and c_{2B}, c_{3A} and c_{3B}, and c_{9A} and c_{9B}, respectively. For example, the second operation of an A-base should be processed by t_5, and the forth operation of a B-base should be processed by t_{20}. Thus, if there is a token in b_1 representing an A-base with color f_{AB2}, then t_5 with color c_5 can fire, but if it is representing a B-base with color f_{BB4}, then t_{20} with color c_{20} can fire. Transition t_3 has two colors: When there is a token in b_0 representing an A-base with color f_{AB1} and an A-part in b_4 with color f_{AP1}, t_3 can fire with color c_{3A}. When there is a token in b_0 representing a B-base with color f_{BB3} and a B-tray in b_4 with color f_{BP2}, t_3 can fire with color c_{3B}. Note that the tokens in p_0 should have colors. Sometimes we are concerned only with the number of tokens in a place; for example, in Figure 13.12, α in p_0 indicates there are α tokens regardless of their colors. However, when a transition is to be fired, the color of a token should be taken into account.

As a process proceeds, the color of a token may change. However, such change occurs only when an assembly transition fires. For example, the colors of tokens in Figure 13.12 can change as follows. In p_0, color f_{AB1} is used for a token representing an A-base enabling t_1, f_{AP1} for an A-part enabling t_2, f_{AP3} for an A-part enabling t_9, f_{BB1} for a B-base enabling t_{12}, f_{BP1} for a B-part enabling t_9, and f_{BP2} for a B-part enabling t_2. When a token with color f_{AB1} enters b_0 by firing t_1, and a token with color f_{AP1} comes to b_4 by firing t_2 with color c_{2A}, these two tokens then enable t_3 with color c_{3A}. After firing t_3, their colors change into f_{AB2} and f_{AP2}, respectively. When a token with color f_{AB2} comes to b_3 and another with f_{AP2} comes to b_5, and a token with color f_{AP3} comes to b_5 by firing t_9 with color c_{9A}, these three tokens enable t_{10}. When t_{10} fires, two tokens are taken away from b_5, one with color f_{AP2} and the other with f_{AP3}. Number 3 of arc (t_{10}, p_0) indicates that three tokens with colors $f_{AB2}, f_{AP2},$ and f_{AP3} are taken to p_0. In this way, the CROPN is kept conservative. When a token with color f_{BB1} comes to b_3 by firing t_{12}, and a token with color f_{BP1} comes to b_5 by firing t_9 with color c_{9B}, these two tokens together enable t_{14}. After firing t_{14}, the color of the former changes into f_{BB2}. When the token with color f_{BB2} comes to b_2 by firing t_{15}, and a token with color f_{BP2} comes to b_4 by firing t_2 with color c_{2B}, these two tokens then enable t_{16}. After firing t_{16}, their colors change to f_{BB3} and f_{BP3}, respectively. When a token with color f_{BB3} comes to b_0 by firing t_{17}, this token, together with a token with color f_{BP3} in b_4, enables t_3 with color c_{3B}. After firing t_3 with color c_{32}, the color of the former changes into f_{BB4}, and the token moves to b_1 and enables t_{20}. Thus, the processes are accurately described.

13.5   REALIZABLE RESOURCE REQUIREMENT

In the part production process, each part can be processed independently. Hence, one can allow only one part in the system, and the parts to be processed in the system can be completed one by one. In this way, the production process is always feasible.

Unlike the part production process, in FAS the base components and parts to be mounted onto the former must be released into the system concurrently. This gives rise to a problem: Given a set of product types that should be processed in the system concurrently (the set of buffers with their capacity, the FAS layout, and product routes) can each operation be executed at least once? This question is called a realizability problem in Roszkowska (2004), and it is shown that it is NP-complete to answer this question in the strong sense. This implies that to find a minimal resource requirement is NP-complete. However, a realizable resource requirement is needed as a necessary condition for deadlock control. We intend to find a simple algorithm to compute a realizable resource requirement.

The hardship of the problem is that the use of resources is dependent on the sequences in which the operations are executed, or the firing sequences of the transitions in our CROPN. This is a combinatorial problem. However, for each product the sequence of operations is prescribed according to its process plan. To complete its assembly, its base component must first be released into the system. Nevertheless, a resource requirement can be reduced if a part to be mounted onto a base component is released into the system just before FAS needs it.

Definition 13.7: Assume that in the CROPN of FAS there is t such that $^\bullet t = \{p_0\}$, where p_0 represents the central storage, $p_e \in t^\bullet$, and $t_a \in p_e^\bullet$, where t_a is an assembly transition with color c. Firing t_a requires tokens in p_i, $i = 1, 2, \ldots, k$, $k \geq 1$, with color a_i, and tokens with color h_e that move into p_e by firing t. If in marking M there exists an i such that $M(p_i)(a_i) < I(p_i, t_a)(a_i, c)$ and transition t fires, then we define t's firing to be premature.

For example, to perform the final assembly of A, there should be an A-base in b_3, an A-part from b_4, and another A-part from p_0. If the former items are not ready at b_3 and b_4, firing t_9 to release an A-part into b_5 is of no use for the assembly process but it occupies buffer spaces. Thus, such premature firing of t_9 should be avoided. With the definition, we present an algorithm to find the realizable resource requirement for the execution of the assembly processes. Assume that there are β types of products to be assembled in the system, and product $i \in \{1, 2, \ldots, \beta\}$ needs d_i parts to be mounted onto its base component. Thus, if FAS is required to assemble one product for each type, there are totally $\alpha = \beta + \Sigma_{i=1}^{\beta} d_i$ parts, including β base components.

Algorithm 13.1: Find the realizable resource requirements for assembling one product of each type concurrently. In the algorithm, R[] is an m-vector, with R[i] being the resource requirement modeled by p_i, T_E is the set of enabled transitions at a marking, and $T_{EF} \subseteq T_E$ is a set of transitions whose firing is premature.

Step 1: Initialization: Let $K(p) = \infty$ for all $p \in P = \{p_0, p_1, p_2, \ldots, p_m\}$ in the CROPN; put α tokens with their colors in place p_0 that represents the central storage, and set the current marking $M = M_0$ with all tokens in p_0; and let R[] = 0 be an m-vector with integer values.

Step 2: Find a set of transitions $T_E \subseteq T$ such that $\forall t \in T_E$ is enabled at the current marking M. If $T_E = \varnothing$, stop; otherwise, go to the next step.

Step 3: In T_E find the set of transitions T_{EF} such that $\forall t \in T_{EF}$, whose firing is premature, and setting $T_E = T_E - T_{EF}$, do the following:

1. Form a firing sequence f of the transitions in T_E. The order can be set arbitrarily.
2. Calculate $M' = M[>f$, and let $M = M'$.
3. For $i = 1$ to m, if $M(p_i) > R[i]$, $R[i] = M(p_i)$.
4. Go to step 2.

When the algorithm stops, vector R[] returns the realizable resource requirement.

Example 13.4

Find the realizable resource requirements in the CROPN shown in Figure 13.12 by using Algorithm 13.1. We do not care about the marking in p_0 representing the central storage. Hence, we drop it in the calculation; i.e., we consider the marking in b_0 to b_5 only. At the initial marking, (0, 0, 0, 0, 0, 0) with $M(p_0) = \alpha$, first we find the set of transitions that are enabled, i.e., $T_E = \{t_1, t_2, t_9, t_{12}\}$. Since firing t_2 and t_9 is premature, they are removed from T_E. A firing sequence $f_1 = t_1 t_{12}$ is formed. After its firing, the marking becomes (1, 0, 0, 1, 0, 0) and $R[]$ is updated as $R[0] = R[3] = 1$. In the next step, the set of enabled transitions is $\{t_2, t_9\}$. Firing either is not premature. Note that t_1 and t_{12} are not enabled, for only one product for each type is put in p_0 at M_0. Hence $f_2 = t_2 t_9$. After firing t_2 and t_9, the marking becomes (1, 0, 0, 1, 1, 1) and $R[]$ is updated as $R[4] = R[5] = 1$. Then, the set of enabled transactions is $\{t_3, t_{14}\}$. In this way, the algorithm can be repeated until the assembly for the two products is completed, and a firing sequence $t_1 t_{12} t_2 t_9 t_3 t_{14} t_5 t_6 t_{15} t_8 t_2 t_9 t_{16} t_{10} t_{17} t_3 t_{20}$ is formed. Based on it, the resource requirement for b_0, b_1, b_2, b_3, b_4, and b_5 is 1, 1, 2, 1, 1, and 2, respectively.

Based on Algorithm 13.1, we know that for the set of products to be processed in the system, all the operations can be executed at least once. Thus, according to the definition of realizability in Roszkowska (2004), the assembly process is realizable. We have the following property. In fact, for this example, it reaches the minimal resource requirement, for it is easy to verify that if a single space is removed, the operations cannot be completed.

Property 13.1: For the given set of product types, the assembly process is realizable with the buffer capacity obtained by Algorithm 13.1.

Assume that there are m places (buffers) and n transitions in the CROPN for FAS. Then we have the result for the computational complexity of Algorithm 13.1 as follows:

Property 13.2: The computational complexity to find the realizable buffer capacity requirement by Algorithm 13.1 is $O(\alpha n(n + m))$.

Proof: The computation mainly involves finding the enabled transitions and checking the marking. To find enabled transitions in ROPN, we need to examine each transition to see if it is enabled. There are n transitions, and we need to do that at most n times. Hence, we need at most n^2 times of computation. To check the marking, each time we need to make a comparison for m places, and we need to do that at most n times. Thus, the number of times to do that is $n \times m$. There are a total of α tokens in the system. Therefore, the computational complexity is $O(\alpha n(n + m))$. ∎

Property 13.2 implies that the algorithm is polynomial. Notice that in CROPN m is the number of buffers, and the number of transitions is dependent on the number of places. At worst, there are two transitions between any two places: $n = m(m - 1)$.

The solution found by Algorithm 13.1 is shown to be feasible for the execution of a single product for each type of products. However, we do not know if it is feasible when there are multiple products for each type. By using Algorithm 13.1, we can find $R_i[]$ by releasing one product of type i into the CROPN for product type i. For example, a base and two parts are put into p_0 in the CROPN shown in Figure 13.11a, and $R_A[]$ can be found for the system shown in Figure 13.2. Based on $R_i[]$, $i \in \{1, 2, \ldots, \beta\}$, the resource requirement for deadlock avoidance with multiple products in the system can be calculated as follows:

Definition 13.8: Place p is said to be an assembly place if $\exists t \in p^\bullet$ such that t is an assembly transition. Further, if t is for the assembly of product type i, it is said that i is a product type assembled in p.

For an assembly place p, let $J(p)$ denote the set of product types that are assembled in p and $\bar{J}(p)$ denote the set of product types that use p in the assembly process but are not assembled in p. Clearly, we have $J(p) \cap \overline{J}(p) = \varnothing$. If $J(p) = \varnothing$, p is said to be a nonassembly place. Let $Z[k] = \sum_{i \in J(p_k)} R_i[k]$ and $W[k] = \max_{j \in \overline{J}(p_k)} R_j[k]$ then $R[k] = Z[k] + W[k]$ is the minimal resource requirement for deadlock control with multiple products in the system, where k is the kth buffer place.

Example 13.5

Find the realizable resource requirements for deadlock control in the CROPN shown in Figure 13.12. From the CROPNs shown in Figure 13.11a and b, we obtain $R_A[] = (1, 1, 1, 1, 1, 2)$ and $R_B[] = (1, 1, 1, 1, 1, 1)$. Places $b_0, b_2, b_3, b_4,$ and b_5 are assembly places, and $J(b_0) = J(b_3) = J(b_4) = J(b_5) = \{A, B\}$, $\overline{J}(b_0) = \overline{J}(b_3) = \overline{J}(b_4) = \overline{J}(b_5) = \varnothing$, $J(b_2) = \{B\}$, $\overline{J}(b_2) = \{A\}$. Place b_1 is a nonassembly place and $\overline{J}(b_1) = \{A, B\}$. Thus:

$Z[0] = R_A[0] + R_B[0] = 2$, $Z[3] = R_A[3] + R_B[3] = 2$, $Z[4] = R_A[4] + R_B[4] = 2$, $Z[5] = R_A[5] + R_B[5] = 3$, $W[0] = W[3] = W[4] = W[5] = 0$

$Z[2] = R_B[2] = 1$, $W[2] = R_A[2] = 1$

$Z[1] = 0$, $W[1] = \max(R_A[1], R_B[1]) = 1$

Therefore, we have $R[] = (2, 1, 2, 2, 2, 3)$, which is smaller than $(2, 1, 2, 2, 3, 3)$ that is obtained in Roszkowska (2004).

13.6   DEADLOCK AVOIDANCE CONTROL POLICY

In this section we focus on the deadlock avoidance problem to develop a deadlock avoidance policy for FAS based on the CROPN obtained above. It is shown by Roszkowska (2004) that the problem to find a maximally permissive deadlock avoidance policy for the FAS is NP-hard. Thus, we pursue a nonmaximally permissive control policy. We assume that the buffer capacity in the system is greater than or equal to R[] obtained in the last section, as done in Roszkowska (2004).

Before we develop a control policy, we examine how deadlock occurs in FAS. It is shown that in part production processes, deadlock occurs because of circular wait, or in ROPN, it must occur in so-called production process circuits (PPCs), as defined in Chapter 5. Now, let us observe the CROPN shown in Figure 13.12 for assembly processes. The base component flow, in some sense, is similar to the part production processes. For the base component flow in Figure 13.12, there are two circuits: $v_1 = \{b_0, t_3, b_1, t_5, b_2, t_{17},$ $b_0\}$ and $v_2 = \{b_2, t_8, b_3, t_{15}, b_2\}$. Deadlocks can occur in such circuits, just like the deadlocks in part production processes. In fact, deadlock situation 1 shown in Section 13.2 results from such circuits. Notice that the parts to be mounted onto the base components flow separately. Like the base component flow, the part flow can be deadlocked as well.

We must coordinate the assembly processes well to avoid deadlock. Consider the simple subnet shown in Figure 13.13. Assume that t_3 is an assembly transition for A-products. It requires an A-base component and A-part in p_3 and p_4, respectively. t_4 is an assembly transition for B-products and requires a B-base component and B-part

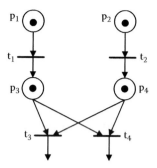

FIGURE 13.13 PN subnet for assembly operations.

in p_3 and p_4, respectively. If at some time p_3 has A-base components and p_4 B-parts, although there may be B-base components in p_1 and A-parts in p_2, assembly cannot be done and a deadlock occurs. Hence, to avoid deadlock in FAS, we should avoid deadlock resulting from base component flow, part flow, and assembly.

Consider the FAS shown in Figure 13.1. There are input and output buffers in workstation 1. The buffers belonging to a workstation can be seen as a special set. Let B_i denote the set of buffers in workstation i. It may contain only one buffer. B_i is called a buffer group. We also notice that there is some $t \in p_0{}^\bullet$, and $t^\bullet$ is an input place of an assembly transition such that firing t releases a part into the system. Such a set of transitions is denoted by T_{tray}. For example, in the CROPN shown in Figure 13.12, t_2 and t_9 are such transitions. We present the deadlock avoidance control policy for FAS, considered here as follows:

WZ-policy: A transition $t \in T$ in CROPN for FAS at marking M is enabled, and its firing changes M to M'. Then t can fire only if all of the following conditions are met:

1. Assume that there are k groups of buffers B_{1-k} for the base components. Then in M' there are at least $k - 1$ free buffer spaces such that at most, one group B_i is full.
2. Assume that there are k buffers b_{1-k} for the parts to be mounted onto the base components. Then in M' there are at least $k - 1$ free buffer spaces such that at most, one buffer is full.
3. Assume transition $t \notin T_{tray}$, and its firing moves V_h tokens of type-h product into p_i. Let $M(p_i)(k)$ denote the number of tokens representing type k product in p_i at marking M. Further assume transition $t_1 \in T_{tray}$, and its firing moves U_h tokens of type-h product into p_i. Then
   a. $K(p_i) - W[i] - \Sigma_{k \in AS(i), k \neq h} \max(M(p_i)(k), R_k[i]) - M(p_i)(h) - U_h \geq V_h$, if $h \in J(i)$, where $R_k[]$ and $W[i]$ are obtained from the last section
   b. $K(p_i) - Z[i] - M(p_i)(h) \geq V_h$, if $h \in \bar{J}(i)$
4. Assume $t \in T_{tray}$, its firing moves U_h tokens of type-h product with color c_1 into p_i, and these tokens, together with V_h tokens of type-h product in p_j and Y_h tokens with color c_2 in p_i, enable assembly transition t_a. Then $M(p_j)(h) \geq V_h$, $M(p_i)(c_2) \geq Y_h$, and $M(p_i)(c_1) = 0$.

Conditions 1 and 2 guarantee that the base component and part flows will not be deadlocked, respectively. Condition 3 guarantees that the state shown in Figure 13.13 can never occur. For example, in the CROPN shown in Figure 13.12, an A-base is needed in b_3 to perform the last assembly of A-product. According to condition 3, if there is no A-base in b_3 at marking M, a space in b_3 should be reserved for an A-product. However, if there is an A-base in it, no matter if it is a raw part or the operation for the part has been completed, no space needs to be reserved. If there is no A-base in b_3, there must be at least one free space, and an A-base can be moved into b_3 only if conditions 1 and 2 are satisfied.

Condition 4 avoids any premature firing such that the buffer spaces are effectively used. For example, to enable t_{10} in Figure 13.12, we need an A-base in b_3, an A-part from b_4 in b_5, and an A-part from p_0. According to condition 4, only when there are an A-base in b_3 and an A-part from b_4 in b_5 can t_9 then fire to release an A-part to b_5 from p_0, since multiple A-parts from p_0 occupying spaces in b_5 are not meaningful before A-base and A-part are ready.

Notice that condition 3 does not make a restriction on a nonassembly place. A nonassembly place needs to meet conditions 1 and 2 only.

Lemma 13.1: An initially marked subnet of CROPN containing n places is live if at any marking M there are at least $n - 1$ free spaces available such that at most, one place is full.

Proof: We only need to show that if the subnet has only $n - 1$ free spaces and at the same time the distribution condition is satisfied, then the subnet is live, because if there are more free spaces, the problem is easier. We use a three-place subnet to show it, and for more place situations, it can be shown similarly.

In Figure 13.14, a subnet contains three places. It is the most complex subnet containing three places. According to the assumption, it needs only two free spaces for the subnet to be live. Assume that the first space is in p_1 and the other in p_2. According to our control policy, t_{31} or t_{32} can fire, for by firing one of them, the condition in the lemma is still satisfied. Because there are tokens in p_3, there must be a token that enables t_{31} or t_{32}, or both. We can select a token in p_3 to fire t_{31} or t_{32}. Assume that t_{31}

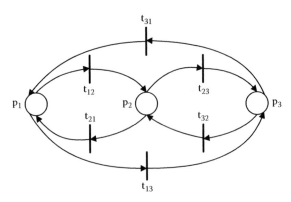

FIGURE 13.14 A subnet for the proof of Lemma 13.1.

fires. Then a token is moved into p_1 from p_3, and at the same time the space in p_1 is moved into p_3. This time we can fire t_{12} or t_{13}. Assume that t_{12} fires. A token is moved into p_2, and we can fire t_{21} or t_{23}. In this way, after any marking is reached, we can still fire any transition again. This implies that the subnet is live. ■

It is easy to prove the following lemma.

Lemma 13.2: If in a subnet of CROPN a place is replaced by a direct place path and the condition given in Lemma 13.1 is still satisfied by treating this direct place path as a place with the path capacity, then the subnet is still live.

With Lemmas 13.1 and 13.2, now we can show that the assembly process is deadlock-free under the control of W-policy.

Theorem 13.1: The CROPN of an FAS is deadlock-free if WZ-policy is applied.

Proof: We know from the assumption that a group of buffers B in our problem is a single place or a direct place path, so it follows from conditions 1 and 2, and Lemmas 13.1 and 13.2, that the base component flow and the other part flow are deadlock-free. Thus, the system can be deadlocked only from the assembly process.

By condition 3, if a place corresponding to buffer b_i is an assembly place and V_k spaces are required for b_i to assemble one product of type k, then at any marking M there are at least V_k parts of type k in b_i, or $G_k - M(b_i)(k)$ spaces are reserved for the part type. In this way, assembly deadlock is avoided. By condition 4, the tokens moved to a place by firing a transition in T_{tray} do not occupy any unnecessary spaces. From the proof of Lemma 13.1, we know that when tokens in a place can be used to fire the transitions, we can select any token in the place. Thus, each type of product in the assembly process has a path itself, and no deadlock will result from assembly operations. At the same time, the reserved spaces are calculated based on $R[]$, or the reservation is feasible. Therefore, the system is deadlock-free. ■

Although the WZ-policy is a group of sufficient conditions, it is not too conservative, for we require only $n - 1$ free space for the base component and part transportation, respectively, where n is the total number of groups of buffers. In general, in FAS each buffer has multiple spaces. Hence, the restriction imposed due to this policy is not too much. We do not impose much restriction on the number of parts in the system; rather, we restrict the ratio of parts among the types. In FAS, the number of parts, not the ratio of parts of different types, greatly affects the productivity. Therefore, our policy is acceptable.

Theorem 13.2: The WZ-policy is less conservative than the R-policy (Roszkowska, 2004).

Sketch of proof: The R-policy requires meeting at least one of two conditions, as follows. Condition I is for space reservation, and this condition has two parts:

Condition I-a: If an assembly operation is executed in buffer 1, after the assembly the part is moved into buffer 2 immediately. Then, a space in buffer 2 should be reserved at any marking.

Condition I-b: If a place (buffer) b_k is an assembly place and product type $j \in J(k)$ is an assembly product type in b_k, then a space in b_k should be reserved for j if there is no part of type j in b_k waiting for assembly, even if a token representing product type j cannot enter this place.

Condition II requires that if firing a transition moves a token into place b_k, then there must be at least one free space in b_k after the token enters b_k.

Now we examine the WZ-policy. For any place in CROPN, assembly place or not, no free space is required to be reserved at *any marking*, while the R-policy does. It requires conditions 1 and 2 to be met, allowing that places can be full at some marking. Thus, it is less conservative than conditions I-a and II in the R-policy. Assume that b_k is an assembly place and product type $j \in J(k)$ is an assembly product type in b_k. Then, by WZ-policy, if there is a j-part in b_k, no matter whether it is one that waits for assembly or has just been completed, it needs not reserve a space in b_k for type j product. Furthermore, if there is no j-part in b_k, a j-part can enter b_k immediately if conditions 1 and 2 are met. Thus, the WZ-policy is also less conservative than condition I-b in the R-policy. This completes the proof. ∎

We will demonstrate this conclusion in the next section by an example. In implementing the WZ-policy, for each transition t, only the number of tokens and their type in $t^\bullet$ need to be checked. Thus, we immediately have the following result:

Theorem 13.3: The complexity for implementing the W-policy is polynomial, i.e., $O(|P|\beta)$, where P is the place set in CROPN and β is the number of product types.

13.7 ILLUSTRATIVE EXAMPLE

Example 13.6

Consider the FAS shown in Figure 13.1 (Roszkowska, 2004). Assume that the capacity for b_{0-5} is 2, 4, 4, 3, 6, and 5, respectively. Apply the W-policy to it for deadlock avoidance.

The CROPN is shown in Figure 13.12. By using the R-policy, a marking is reached as shown in Figure 13.15, where a dot denotes a part for product A and a square denotes a part for product B. At this marking, transition t_{12} is forbidden by the R-policy. This implies that no more jobs can be released into the system.

Now we consider the system by the WZ-policy. We obtain that $R_A[] = (1, 1, 1, 1, 1, 2)$ and $R_B[] = (1, 1, 1, 1, 1, 2)$. Places b_0 and b_{2-5} are assembly places. In the marking we have $M(b_0)(A) = 1$, $M(b_2)(A) = 3$, $M(b_3)(A) = 1$, $M(b_4)(A) = 3$, and $M(b_5)(A) = 3$. Thus, for product A, condition 3 is satisfied. For product B, we have $M(b_0)(B) = 1$, $M(b_2)(B) = 0$, $M(b_3)(B) = 1$, $M(b_4)(B) = 1$, and $M(b_5)(B) = 1$. Notice that there is a free space in b_2, or condition 3 is also satisfied for product B. Places b_0 and b_1 form a direct place path. There are three groups of places for the material flow process of base components $B_1 = \{b_0, b_1\}$, $B_2 = \{b_2\}$, and $B_3 = \{b_3\}$. By the W-policy, two free spaces are necessary, and they should be distributed in two different groups of places. In the current marking, the free spaces in b_{0-3} are 0, 2, 1, and 1, respectively. Hence condition 1 is satisfied. The free spaces in b_4

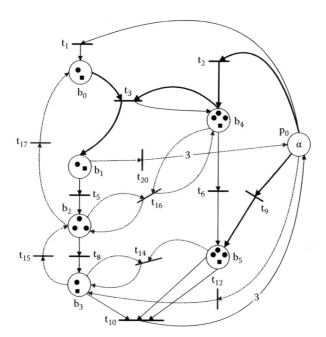

FIGURE 13.15 The CROPN for Example 13.6.

and b_5 are 1 and 1, respectively. We require only one free space, and it can be located in either b_4 or b_5, so condition 2 is also satisfied. This implies that this marking is legal for the W-policy and can be reached. However, there are more free spaces in the buffers than what is required by the W-policy, so more parts can be released into the system if other conditions are satisfied. In fact, at this marking, t_{12} is enabled and firing it does not violate the W-policy. It is easy to check that there will be no deadlock by firing t_{12} if this system is controlled by the W-policy. This implies that more jobs can be released into the system than with the R-policy.

In fact, we have $\sum_{i=0}^{3} K(b_i) = 13$. By condition 1, two free spaces are needed for deadlock-free operations of base component flow. This implies that at most eleven products can be released into the system for assembly. This is shown to be true by the token evolution from the initial state, as follows.

Assume that initially $M_0(p_0) > 0$ and, for any i, $M_0(b_i) = 0$; the token evolution from M_0 is presented in Table 13.1, where the number in the brackets after a place represents the capacity of the place. In the table, ① is used for a token of A-base with color f_{AB1}, ② for a token of A-base with color f_{AB2}, ③ for A-part with color f_{API}, ④ for A-part with color f_{AP2}, and ⑤ for A-part with color f_{AP3}; ❶ for B-base with color f_{BB1}, ❷ for B-base with color f_{BB2}, ❸ for B-base with color f_{BB3}, ❹ for B-base with color f_{BB4}, ❺ for B-part with color f_{BP1}, ❻ for B-part with color f_{BP2}, and ❼ for B-part with color f_{BP3}. It is shown that a total of eleven products (eleven base components and their associated parts) are released into the system for assembly. We have seen that by the R-policy, only nine products can be released into the system. This demonstrates the conclusion in the last section: the W-policy is less conservative than the R-policy.

TABLE 13.1

The Token Evolution from M_0

Marking	Type	$b_0(2)$	$b_1(4)$	$b_2(4)$	$b_3(3)$	$b_4(6)$	$b_5(5)$
M_1	A	①				③	
	B				❶		❺
M_2	A	①	②			③④	
	B				❷❶		❺
M_3	A	①	②	②		③④	④
	B			❷	❷❶	❻	❺
M_4	A	①	②	②②		③④	④④
	B			❸❷	❷❶	❼❻	❺
M_5	A	①	②②	②	②	④④	④④④⑤
	B	❸		❸❷	❷❶	❼❼❻	❺
M_6	A	①	②②	②②	②	④④③	④④④⑤
	B	❸	❹	❸	❷❷	❼❼	

13.8   INDUSTRIAL CASE STUDY

This section presents an industrial case study to show the application of the deadlock control policy presented in this chapter. This system is the main part of an assembly line for the core mechanisms of recorders. Its schematic structure is shown in Figure 13.16. It contains nine workstations, and each workstation w_i has two buffers, a_i and b_i, for bases and parts (trays), respectively, with capacity $K(a_i) = K(a_9) = K(b_j) = 4$, $K(a_{i+4}) = 5$, $i \in N_4$, $j \in N_9$. In the system, the bases are delivered among buffers a_i and the central buffer. Trays are delivered among buffers b_i and the central buffer by a number of robots. Because robots have no contribution to deadlock in our method, they are omitted in Figure 13.16.

There are two types of recorders for assembly: one-way and two-way. They have different structures and require different models. Their assembly processes are shown in Figure 13.17. All of NR rotary-rod and NR rotary-rod-spring, REC button and REC spring, anti-erasion-claw and anti-erasion-claw-spring, and each part of STOP buttons, lock-plates, and slides are put in trays, respectively. Thus, the material

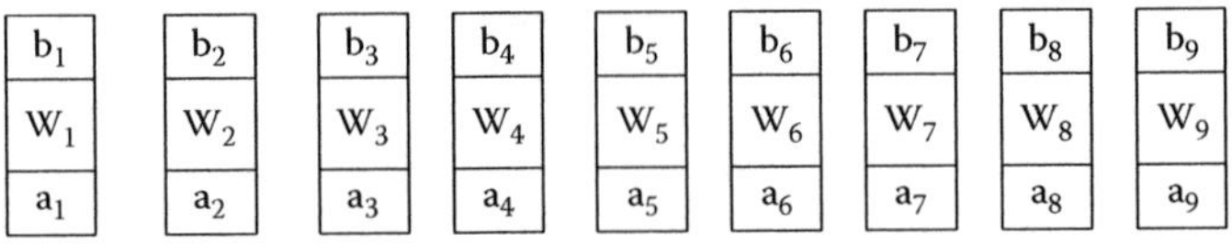

b_1	b_2	b_3	b_4	b_5	b_6	b_7	b_8	b_9
W_1	W_2	W_3	W_4	W_5	W_6	W_7	W_8	W_9
a_1	a_2	a_3	a_4	a_5	a_6	a_7	a_8	a_9

FIGURE 13.16 Schematic structure of the assembly line for the case study.

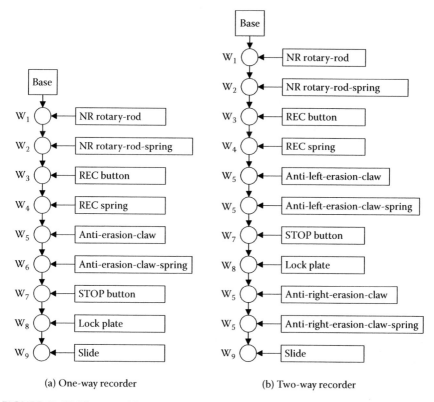

(a) One-way recorder　　　　　　　　(b) Two-way recorder

FIGURE 13.17 The assembly processes.

flows in the assembly processes of one-way and two-way recorders are shown in Figure 13.18, where O-Base and T-Base stand for bases and O-Tray and T-Tray stand for trays of one-way and two-way recorders, respectively.

By using the modeling method developed in this chapter, the CROPN model for this problem is obtained as shown in Figure 13.19, where a_i's and b_i's correspond to the buffers for base and tray, and p_0, the central buffer; x_i's and y_i's stand for the events of delivering bases and trays; w_i's, the assembly operations; and z_i's, the events of delivering materials between the central buffer and other buffers, respectively. This model has only nineteen places and twenty-two transitions. Thus, this industrial problem can be easy to handle in terms of the model size.

In the CROPN for the case problem, all the places but p_0 are assembly places and $T_{\text{tray}} = \{z_2, z_3, z_4, z_5, z_6, z_7\}$. We find $R_i[] = (1, 1, 1, 1, 1, 1, 1, 1, 1, 1, 1, 1, 1, 1, 1, 1, 1, 1, 1)^T$ for any product type i. This means at least n spaces are needed in each buffer a_i or b_i if n different product types are to be assembled concurrently. There is only one circuit, $v = \{a_5, x_5, a_6, x_6, a_7, x_7, a_8, x_9, a_5\}$, that does not contain place p_0. Note that p_0 models the central buffer that can be treated as an infinite capacity buffer, or $K(p_0) = \infty$. Thus, to avoid deadlock in this system by the proposed approach, control the transition firings such that (1) among the places a_5, a_6, a_7, and a_8, only one place can be full at any time, and all other places can be full when other conditions are met; (2) if in place a_i or

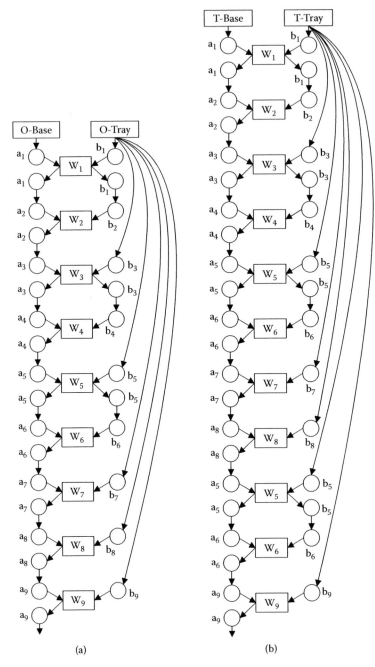

FIGURE 13.18 The material flows in the assembly processes of (a) one-way and (b) two-way recorders.

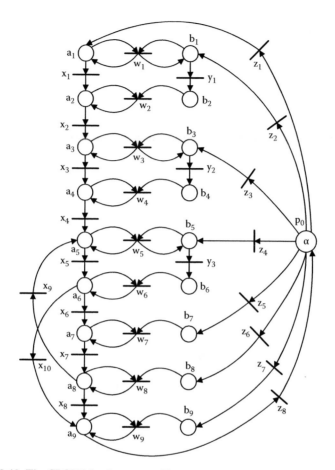

FIGURE 13.19 The CROPN for the case problem.

b_i there is no base or part for product type j, a space is kept in this place for type j that is identified by colors; and (3) there is no premature firing for any transition in T_{tray}. In practice, there are at most two or three different product models in assembling at a time. Noticing that the capacity in places a_5, a_6, a_7, and a_8 is five, and four for other places, the control policy can be easily implemented and the resources can be efficiently utilized.

It should be noted that if the R-policy is used for this problem, all the places cannot be full at any marking, so at most three different models can be assembled concurrently in the system. However, by our method, only three of the four places, a_5, a_6, a_7, and a_8, need one reserved space; all other places can be full. Thus, four different models can be processed concurrently in the system.

13.9 SUMMARY

In the past decades, the problem of deadlock avoidance in automated manufacturing systems for sequential processes or part production processes has been studied extensively, and many results were obtained. Relatively, the problem of deadlock avoidance for FAS has received less attention. In FAS, the base components are transported with pallets, but the parts to be mounted onto them are transported via trays. When an assembly operation is performed by using some parts in a tray, the tray with the others still occupies a buffer space. In this way, an assembly/disassembly material flow is formed. In such systems, deadlock can occur in the base component flow, part flow, and assembly operations. Thus, it is a great challenge to avoid deadlock in such systems.

In this chapter, we study the problem of deadlock avoidance in assembly/disassembly processes. We propose using resource-oriented Petri nets to capture the discrete event dynamics of FAS concisely. Based on the PN model, a deadlock control policy is proposed. It is shown that the control policy is computationally efficient and better than the one in the literature.

In Hsieh (2004), a joint assembly system is studied where the resources are machines. A machine performs operations and holds parts at the same time. Hsieh (2004) did not present a closed-loop control policy but focused on mainly the impact caused by machine failure. Such a system can be modeled by the method presented here with some modifications. The control policy can also be applied.

REFERENCES

Fanti, M. P., B. Maione, and B. Turchiano. 1997. Event control for deadlock avoidance in assembly systems. In *Proceedings of IEEE Conference on Systems, Man, and Cybernetics,* Orlando, FL, 3756–61.

Fanti, M. P., B. Maione, and B. Turchiano. 1998. Design and implementation of supervisors avoiding deadlocks in flexible assembly systems. In *Proceedings of IEEE-SMC Multiconference on Computational Engineering System Applications*, Hammamet, Tunisia, 667–672.

Hsieh, F.-S. 2004. Fault-tolerant deadlock avoidance algorithm for assembly processes. *IEEE Transactions on Systems, Man, and Cybernetics A* 34:65–79.

Paik, C., and D. Tcha. 1995. Throughput equivalencies in fork/join queueing networks with finite buffers and general service times. *International Journal of Production Research* 33:695–703.

Roszkowska, E. 2004. Supervisory control for deadlock avoidance in compound processes. *IEEE Transactions on System, Man, and Cybernetics A* 34:52–64.

Roszkowska, E., and R. Wojcik. 1993. Problems of process flow feasibility in FAS. In *CIM in process and manufacturing industries*, 115–20. Oxford: Pergamon.

Wu, N. Q. 1997. Avoiding deadlocks in automated manufacturing systems with shared material handling system. In *Proceedings of 1997 IEEE International Conference on Robotics and Automation*, Albuquerque, NM, 2427–33.

Wu, N. Q., M. C. Zhou, and Z. W. Li. 2008. Resource-oriented Petri net for deadlock avoidance in flexible assembly systems. *IEEE Transactions on System, Man, and Cybernetics A* 38:56–69.

Bibliography

Abdallah, I., and A. Elmaraghy. 1998. Deadlock prevention and avoidance in FMS: A Petri net based approach. *International Journal of Advanced Manufacturing Technology* 14:704–15.

Aho, A. V., J. E. Hopcroft, and J. D. Ullman. 1987. *Data structures and algorithms*. Reading, MA: Addison-Wesley.

Al-Jaar, R. Y., and A. A. Desrochers. 1990. Performance evaluation of automated manufacturing systems using generalized stochastic Petri nets. *IEEE Transactions on Robotics and Automation* 6:621–39.

Bader, M. E., R. P. Hall, and G. Strasser. 1990. Integrated processing equipment. *Solid State Technology* 33:149–54.

Banaszak, Z. A., and B. H. Krogh. 1990. Deadlock avoidance in flexible manufacturing systems with concurrently competing process flows. *IEEE Transactions on Robotics and Automation* 6:724–34.

Bozer, Y. A., and M. M. Srinivasan. 1989. Tandem configuration for AGV systems offer flexibility and simplicity. *Industrial Engineering* 21:23–27.

Bozer, Y. A., and M. M. Srinivasan. 1991. Tandem configuration for automated guided vehicle systems and the analysis of single vehicle loops. *IIE Transactions* 23:72–82.

Bozer, Y. A., and M. M. Srinivasan. 1992. Tandem AGV systems: A partitioning algorithm and performance comparison with conventional AGV systems. *European Journal of Operational Research* 63:173–92.

Burggraaf, P. 1995. Coping with the high of wafer fabs. *Semiconductor International* 38:45–50.

Byeon, E.-S., S. D. Wu, and R. H. Storer. 1998. Decomposition heuristics for robust job-shop scheduling. *IEEE Transactions on Robotics and Automation* 14:303–13.

Chen, S.-J., and L. Lin. 1999. Reducing total tardiness cost in manufacturing cell scheduling by a multi-factor priority rule. *International Journal of Production Research* 37:2939–56.

Cho, H., T. K. Kumaran, and R. A. Wysk. 1995. Graph-theoretic deadlock detection and resolution for flexible manufacturing systems. *IEEE Transactions on Robotics and Automation* 11:413–21.

Chu, F., and X. L. Xie. 1997. Deadlock analysis of Petri nets using siphons and mathematical programming. *IEEE Transactions on Robotics and Automation* 13:793–804.

Coffman, E. G., Jr., M. J. Elphick, and A. Shoshani. 1971. System deadlocks. *ACM Computing Surveys* 3:67–78.

Cormen, T. H., C. E. Leiserson, and R. E. Rivest. 1990. *Introduction to algorithms*. New York: MIT Press/McGraw-Hill.

Dowsland, K. A., and A. M. Greaves. 1994. Collision avoidance in bi-directional AGV systems. *Journal of Operations Research Society* 45:817–26.

Egbelu, P. J., and J. M. A. Tanchoco. 1986. Potential for bidirectional guided-path for automated vehicles based systems. *International Journal of Production Research* 24:1075–97.

Ezpeleta, J., J. M. Colom, and J. Martinez. 1995. A Petri net based deadlock prevention policy for flexible manufacturing systems. *IEEE Transactions on Robotics and Automation* 11:171–84.

Fanti, M. P., B. Maione, S. Mascolo, and B. Turchiano. 1997. Event-based feedback control for deadlock avoidance in flexible production systems. *IEEE Transactions on Robotics and Automation* 13:347–63.

Fanti, M. P., B. Maione, and B. Turchiano. 1997. Comparing digraph and Petri net approaches to deadlock avoidance in FMS modeling and performance analysis. *IEEE Transactions on Systems, Man, and Cybernetics A* 30:783–98.

Fanti, M. P., B. Maione, and B. Turchiano. 1997. Event control for deadlock avoidance in assembly systems. In *Proceedings of IEEE Conference on Systems, Man, and Cybernetics*, 3756–61.

Fanti, M. P., B. Maione, and B. Turchiano. 1998. Design and implementation of supervisors avoiding deadlocks in flexible assembly systems. In *Proceedings of IEEE-SMC Multiconference on Computational Engineering Systems Application*, Hammamet, Tunisia, 667–672.

Fanti, M. P., G. Maione, and B. Turchiano. 2002. Design of supervisors to avoid deadlock in flexible assembly system. *International Journal of Flexible Manufacturing Systems* 14:157–75.

Fanti, M. P., and M. C. Zhou. 2004. Deadlock control methods in automated manufacturing systems. *IEEE Transactions on Systems, Man, and Cybernetics A* 34:5–21.

Gligor, V. D., and S. H. Shuttuck. 1980. On deadlock detection in distributed systems. *IEEE Transactions on Software Engineering* 6(5).

Hammond, G. (1986). *AGVS at work*. New York: Springer-Verlag.

Holt, R. C. 1972. Some deadlock properties of computer systems. *ACM Computing Surveys* 6:79–96.

Hsieh, F.-S. 2004. Fault-tolerant deadlock avoidance algorithm for assembly processes. *IEEE Transactions on Systems, Man, and Cybernetics A* 34:65–79.

Hsieh, F., and S. Chang. 1994. Dispatching-driven deadlock avoidance controller synthesis for flexible manufacturing systems. *IEEE Transactions on Robotics and Automation* 10:196–209.

Huang, J., U. S. Palekar, and S. G. Kapoor. 1993. A labeling algorithm for the navigation of automated guided vehicles. *Journal of Engineering for Industry* 115:315–21.

Jeng, M. D., and F. DiCesare. 1995. Synthesis using resource control nets for modeling shared-resource systems. *IEEE Transactions on Robotics and Automation* 11:317–27.

Jensen, K. 1981. Colored Petri nets and the invariant method. *Theoretical Computer Science* 14:317–36.

Jensen, K. 1986. Colored Petri nets. *Lecture Notes in Computer Science*, vol. 254, part 1, 248–99. Berlin: Springer-Verlag.

Johri, P. K. 1993. Practical issues in scheduling and dispatching in semiconductor wafer fabrication. *Journal of Manufacturing Systems* 12:74–485.

Joshi, S. B., E. G. Mettala, J. S. Smith, and R. A. Wysk. 1995. Formal models for control of flexible manufacturing cells: Physical and system models. *IEEE Transactions on Robotics and Automation* 11:558–70.

Kim, C. O., and S. S. Kim. 1997. An efficient real-time deadlock-free control algorithm for automated manufacturing systems. *International Journal of Production Research* 35:1545–60.

Kim, C. W., and J. M. A. Tanchoco. 1991. Conflict-free shortest bi-directional AGV routing. *International Journal of Production Research* 29:2377–91.

Kim, C. W., and J. M. A. Tanchoco. 1993. Operational control of bi-directional automated guided vehicle system. *International Journal of Production Research* 31:2123–38.

Kocay, W., and D. L. Kreher. 2005. *Graphs, algorithms and optimization*. Boca Raton, FL: Chapman & Hall/CRC.

Koff, G. A. 1987. Automatic guided vehicle: Application, control, and planning. *Material Flows* 4:3–16.

Krishnamurthy, N. N., R. Batta, and M. H. Karwan. 1993. Developing conflict-free routes for automated guided vehicles. *Operations Research* 41:1077–90.

Lawley, M. A. 1999. Deadlock avoidance for production systems with flexible routing. *IEEE Transactions on Robotics and Automation* 15:1–13.

Lawley, M. A., S. Reveliotis, and P. Ferreira. 1998. The application and evaluation of banker's algorithm for deadlock-free buffer space allocation in flexible manufacturing systems. *International Journal of Flexible Manufacturing Systems* 10:73–100.

Lawley, M. A., S. Reveliotis, and P. Ferreira. 1998. A correct and scalable deadlock avoidance policy for flexible manufacturing policy. *IEEE Transactions on Robotics and Automation* 14:796-809.

Lee, C.-C., and J. T. Lin. 1995. Deadlock prediction and avoidance based on Petri nets for zone-control automated guided vehicle systems. *International Journal of Production Research* 33:3249–265.

Lee, D. Y., and F. DiCesare. 1994. Scheduling FMS using Petri nets and heuristic search. *IEEE Transactions on Robotics and Automation,* 10:123–132.

Li, Z. W., and M. C. Zhou. 2004. Elementary siphons of Petri nets and their application to deadlock prevention in flexible manufacturing systems. *IEEE Transactions on Systems, Man, and Cybernetics A* 34:38–51.

Li, Z. W., and M. C. Zhou. 2008. Control of elementary and dependent siphons in Petri nets and their applications. *IEEE Transactions on Systems, Man, and Cybernetics A* 38:133–48.

Li, Z. W., M. C. Zhou, and N. Q. Wu. 2008. A survey and comparison of Petri net-based deadlock prevention policy for flexible manufacturing systems. *IEEE Transactions on Systems, Man, and Cybernetics C* 38:173–88.

Malmbog, J. 1990. A model for the design of zone-control automated guided vehicle systems. *International Journal of Production Research* 28:1741–758.

Marsan, M. A., G. Conte, and G. Balbo. 1984. A class of generalized stochastic Petri nets for the performance evaluation of multiprocessor systems. *ACM Transactions on Computer Systems* 2:93–122.

Murata, T. 1989. Petri nets: Properties, analysis and applications. *Proceedings of the IEEE* 77:541–80.

Narahari, Y., and N. Viswanadham. 1985. A Petri net approach to modeling and analysis of FMS. *Annals of Operations Research* 3:449–72.

Paik, C., and D. Tcha. 1995. Throughput equivalencies in fork/join queueing networks with finite buffers and general service times. *International Journal of Production Research* 33:695–703.

Park, J., and S. A. Reveliotis. 2002. Liveness-enforcing supervision of resource allocation systems with uncontrollable behavior and forbidden states. *IEEE Transactions on Robotics and Automation* 18:234–39.

Park, S. C., N. Raman, and M. J. Shaw. 1997. Adaptive scheduling in dynamic flexible manufacturing systems: A dynamic rule selection approach. *IEEE Transactions on Robotics and Automation* 13:486–502.

Perkinson, T. L., R. S. Gyurcsik, and P. K. MacLarty. 1996. Single-wafer cluster tool performance: An analysis of the effects of redundant chambers and revisitations sequences on throughput. *IEEE Transactions on Semiconductor Manufacturing* 9:384–400.

Perkinson, T. L., P. K. MacLarty, R. S. Gyurcsik, and R. K. Cavin III. 1994. Single-wafer cluster tool performance: An analysis of throughput. *IEEE Transactions on Semiconductor Manufacturing* 7:369–73.

Proth, J.-M., and X. L. Xie. 1996. *Petri nets: A tool for design and management of manufacturing systems.* New York: John Wiley & Sons.

Ramaswamy, S. E., and S. B. Joshi. 1996. Deadlock-free schedules for automated manufacturing workstations. *IEEE Transactions on Robotics and Automation* 12:391–400.

Reveliotis, S. A. 1999. Accommodating FMS operational contingencies through routing flexibility. *IEEE Transactions on Robotics and Automation* 15:3–19

Reveliotis, S. A. 2000. Conflict resolution in AGV systems. *IIE Transactions* 32:647–59.

Reveliotis, S. A., and P. M. Ferreira. 1996. Deadlock avoidance policies for automated manufacturing cells. *IEEE Transactions on Robotics and Automation* 12:845–57.

Rostami, S., B. Hamidzadeh, and D. Camporese. 2001. An optimal periodic scheduler for dual-arm robots in cluster tools with residency constraints. *IEEE Transactions on Robotics and Automation* 17:609–18.

Roszkowska, E. 2004. Supervisory control for deadlock avoidance in compound processes. *IEEE Transactions on System, Man, and Cybernetics A* 34:52–64.

Roszkowska, E., and R. Wojcik. 1993. Problems of process flow feasibility in FAS. In *CIM in process and manufacturing industries*, 115–20. Oxford: Pergamon

Silberschatz, A., and P. G. Galvin. 1994. *Operating system concepts*. 4th ed. Reading, MA: Addision-Wesley.

Singer, P. 1995. The driving forces in cluster tools development. *Semiconductors International* 38:113–18.

Sinriech, D., and J. M. A. Tanchoco. 1995. An introduction to the segmented flow approach for discrete material flow systems. *International Journal of Production Research* 33:3381–410.

Sinriech, D., J. M. A. Tanchoco, and Y. T. Herer. 1996. The segmented bi-directional single-loop topology for material flow systems. *IIE Transactions* 28:4–54.

Venkatesh, S., R. Davenport, P. Foxhoven, and J. Nulman. 1997. A steady-state throughput analysis of cluster tools: Dual-blade versus single-blade robots. *IEEE Transactions on Semiconductor Manufacturing* 10:418–23.

Viswanadham, N., and Y. Narahari. 1987. Colored Petri net models for automated manufacturing systems. Paper presented at Proceedings of IEEE International Conference on Robotics and Automation, Raleigh, NC.

Viswanadham, N., Y. Narahari, and T. L. Johnson. 1990. Deadlock prevention and deadlock avoidance in flexible manufacturing systems using Petri net models. *IEEE Transactions on Robotics and Automation* 6:713–23.

Wu, N. Q. 1997. Avoiding deadlocks in automated manufacturing systems with shared material handling system. In *Proceedings of 1997 IEEE International Conferenc on Robotics and Automation*, Albuquerque, NM, 2427–33.

Wu, N. Q. 1999. Necessary and sufficient conditions for deadlock-free operation in flexible manufacturing systems using a colored Petri net model. *IEEE Transactions on Systems, Man, and Cybernetics C* 29:192–204.

Wu, N. Q., and W. Q. Zeng. 2002. Deadlock avoidance in AGV system using colored Petri net model. *International Journal of Production Research* 40:223–38.

Wu, N. Q., and M. C. Zhou. 2000. Resource-oriented Petri nets for deadlock avoidance in automated manufacturing. In *Proceedings of 2000 IEEE International Conference on Robotics and Automation,* San Francisco, CA, 3377–82.

Wu, N. Q., and M. C. Zhou. 2001. Avoiding deadlock and reducing starvation and blocking in automated manufacturing systems. *IEEE Transactions on Robotics and Automation* 17:657–68.

Wu, N. Q., and M. C. Zhou. 2001. Resource-oriented Petri nets in deadlock avoidance of AGV systems. In *Proceedings of 2001 IEEE International Conference on Robotics and Automation*, Seoul, Korea, 64–69.

Wu, N. Q., and M. C. Zhou. 2002. Deadlock avoidance in semiconductor track systems. In *Proceedings of 2002 IEEE International Conference on Robotics and Automation*, Washington, DC, 193–98.

Wu, N. Q., and M. C. Zhou. 2003. AGV routing for conflict resolution in AGV systems. In *Proceedings of 2003 IEEE International Conference on Robotics and Automation*, Taipei, Taiwan, 1428–33.

Wu, N. Q., and M. C. Zhou. 2004. Modeling and deadlock control of automated guided vehicle systems. *IEEE Transactions on Mechatronics* 9:50–57.

Wu, N. Q., and M. C. Zhou. 2005. Modeling and deadlock avoidance of automated manufacturing systems with multiple automated guided vehicles. *IEEE Transactions on Systems, Man, and Cybernetics B* 35:1193–202.

Wu, N. Q., and M. C. Zhou. 2007. Deadlock modeling and control of semiconductor track systems using resource-oriented Petri nets. *International Journal of Production Research* 45:3439–56.

Wu, N. Q., and M. C. Zhou. 2007. Deadlock resolution in automated manufacturing systems with robots. *IEEE Transactions on Automation Science and Engineering* 4:474–80.

Wu, N. Q., and M. C. Zhou. 2007. Real-time deadlock-free scheduling for semiconductor track systems based on colored timed Petri nets. *OR Spectrum* 29:421–43.

Wu, N. Q., and M. C. Zhou. 2007. Shortest routing of bi-directional automated guided vehicles avoiding deadlock and blocking. *IEEE/ASME Transactions on Mechatronics* 12:63–72.

Wu, N. Q., M. C. Zhou, and G. Hu. 2007. On the Petri net modeling of automated manufacturing systems. In *Proceedings of 2007 IEEE International Conference on Networking, Sensing and Control*, London, 228–33.

Wu, N. Q., M. C. Zhou, and Z. W. Li. 2008. Resource-oriented Petri net for deadlock avoidance in flexible assembly systems. *IEEE Transactions on System, Man, and Cybernetics A* 38:56–69.

Wysk, R. A., N. S. Yang, and S. Joshi. 1991. Detection of deadlocks in flexible manufacturing cells. *IEEE Transactions on Robotics and Automation* 7:853–59.

Xie, X., and M. D. Jeng. 1999. ERCN-merged nets and their analysis using siphons. *IEEE Transactions on Robotics and Automation* 15:692–703.

Xing, K. Y., B. S. Hu, and H. X. Chen. 1996. Deadlock avoidance policy for Petri net modeling of flexible manufacturing systems with shared resources. *IEEE Transactions on Automatic Control* 41:289–95.

Xiong, H. H., and M. C. Zhou. 1999. A Petri net method for deadlock-free scheduling of flexible manufacturing systems. *International Journal of Intelligent Control and Systems* 3:277–95.

Yeh, M.-S., and W.-C. Yeh. 1998. Deadlock prediction and avoidance for zone-control AGVS. *International Journal of Production Research* 36:2879–89.

Yim, D.-S., J.-I. Kim, and H.-S. Woo. 1997. Avoidance of deadlocks in flexible manufacturing systems using a capacity-designated directed graph. *International Journal of Production Research* 35:2459–75.

Yoon, H. J., and D. Y. Lee. 2001. Identification of potential deadlock set in semiconductor track systems. In *Proceedings of 2001 IEEE International Conference on Robotics and Automation*, Seoul, Korea, pp. 1820–25.

Yoon, H. J., and D. Y. Lee. 2004. Deadlock-free scheduling of photolithography equipment in semiconductor fabrication. *IEEE Transactions on Semiconductor Manufacturing* 17:42–54.

Zeng, L., H.-P. Wang, and S. Jin. 1991. Conflict detection of automated guided vehicles: A Petri net approach. *International Journal of Production Research* 29:865–79.

Zhou, M. C., and F. DiCesare. 1991. Parallel and sequential mutual exclusions for Petri net modeling of manufacturing systems with shared resources. *IEEE Transactions on Robotics and Automation* 7:515–27.

Zhou, M. C., and F. DiCesare. 1993. *Petri net synthesis for discrete event control of manufacturing systems*. Boston: Kluwer Academic Publications.

Zhou, M., F. DiCesare, and A. Desrochers. 1992. A hybrid methodology for synthesis of Petri nets for manufacturing systems. *IEEE Transactions on Robotics and Automation* 18:350–61.

Zhou, M. C., and M. P. Fanti, ed. 2004. *Deadlock resolution in computer-integrated systems*. New York: Marcel Dekker.

Zhou, M. C., and K. Venkatesh. 1998. Modeling, simulation and control of flexible manufacturing systems: Petri net approach. Singapore: World Scientific.

Zuberek, W. M. 2001. Timed Petri nets in modeling and analysis of cluster tools. *IEEE Transactions on Robotics and Automation* 17:562–75.

Index

A

Abbreviations, *xxv–xxvi*
AC, *see* Asymmetric choice (AC) net
"A Coordination Theory of Intelligent
　　Machines," 8
ACROPN, *see* Augmented colored ROPN
　　(ACROPN)
Active circuits, 149, *see also* Circuits
Adidas, 2, 5
Advances in Petri Nets, 6
Agile manufacturing, 4
Agility, 4
AGVs, *see* Automated guided vehicle systems
　　(AGVs)
Al-Jaar, Robert, 7
AMS, *see* Automated manufacturing systems
　　(AMS)
Analysis by time marked graph, 199,
　　201–203
"A Petri Net Method for Schedulability and
　　Scheduling Problems in Single-Arm
　　Cluster Tools with Wafer Residency
　　Time Constraints," *xviii*
"Applications of Petri Nets in Manufacturing
　　Systems: Modeling, Control and
　　Performance," 7
"Approximation Methods for Stochastic Petri
　　Nets," 9
Assembly/disassembly systems, modeling and
　　control
　　case study, 262–263, 265
　　deadlock avoidance policy, 256–260
　　flexible assembly system, 240–241
　　fundamentals, 239, 265–266
　　illustrative example, 260–261
　　individual products, 247–250
　　modeling by colored ROPN, 246–253
　　realizable resource requirement,
　　　253–256
　　resource models, 246–247
　　R-policy, 242–245
　　whole system, ROPN, 250–253
Asymmetric choice (AC) net, 21
"A Theory for the Synthesis and Augmentation of
　　Petri Nets in Automation," 7
"A Transformation Theory for Petri Nets and
　　Their Applications to Manufacturing
　　Automation," 8

Augmented colored ROPN (ACROPN),
　　109–112
Augmented marked graph, 52
Augmented state machines, 68–69
Automated flexible assembly systems, *xviii, see
　　also* Flexible assembly system (FAS)
Automated guided vehicle systems (AGVs),
　　control and routing
　　bidirectional paths, 140–154
　　colored ROPN, 135–136, 140–143, 148–150,
　　　154–158
　　computational complexity, 139–140
　　cycles, 143–148
　　deadlock avoidance, 136–139, 143–150
　　deadlock problem treatment, *xvii*
　　effect on deadlock, 85
　　examples, 150–154, 163–165
　　fundamentals, 133–134, 169
　　illustrative examples, 163–165
　　modeling, 140–143
　　performance comparison, 165–168
　　problem description, 155–158
　　rerouting, 158–162
　　resource-oriented Petri nets, *xvi*
　　route expansion, 162–163
　　routing, colored ROPN, 154–158
　　unidirectional paths, 135–140
Automated guided vehicle systems (AGVs), FMS
　　control
　　deadlock avoidance, 178–182
　　example, 182–183
　　fundamentals, 171–173, 183
　　illustrative example, 182–183
　　system modeling, colored ROPN, 173–178
Automated manufacturing systems (AMS)
　　colored ROPN, 62
　　complexity, resource sharing, 43
　　deadlock, 246
　　deadlock control policy, 189–190, 192
　　dependent PPCs, interactive subnets, 126
　　hardware entities, 2
　　performance improvement, examples, 128
　　Petri net modeling, 2–5
　　resource-oriented Petri net, 60
Automation, historical developments, 6–10, 12
"Avoiding Deadlock and Reducing Starvation
　　and Blocking in Automated
　　Manufacturing Systems Based on a
　　Petri Net Model," *xvi*

"Avoiding Deadlocks in Automated
      Manufacturing Systems with Shared
      Material Handling System," *xv*

B

Base and base flow, 240–241
Bibliography, 267–271
Bidirectional paths, AGVs
      colored ROPN, 140–143, 148–150
      conflict detection, 134
      deadlock avoidance, 143–150
      examples, 150–154
      modeling, 140–143
      resource-oriented Petri nets, *xvi*
      unidirectional paths comparison, 134, 140–141
Billington, Jonathan, 10
Blocking, deadlock avoidance
      automated manufacturing systems, 246
      control law, 118–121, 128
      defined, 115
      dependent PPCs, interactive subnets, 123–127
      examples, 128–131
      fundamentals, 115–116, 131
      performance improvement, examples,
            128–131
      relaxed control policy, 118–121
      route expansion, AGV systems, 162, 165
      routing, AGV systems, 155, 157
      simple example, 116–118
Boeing 747, 5
Boundedness, 22–24
Buffers, buffer groups, and buffer space
      deadlock avoidance, 257
      dependent PPCs, interactive subnets, 127
      effect on deadlock, 85
      flexible manufacturing system modeling, 174
      inhibitor arcs, 29
      robots-as-temporary-buffer policy, 191
      R-policy, 243–244
      system modeling, colored ROPN, 87–88

C

Cao, Tiehua, 9
Case study, 262–263, 265
Characteristics, PN modeling
      process-oriented, 52–54
      resource-oriented, 68–69
Choice, 121
Ciardo, G., 7
Circuits, *see also* Production process circuits
      (PPCs)
      AGV systems, colored ROPN, 137–139,
            150–154
      colored resource-oriented Petri nets, 148–150
      computational complexity, 140

Circular wait, 84
Closed-loop control, 154
Cluster tools
      analysis by time marked graph, 199, 201–203
      deadlock analysis, 209–211, 213–214
      dual-blade robot cluster tool analysis,
            213–216
      fundamentals, 197–199
      modeling by colored CROPN, 204–209
      revisitation, 211–216
      single-blade robot cluster tool analysis,
            209–214
      throughput analysis, 211–217
      zero revisitation, 211–212, 215–216
Colored Petri net (CPN)
      conflict detection, AGV bidirectional paths,
            134
      definitions, 33–37
      fundamentals, 31, 41
      P-invariant, 38–40
      simple example, 31, 33
      transition enabling and firing rules, 37–38
Colored resource-oriented Petri nets (CROPN)
      AGV systems, 137–148, 154–158, 163,
            173–178
      assembly/disassembly systems, modeling,
            246–253
      bidirectional paths modeling, 140–143
      circuits, 148–150
      cluster tools, modeling, 204–209
      cycle chains, 143–150
      cycles, 143–148
      deadlock avoidance, 137–139, 143–150,
            178–182
      dependent PPCs, interactive subnets, 121,
            125–126
      examples, 163–165
      flexible manufacturing system modeling,
            173–178
      FMS control, 173–178, 189–193
      free-choice Petri nets, 62
      implementation, 105–107
      individual products, 247–250
      modeling cluster tools, 203–209
      multiple AGVs, 173–178
      multiple robots, 189–193
      part production processes, ROPN modeling,
            62–65
      performance comparison, routing, 165–168
      problem description, routing, 155–158
      production process circuits, 180
      realizable resource requirement, 253–256
      relaxed control policy, 119–121
      rerouting, 158–162
      resource models, 246–247
      robots, 189–193
      route expansion, 162–163

routing, 154–158, 163–165
simplified, overall system liveness, 103–104
unidirectional paths, AGV systems, 135–136
whole system ROPN, 250–253
Colored-timed resource-oriented Petri nets
(CTROPN)
dual-blade robot cluster tool, 213–214
fundamentals, 208–209
single-blade robot cluster tool, 208–210
Combinatorial problems, 253
"Communication with Automata," 6
Computational complexity
track systems, 229
unidirectional paths, 139–140
Computationally retractable algorithm, 172–173
Computation Structure Group (MIT), 6
Conditions, 15, *see also* Places
Conflict cycle chain, AGV systems
rerouting, 158–159
route expansion, 164
Conflict detection, 134
Conservativeness
colored Petri net, 38–38
deadlock avoidance policies, 118
Petri net properties, 24
process-oriented Petri nets, 52
process- *vs.* resource-oriented Petri nets,
72–73
Consistent tokens, 225–227, *see also* Tokens
Contemporary manufacturing systems, 3
Control, assembly/disassembly systems
case study, 262–263, 265
deadlock avoidance policy, 256–260
flexible assembly system, 240–241
fundamentals, 239, 265–266
illustrative example, 260–261
individual products, 247–250
modeling by colored ROPN, 246–253
realizable resource requirement, 253–256
resource models, 246–247
R-policy, 242–245
whole system, ROPN, 250–253
Control law, *see also* Deadlock avoidance
deadlock avoidance, reducing starvation and
blocking, 118–121, 128
interactive subnets, 102
Convertibility, 5
Coverability tree method, 22, 26
CPN, *see* Colored Petri net (CPN)
CROPN, *see* Colored resource-oriented Petri nets
(CROPN)
CTROPN, *see* Colored timed resource-oriented
Petri nets (CTROPN)
Customized flexibility, 4–5
Cycle chain, AGV systems
colored ROPN, 150–154
cycles, deadlock avoidance, 143–148

Cycles, AGV deadlock avoidance, 143–148
Cycling tokens, 92

D

Deadlock, *see also* Liveness
automated guided vehicle systems, 133–134
cause of, 171
conditions for, 84
detection and recovery, 83
dual-blade robot cluster tool analysis, 214–215
existence, 89, 91–93
liveness, 25, 74
prevention, 83
process-oriented Petri net conservativeness,
52
reconfigurable manufacturing systems, 84–86
resolution problem, 171–173
resource sharing, 47
ROPN characteristics, 69
R-policy, 244
single-blade robot cluster tool analysis,
209–211
situations, 107, 109
Deadlock avoidance
AGV systems, 136–139
assembly/disassembly systems, modeling and
control, 256–260
colored ROPN, 93–102
cycles, 143–148
FMS control, 178–182, 186–192
fundamentals, 83, 93–95
interactive subnets, 95–102
multiple AGVs, 178–182
multiple PPCs, 98–102
multiple robots, 186–192
one PPC, 95
policy, 93–102
ROPN modeling, 109–112
routing, AGV systems, 154
shared material handling system, 107–112
situations, 107, 109
starvation and blocking
control law, 118–121, 128
dependent PPCs, interactive subnets,
123–127
examples, 128–131
fundamentals, 115–116, 131
performance improvement, examples,
128–131
relaxed control policy, 118–121
simple example, 116–118
subnets, 95–98
system modeling, colored ROPN, 88–89
track systems
fundamentals, 218
illustrative example, 229–230

implementation, 228–229
modeling by ROPN, 219–220
semiconductor track system, 217–219
strongly connected subnet, 220, 224–228
two PPCs, 95–98
unidirectional paths, AGVs, 136–139
Deadlock-free scheduling, track system
dispatching rules, 231, 233–234
fundamentals, 230–231
illustrative example, 234–236
"Deadlock Modeling and Control of
Semiconductor Track Systems Using
Resource-Oriented Petri Nets," *xviii*
"Deadlock Prevention and Deadlock Avoidance
in Flexible Manufacturing Systems
Using Petri Net Models, *xv*
"Deadlock Resolution in Automated
Manufacturing Systems with Robots,"
xvii
Dependent PPCs, interactive subnets,
123–127
Deposit locations, 134
DES, *see* Discrete event systems (DES)
"Design and Performance Prediction of
Computer Resources for Real-Time
Computer Integrated Manufacturing
Systems," 8
Desrochers, Alan A., 7–9
Destination marking, 141–142
Diagnosability, 5
DiCesare, Frank, 7–9, 12
Digital Equipment Corporation (now HP), 7
Digraphs, 15
Directed bipartite graphs, 15
Disassembly systems, *see* Assembly/disassembly
systems, modeling and control
Discrete event systems (DES), 13, 44–47
Discrete time models, 1
Dispatching rules, 231, 233–234
Dual-blade robots
analysis by timed marked graph, 201–203
cluster tools, 198–199, 203–204, 207–208
deadlock analysis, 213–214
modeling, CROPN, 203–204, 207–208
revisitation, 214–216
throughput analysis, 214–216
zero revisitation, 215–216
Dugan, J., 7

E

EFC, *see* Extend free-choice (EFC) net
Electronic circuits, 1
Elementary circuits, 18–19
Elementary production process circuits (PPCs),
91, *see also* Production process
circuits (PPCs)

Enabling and firing rules, *see also*
Transitions and transition enabling
and firing
colored Petri net, 37–38
finite capacity Petri nets, 18
macro-transitions, 206–207
Petri nets, 16–17
Engineering Research Center for
Reconfigurable Manufacturing
Systems (RMSs), 4
Events, 15, *see also* Transitions and transition
enabling and firing
Examples
AGV systems, multiple, 182–183
assembly/disassembly systems, modeling and
control, 260–261
colored Petri net, 31, 33
deadlock avoidance, 116–118, 128–131,
230–231
deadlock-free scheduling, track system,
234–236
flexible and reconfigurable manufacturing
systems, control, 104–105
FMS control, 182–183, 193
multiple AGVs, 182–183
multiple robots, 193
process- *vs.* resource-oriented Petri nets,
74–80
reducing starvation and blocking, 116–118
robots, multiple, 193
routing, colored ROPN, 163–165
track systems, 230–231
Existence, deadlock, 89, 91–93
Expansion flexibility, 3, *see also* Flexibility
Extended resource control net ERCN-merged
net, 8, 52
Extend free-choice (EFC) net, 21

F

FAS, *see* Flexible assembly system (FAS)
FC, *see* Free-choice (FC) net
Feng, Chengche, 9
Finite capacity
macro-transitions, 206
Petri nets, 18
resource-oriented Petri net, 57, 69
system modeling, colored ROPN, 89
transition firing, 188
Firing rules, *see also* Transitions and transition
enabling and firing
colored Petri net, 37–38
Petri nets, 16–17
Fixtures, effect on deadlock, 85
Flexibility
expansion, 3
track systems, 219, 229

Flexible assembly system (FAS), *see also*
    Assembly/disassembly systems,
        modeling and control
    deadlock resolution, *xviii*
    fundamentals, 240–241
Flexible manufacturing systems (FMS)
        control
    automated guided vehicle systems, 133
    CROPN, system modeling, 87–89
    deadlock
        existence, 89, 91–93
        reconfigurable manufacturing systems,
            84–86
        situations, 107, 109
    deadlock avoidance
        fundamentals, 93–95
        interactive subnets, 95–102
        maximally permissive policy,
            116–118
        multiple PPCs, interactive subnet formed
            by, 98–102
        one PPC, subnet formed by, 95
        policy, 93–102
        ROPN modeling, 109–112
        shared material handling system,
            107–112
        situations, 107, 109
        subnets, 95–98
        two PPCs, interactive subnet formed by,
            95–98
    fundamentals, 83–84, 112–113
    illustrative example, 104–105
    implementation, 105–107
    liveness, overall system, 102–104
    maximally permissive deadlock avoidance
        policy, 116–118
    multiple AGVs
        deadlock avoidance, 178–182
        example, 182–183
        fundamentals, 171–173, 183
        illustrative example, 182–183
        system modeling, colored ROPN, 173–178
    reconfigurable manufacturing systems
        comparison, 5
Flexible manufacturing systems (FMS) control,
        multiple robots
    deadlock control policy, 186–192
    fundamentals, 185, 194
    illustrative example, 193
    motivation through example, 185–186
Fork/join pattern, 240–241
Free-choice (FC) net
    colored ROPN, 62
    liveness, 27
    subclass of PN, 20
Free tokens, 227–230, *see also* Tokens
Full circuits, AGV system rerouting, 158

G

General Motors (GM), 7
Giua, Alessandro, 8, 10
GM (General Motors), 7
GreatSPN, 7

H

Hierarchical models, 176, 231
Hold and wait, 84
Hybrid systems, *xviii*, 9

I

IBM company, 7
IBS, *see* Ill-behaved siphon (IBS)
IITs, *see* Interactive input transitions (IITs);
        Intercircuit input transitions (IITs)
Ill-behaved siphon (IBS)
    flexible manufacturing system modeling, 178
    liveness, 27, 73–74, 76–80
    process-oriented Petri net conservativeness,
        52–54
    resource sharing, ROPN, 68
Illustrative examples, *see* Examples
Implementation
    deadlock avoidance, track system, 229–230
    flexible and reconfigurable manufacturing
        systems, control, 105–107
Incidence matrix
    colored Petri net, 38–38
    Petri net properties, 24
Inconsistent tokens, 225–227, *see also* Tokens
INDEXER, infinite capacity, 219
Individual products, 247–250
Industrial case study, 262–263, 265
Infinite capacity state machine net, 26–27
Ingersoll-Rand factory, 3
Inhibitor arcs, 28–30
Integrability, 4
Intelligent Task Planning Using Petri Nets, 9
Interactive circuits, 152–153
Interactive cycle chain, AGV systems
    colored ROPN, 150–154
    cycles, deadlock avoidance, 146–148
    rerouting, AGV systems, 159–161
Interactive input transitions (IITs)
    control law, 102
    interactive subnets, 96–98
    liveness, overall system, 103
Interactive output transitions (IOTs), 96, 98–99
Interactive subnets
    AGV systems, colored ROPN, 138–139
    cycle chain, AGV systems, 146–148
    deadlock avoidance, 94–96
    dependent PPCs, 123–127

multiple PPCs, 98–102
one PPC, 95, 98–102
two PPCs, 95–98
Intercircuit input transitions (IITs), 120–121
IOTs, *see* Interactive output transitions
    (IOTs)

J

Jeng, MuDer, 8, 10
Jensen, K., 7
Johnson & Johnson, 7
Joint assembly system, 266
Joshi, Jagdish S., 8
Jungnitz, Hauke J., 9

K

K-boundedness, 23
Kim, Jongwook, 9
Kirchhoff's laws, 1
Knots, 106–107
Koh, I., 8

L

Lawley, policy of, 235
Lean manufacturing, 4
Leaving tokens, deadlock existence, 92, *see also*
    Tokens
Lee, Doo Yong, 9
Lee, Yoon and, policy of, 235–236
Lewis, Frank L., 10
Li, Zhiwu, 10
Lin, Chuan, 10
Linear capacitors, 1
Liveness, *see also* Deadlock
    AGV systems, 137–139, 147, 150
    augmented colored ROPN, 110–112
    circuits, 138–139
    colored ROPN, 93–102, 150
    cycle chain, 147
    deadlock avoidance policy, 178–179
    deadlock existence, 93
    dependent PPCs, interactive subnets, 122
    overall system, 102–104
    Petri net properties, 25–27
    POPN conservativeness, 52
    process- *vs.* resource-oriented Petri nets,
        73–74
    relaxed control policy, 120
    resource-oriented Petri nets, *xvi*
    resource sharing, ROPN, 68
    track systems, 230–231
LL, *see* Loadlocks (LL)
Loadlocks (LL), 199, 201–203
Location, deposits, 134

M

Machine flexibility, 3
Machines, effect on deadlock, 85
Macro-transitions, *see also* Transitions and
        transition enabling and firing
    automated guided vehicle system, 176–178
    cluster tools modeling, CROPN, 205–207
    deadlock avoidance policy, 179–182
*Manufacturing Systems Control Design: A
        Matrix Based Approach,* 10
Marked graph (MG)
    analysis, 199, 201–204
    liveness, 27
    subclass of PN, 20–21
    timed PN, 28
Market flexibility, 3
Marking change, 27, 37
Material flow, 240–241
Material handling and material handling systems
        (MHS)
    automated guided vehicle systems, 133
    deadlock avoidance, 85, 107–112
    deadlock situations, 107, 109
    flexibility, 3
    fundamentals, 107
    processes, resource-oriented PN modeling,
        65–66
    resource sharing, 50–52
Mathematical modeling, 1
Matrix format, output function
    colored PN, 34–37
    colored ROPN, 64–65
Maximally permissive deadlock control policy
    example, 116–118
    M-policy, 181
    optimal schedule impact, 115–116
    productivity impact, 112
    resolution problem, 172–173
Mergers
    dependent PPCs, interactive subnets, 121
    extended resource control net ERCN-merged
        net, 52
    part production processes, ROPN modeling,
        60, 62
MG, *see* Marked graph (MG)
MHS, *see* Material handling and material
        handling systems (MHS)
miAdidas, 2
Missions, time fulfilling, 165–168
Modeling
    assembly/disassembly systems, 246–253
    bidirectional paths, 140–143
    case study, 262–263, 265
    cluster tools, 203–208
    colored ROPN, 246–253
    deadlock avoidance, 220–221, 256–260

discrete event systems, 44–47
flexible assembly system, 240–241
FMS control with multiple AGVs, 173–178
fundamentals, 1–2
individual products, 247–250
process-oriented Petri net modeling,
    44–47
realizable resource requirement, 253–256
resource models, 246–247
R-policy, 242–245
track system, 219–220
unidirectional paths, 135–136
whole system, ROPN, 250–253
*Modeling, Simulation, and Control of Flexible
    Manufacturing Systems: A Petri Net
    Approach,* 8
"Modeling and Deadlock Avoidance of
    Automated Manufacturing Systems
    with Multiple Automated Guided
    Vehicles," *xvii*
"Modeling and Deadlock Control of Automated
    Guided Vehicle Systems," *xvii*
Modularity, 4
Motivation through example, 185–186
Motorola, 4
M-policy, 180–183
Multiple AGVs, *xvii, see also* Automated guided
    vehicle systems (AGVs)
Multiple PPCs, 98–102, *see also* Production
    process circuits (PPCs)
Multishared transition, 252, *see also* Transitions
    and transition enabling and firing
Murata, Tadao, 6, 10
Mutual exclusion, 8, 84

N

Narahari, Y., 7
"Necessary and Sufficient Conditions for
    Deadlock-Free Operation in Flexible
    Manufacturing Systems Using a
    Colored Petri Net Model," *xix*
Nodes, *see also* Zones
    bidirectional paths, AGVs, 140
    reachability, 22
    routing, AGV systems, 155–158
    R-policy, 243
Nonelementary production process circuits
    (PPCs), 91, *see also* Production
    process circuits (PPCs)
Nonmaximally permissive control policy,
    256
Nonpreemption, 84
Non-pseudo-blocked tokens, 228–230, *see also*
    Tokens
NP-hard (nonmaximally permissive control
    policy), 256

O

Objectives, 12–13
Ohm's law, 1
Oil refineries, *xviii*
One PPC, subnet formed by, 95
One-step look-ahead, 139–140
One-way recorders, 262–263, 265
Operation flexibility, 3
Optimal scheduling, 115–116

P

Pallets, effect on deadlock, 85
Panasonic, 4
Parallel mutual exclusion (PME)
    liveness, 76
    resource sharing, 47–49, 68
    ROPN, 68
Partial conservativeness, 24
Part processing
    automated guided vehicle systems, 133
    colored ROPN, 62–65
    deadlock, *xvii,* 85
    deadlock avoidance, 179
    flexible manufacturing system modeling,
        174–175
    liveness, 74
    resource sharing, 48–49
Part production processes, ROPN modeling
    colored ROPN, 62–65
    fundamentals, 58–60
    subnet forming, 60
    subnet merging, 60, 62
    wafers, 220
"Performance and State-Space Analyses of
    Systems Using Petri Nets," 9
Performance comparison, 165–168
"Performance Evaluation of Automated
    Manufacturing Systems Using
    Generalized Stochastic Petri
    Nets," 7
Performance improvement, examples,
    128–131
*Performance Modeling of Automated
    Manufacturing Systems,* 7
Petri Nets: An Introduction, 6
*Petri Nets: A Tool for Design and Management
    of Manufacturing Systems,* 10
"Petri Nets: Properties, Analysis and
    Applications," *xv,* 6
"Petri Nets as Discrete Event Models for
    Supervisory Control," 8
*Petri Nets in Automation and Computer
    Engineering,* 6
Petri Nets in Flexible and Agile Automation,
    8, 10

Petri nets (PN)
   boundedness, 23–24
   conservativeness, 24
   defined, 6, 15
   enabling and firing rules, 16–17
   finite capacity, 18
   fundamentals, 15–20, 30
   incident matrix, 24
   inhibitor arcs, 28–30
   liveness, 25–27
   properties, 22–27
   reachability, 22
   reversibility, 24–25
   special structures, 18–20
   subclass, 20–21
   timed, 27–28
Petri nets (PN), modeling
   automated manufacturing systems, 2–5
   automation historical developments,
      6–10, 12
   defined, 6
   fundamentals, 13–14
   objectives, 12–13
   process, 1–2
   scope, 12–13
Petri Nets Steering Committee, 12
Petri Nets World, 12
Petri Net Synthesis for Discrete Event Control of
   Manufacturing Systems, 8
Petri Net Theory and the Modeling of Systems,
   xv, 6
Petrinetze, 6
Photoresist processes, track systems, 219
Pickup locations, 134
P-invariant, 24, 38–40
Places
   colored ROPN, 64
   incident matrix and conservativeness, 24
   interpretations, 16, 32–33
   liveness, 79
   modeling, automated manufacturing systems,
      44–45, 47
   resource-oriented Petri net, 57–58
   resource sharing, ROPN, 67
   special structures in Petri nets, 18
   time durations, 207
   track systems, 218
PLCs, *see* Programmable logic controllers
   (PLCs)
PM, *see* Processing modules (PMs)
PME, *see* Parallel mutual exclusion (PME)
PN, *see* Petri nets (PN)
Policy, *see* Deadlock avoidance
POPN, *see* Process-oriented Petri net (POPN)
   modeling
Power, process- *vs.* resource-oriented Petri nets,
   71–72

p-path, 18
PPCs, *see* Production process circuits (PPCs)
Preemption, non-, 84
Primitives, 248–249
Problem description, 155–158
Processes, 1–3
Processing modules (PMs), 197, 209
Process-oriented Petri net (POPN) modeling
   characteristics, 52–54
   fundamentals, 43, 54
   material handling, 50–52
   material handling systems, 107, 109
   modeling method, 44–47
   part processing, 48–49
   resource-oriented PN comparison, 80
   resource sharing, 47–52
Process- *vs.* resource-oriented Petri nets
   conservativeness, 72–73
   example, 74–80
   fundamentals, 71, 80–81
   power, 71–72
   size, 71–72
   structure for liveness, 73–74
Production flexibility, 3
Production process circuits (PPCs), *see also*
   Circuits
   augmented colored ROPN, 111
   colored ROPN, 104–105
   control law, complexity in applying, 128
   deadlock avoidance, 94, 256
   deadlock control policy, 186–193
   deadlock existence, 89, 91–93
   dependent, interactive subnets, 123–127
   example, 104–105
   implementation, 105
   interactive subnets, 104–105
   liveness, 74, 79–80
   material handling systems, 109
   M-policy, 180
   relaxed control policy, 118–121
   ROPN characteristics, 68–69
   strong connectedness, 221
Productive researchers, 10
Products, 3
Program flexibility, 3
Programmability, 3
Programmable logic controllers (PLCs), 3
Properties
   boundedness, 23–24
   conservativeness, 24
   incident matrix, 24
   liveness, 25–27
   reachability, 22
   reversibility, 24–25
Proth, J.-M., 10
Pseudo-blocked tokens, 228–231, *see also*
   Tokens

Q

Qualitative properties, 197
Quantitative properties, 197

R

Reachability
  AGV systems, 161–162
  Petri net properties, 22
  rerouting, 161
  route expansion, 162
Realizability problem, 243
Realizable resource requirement, 253–256
"Real-Time Deadlock-Free Scheduling for
    Semiconductor Track Systems Based
    on Colored Timed Petri Nets," *xviii*
Reconfigurability, 4
Reconfigurable manufacturing systems (RMS)
  CROPN, system modeling, 87–89
  deadlock
    existence, 89, 91–93
    fundamentals, 84–86
    situations, 107, 109
  deadlock avoidance
    fundamentals, 93–95
    interactive subnets, 95–102
    multiple PPCs, interactive subnet formed
      by, 98–102
    one PPC, subnet formed by, 95
    policy, 93–102
    ROPN modeling, 109–112
    shared material handling system, 107–112
    situations, 107, 109
    subnets, 95–98
    two PPCs, interactive subnet formed by,
      95–98
  flexible manufacturing systems comparison, 5
  fundamentals, 83–84, 112–113
  illustrative example, 104–105
  implementation, 105–107
  liveness, overall system, 102–104
  resource sharing, 47–48
Recorders, 262–263, 265
Reisig, W., 6
Relaxed control policy, 118–121
Releasing rule, 233, 234
Rensselaer Polytechnic Institute, 7
Rerouting, 158–162
Resource circuits, 53–54
Resource models, 246–247
"Resource-Oriented Petri Net for Deadlock
    Avoidance in Flexible Assembly
    Systems," *xviii*
Resource-oriented Petri net (ROPN) modeling,
    see also Robots, FMS control of
    multiple

assembly/disassembly systems, 250–253
characteristics, 68–69
colored ROPN, 62–65
deadlock avoidance, track system, 220–221
deadlock avoidance with MHS, 109–112
fundamentals, 57, 69
material handling processes, 65–66
part production processes, 58–65
process-oriented PN comparison, 80
resource sharing, 66–68
steps, 57–58
subnet forming, 60
subnet merging, 60, 62
"Resource-Oriented Petri Nets for Deadlock
    Avoidance in Automated
    Manufacturing," *xvi*
"Resource-Oriented Petri Nets in Deadlock
    Prevention and Avoidance," *xix*
Resource sharing
  fundamentals, 47–48
  liveness, 75–76
  material handling, 50–52
  part processing, 48–49
  resource-oriented PN modeling, 66–68
Resource- *vs.* process-oriented Petri nets
  conservativeness, 72–73
  example, 74–80
  fundamentals, 71, 80–81
  power, 71–72
  size, 71–72
  structure for liveness, 73–74
Reversibility, 24–25
Revisitation
  dual-blade robot cluster tool analysis,
    216–217
  single-blade robot cluster tool analysis,
    212–214
  subnet forming, ROPN, 60
R-firing, 188–191
RMS, *see* Reconfigurable manufacturing systems
    (RMS)
RMSs, *see* Engineering Research Center for
    Reconfigurable Manufacturing
    Systems (RMSs)
Robots
  deadlock situations, *xvii*
  material delivery, 247
  modeling, automated manufacturing systems,
    45
Robots, dual-blade
  analysis by timed marked graph, 201–203
  cluster tools, 198–199, 203–208
  deadlock analysis, 213–214
  modeling, CROPN, 203–204, 207–208
  revisitation, 214–216
  throughput analysis, 214–216
  zero revisitation, 215–216

Robots, FMS control of multiple
  deadlock control policy, 186–192
  fundamentals, 185, 194
  illustrative example, 193
  motivation through example, 185–186
Robots, single-blade
  analysis by timed marked graph, 199, 204
  cluster tools, 198–199, 203–204, 207–208
  deadlock analysis, 213–214
  modeling, CROPN, 203–204, 207–208
  revisitation, 214–216
  throughput analysis, 211–214, 214–216
  zero revisitation, 211–212
Robots-as-temporary-buffer (RTB) policy, 191
Roszkowaska studies, see R-policy
Route expansion, 162–163
Routing, 3
Routing, colored ROPN
  examples, 163–165
  flexibility, 63, 85
  fundamentals, 154–155
  illustrative examples, 163–165
  performance comparison, 165–168
  problem description, 155–158
  rerouting, 158–162
  route expansion, 162–163
R-policy
  assembly/disassembly systems, modeling and
      control, 242–245
  realizability, 255
  recorder models, 265
  W-policy comparison, 259–260
  WZ-policy comparison, 261
RTB, see Robots-as-temporary-buffer (RTB)
      policy
Rule-based scheduling/sequencing, 116

S

Safe, boundedness, 23, see also Unsafe state
Sanderson, Arthur C., 9
Saridis, G.N., 8
Scalability, 5
Scheduling, 115–116
"Scheduling and Supervisory Control of Flexible
      Manufacturing Systems Using Petri
      Nets and Heuristic Search," 9
Scope, 12–13
SCOPUS, 6, 10–12
SCS, see Single-capacity systems (SCS)
SDPP, see Shared direct place path (SDPP)
Self-loops
  augmented colored ROPN, 109–110
  individual product models, 249
  material handling processes, modeling, 66
  special structures in Petri nets, 18–19
Semiconductor manufacturing systems (SMS)

AGV systems, colored ROPN, 152
analysis by time marked graph, 199,
    201–203
bidirectional paths, AGVs, 142
cluster tools, 197–217
colored ROPN, 204–209
deadlock analysis, 209–211, 213–214
deadlock avoidance, track system,
    218–231
deadlock-free scheduling, track system,
    231–236
dispatching rules, 231, 233–234
dual-blade robot cluster tool analysis,
    213–216
fundamentals, 197, 236
illustrative example, 229–230, 234–236
implementation, 228–229
modeling, 203–208, 220–221
revisitation, 211–213, 214–217
ROPN, 219–220
semiconductor track system, 217–219
single-blade robot cluster tool analysis,
    208–213
strongly connected subnet, 220, 224–228
throughput analysis, 210–216
zero revisitation, 210–211, 215–216
"Sensitivity Analysis of Discrete Event Dynamic
      Systems by a Petri Net-Based
      Perturbation Method," 9
Sequential mutual exclusion (SME)
  liveness, 76
  resource sharing, 48–49, 68
  ROPN, 68
Shared direct place path (SDPP)
  interactive subnets, 96–99, 101
  relaxed control policy, 119–120
Shared material handling system, 107–112
"Shortest Routing of Bidirectional Automated
      Guided Vehicles Avoiding Deadlock
      and Blocking," xvii
Silva, Manuel, 6, 10
Simplified CROPN, 103–104
Single-blade robots
  analysis by timed marked graph,
      199, 203
  cluster tools, 198–199, 203–205, 207–208
  deadlock analysis, 208–210
  modeling, CROPN, 203–204, 207–208
  revisitation, 211–213
  throughput analysis, 211–213
  zero revisitation, 210–211
Single-capacity systems (SCS), 217, 228
Single-resource allocation system, xvii
Sink places, 18
Siphons, see also Ill-behaved siphon (IBS)
  liveness, 27
  special structures in Petri nets, 18, 20

Situations, deadlock, 107, 109
Size, 71–72
SM, *see* State machine (SM) net
Special structures, 18–20
SPNP, *see* Stochastic Petri net package (SPNP)
Starvation, deadlock avoidance, *see also*
        Flexible manufacturing systems
        (FMS) control
    control law, 102, 118–121, 128
    defined, 115
    dependent PPCs, interactive subnets,
        123–127
    examples, 128–131
    fundamentals, 115–116, 131
    performance improvement, examples,
        128–131
    relaxed control policy, 118–121
    simple example, 116–118
State machine (SM) net
    liveness, 26–27
    subclass of PN, 20–21
Steps, 57–58
Stochastic Petri net package (SPNP), 7, 9
Strong connectedness
    deadlock avoidance, track system, 220,
        224–228
    deadlock existence, 93
    infinite capacity state machine net, 26
    rerouting, AGV systems, 161
    special structures in Petri nets, 19
    track systems, 220, 230
Structure for liveness, *see* Liveness
Structures, special, 18–20
Subclasses, Petri nets, 20–21
Subnets
    multiple PPCs, 98–102
    one PPC, 95–98
    part production processes, ROPN, 60, 62
    resource sharing, ROPN, 67
    track systems, 230
    two PPCs, 95–98
Sun Microsystems, 7

T

"Task Planning and Uncertainty for Robotic
        Systems," 9
"The Modeling, Analysis and Simulation of
        a Discrete Event Dynamic System
        Using Time Petri Net Models," 9
"Theory and Applications of Resource
        Control Petri Nets for Automated
        Manufacturing Systems," 8
Throughput analysis, cluster tools
    dual-blade robots, 214–216
    single-blade robots, 210–213

Timing
    cluster tools modeling, 208–209
    marked graphs, 199, 201–204
    Petri nets, 27–28
TM, *see* Transportation module (TM)
Tokens
    consistent, 225–227
    deadlock existence, 92
    defining colors, 63–64
    free, 227–230
    inconsistent, 225–227
    inhibitor arcs, 29
    interactive subnets, 99
    modeling, automated manufacturing systems,
        44–45
    non-pseudo-blocked, 227–229
    process-oriented Petri net conservativeness,
        52
    pseudo-blocked, 227–230
    resource-oriented Petri net, 57–58, 63
    track systems, 220–221
Tools, effect on deadlock, 85
Track systems, deadlock avoidance
    fundamentals, 217
    illustrative example, 229–230
    implementation, 228–229
    modeling by ROPN, 219–220
    semiconductor track system, 217–219
    strongly connected subnet, 220,
        224–228
Transitions and transition enabling and firing, *see
        also* Macro-transitions
    analysis by timed marked graph, 199–201,
        203
    colored Petri net, 37–38
    colored ROPN, 63–64, 149
    control-enabled, 93
    cycle chain, AGV systems, 148
    deadlock avoidance, 93–94, 179–182
    deadlock existence, 92
    flexible manufacturing system modeling,
        176–178
    FMS with multiple robots, 188–192
    incident matrix and conservativeness, 24
    interactive subnets, 99
    interpretations, 16, 32–33
    liveness, 79
    material handling processes, modeling,
        66
    modeling, automated manufacturing systems,
        44–45
    multiproducts assembly, 252
    postsets, 16
    presets, 16
    process enabling, 99
    reachability, 22
    rerouting, AGV systems, 161

resource-oriented Petri net, 57–58
resource sharing, ROPN, 67
R-policy, 243
special structures in Petri nets, 18
subnets, 95–96
system modeling, colored ROPN, 89
timed marked graph analysis, 199–201
time durations, 207
token color, 252–253
token relationship, 99
track systems, 220
Transportation, deadlock problem treatment, *xvii*
Transportation module (TM)
    cluster tools, 203–204, 207
    colored timed ROPN, 208
    fundamentals, 197, 199
Traps, 20
Trays, 240–241
Trivedi, Kishor S., 7, 10
Two PPCs, interactive subnet formed by, 95–98
Two-way recorders, 262–263, 265

U

Unidirectional paths
    bidirectional path comparison, 134,
        140–141
    colored ROPN, modeling, 135–136
    computational complexity, 139–140
    deadlock avoidance policy, 136–139
    resource-oriented Petri nets, *xvi*
Unions, *see* Mergers
Unsafe state, 182, *see also* Safe, boundedness

V

Van Der Aalst, Wil M.P., 10
Venkatesh, K., 8
Viswanadham, N., 7
Volume flexibility, 3

W

Wafers, process flows, 218–219, *see also*
        Semiconductor manufacturing
        systems (SMS)
Wait, hold and, 84
Wang, Fei-Yue, 8
Watson, J.F., III, 9
Well-behaved siphons, 27
Whole system, ROPN, 250–253
Whole system ROPN, 250–253
Workstation operations, 247–248
W-policy, 257–259, 260
Wu, Naiqi, *xxiii*, 12
WZ-policy, 261

X

Xie, X., 10

Y

Yoon and Lee, policy of, 236–237

Z

Zero preemption, 84
Zero revisitation, cluster tool analysis
    dual-blade robots, 214–215
    single-blade robots, 210–211
Zhou, MengChu, *xxiii–xxiv*, 7–8, 10
Zones, *see also* Nodes
    AGV systems, 135, 137, 140
    bidirectional paths, 140
    computational complexity, 140
    deadlock avoidance, 137
    defined, 134
    routing, 155–158
    R-policy, 244–245
    unidirectional paths, 135